Statistical Methods for Environmental Pollution Monitoring

Statistical Methods for Environmental Pollution Monitoring

Richard O. Gilbert
Pacific Northwest Laboratory

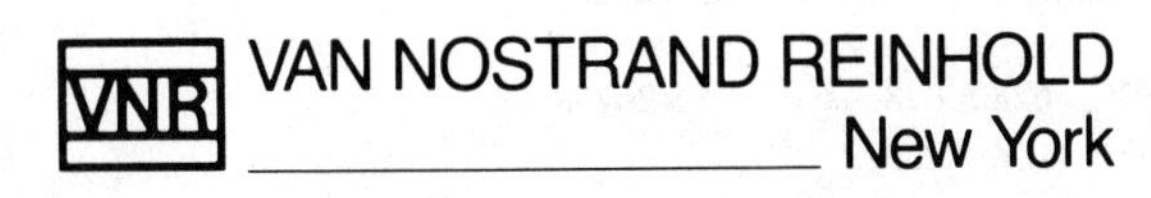

Dedicated to my parents, Mary Margaret and Donald I. Gilbert

Library of Congress Catalog Card Number 86-26758
ISBN 0-442-23050-8

Work supported by the U.S. Department of Energy under Contract No. DE-AC06-76RLO1830.

I(T)P Van Nostrand Reinhold is an International Thomson Publishing company. ITP logo is a trademark under license.

Printed in the United States of America

Van Nostrand Reinhold
115 Fifth Avenue
New York, NY 10003

International Thomson Publishing GmbH
Königswinterer Str. 418
53227 Bonn
Germany

International Thomson Publishing
Berkshire House, 168-173
High Holborn, London WC1V 7AA
England

International Thomson Publishing Asia
221 Henderson Bldg. #05-10
Singapore 0315

Thomas Nelson Australia
102 Dodds Street
South Melbourne 3205
Victoria, Australia

International Thomson Publishing Japan
Kyowa Building, 3F
2-2-1 Hirakawacho
Chiyoda-Ku, Tokyo 102
Japan

Nelson Canada
1120 Birchmount Road
Scarborough, Ontario
M1K 5G4, Canada

16 15 14 13 12 11 10 9 8 7

Library of Congress Cataloging-in-Publication Data

Gilbert, Richard O.
Statistical methods for environmental pollution monitoring.
Bibliography: p.
Includes index.
1. Pollution—Environmental aspects—Statistical methods. 2. Environmental monitoring—Statistical methods I. Title.
TD193.G55 1987 628.5'028 86-26758
ISBN 0-442-23050-8

Contents

Preface

The application of statistics to environmental pollution monitoring studies requires a knowledge of statistical analysis methods particularly well suited to pollution data. This book attempts to help fill that need by providing sampling plans, statistical tests, parameter estimation procedure techniques, and references to pertinent publications. The book is written primarily for nonstatisticians (environmental scientists, engineers, hydrologists, etc.) who have had perhaps one or two introductory statistics courses. Most of the statistical techniques discussed are relatively simple, and examples, exercises, and case studies are provided to illustrate procedures. In addition to being a general reference, this book might be used in an upper undergraduate or lower graduate level applied statistics course or as a supplemental book for such a class.

The book is logically, though not formally, divided into three parts. Chapters 1, 2, and 3 are introductory chapters. Chapters 4 through 10 discuss field sampling designs and Chapters 11 through 18 deal with a broad range of statistical analysis procedures. Some statistical techniques given here are not commonly seen in statistics books. For example, see methods for handling correlated data (Sections 4.5 and 11.12), for detecting hot spots (Chapter 10), and for estimating a confidence interval for the mean of a lognormal distribution (Section 13.2). Also, Appendix B lists a computer code that estimates and tests for trends over time at one or more monitoring stations using nonparametric methods (Chapters 16 and 17). Unfortunately, some important topics could not be included because of their complexity and the need to limit the length of the book. For example, only brief mention could be made of time series analysis using Box-Jenkins methods and of kriging techniques for estimating spatial and spatial-time patterns of pollution, although multiple references on these topics are provided. Also, no discussion of methods for assessing risks from environmental pollution could be included.

I would appreciate receiving comments from readers on the methods discussed here, and on topics that might be included in any future editions. I encourage the reader to examine the references sited in the book since they provide a much broader perspective on statistical methods for environmental pollution than can be presented here.

Financial support for this book was provided by the U.S. Department of Energy, Office of Health and Environmental Research. Dr. Robert L. Watters of that office deserves special mention for his encouragement and support. Pacific Northwest Laboratory provided facilities and secretarial support. David W. Engel wrote the computer code in Appendix B and helped with the trend examples in Chapters 16 and 17. Margie Cochran developed a reference filing system that greatly facilitated the development of the bibliography. Also, Robert R. Kinnison provided encouragement and support in many ways including statistical computing assistance and access to his library. The help of these individuals is very much appreciated. I am also grateful for review comments on drafts of this book by various reviewers, but any errors or omissions that remain are entirely my own responsibility. The encouragement and guidance given by Alex Kugushev during the early stages of this endeavor are also much appreciated. I am deeply grateful to Sharon Popp who has faithfully and with great skill typed all drafts and the final manuscript. I wish to thank the literary executor of the Late Sir Ronald A. Fisher, F.R.S. to Dr. Frank Yates, F.R.S. and to Longman Group Ltd., London (previously published by Oliver and Boyd, Ltd., Edinburgh) for permission to reprint Table III from *Statistical Tables for Biological, Agricultural and Medical Research* (6th Edition, 1974).

Richard O. Gilbert

Statistical Methods for Environmental Pollution Monitoring

1 Introduction

Activities of man introduce contaminants of many kinds into the environment: air pollutants from industry and power plants, exhaust emissions from transportation vehicles, radionuclides from nuclear weapons tests and uranium mill tailings, and pesticides, sewage, detergents, and other chemicals that enter lakes, rivers, surface water, and groundwater. Many monitoring and research studies are currently being conducted to quantify the amount of pollutants entering the environment and to monitor ambient levels for trends and potential problems. Other studies seek to determine how pollutants distribute and persist in air, water, soil, and biota and to determine the effects of pollutants on man and his environment.

If these studies are to provide the information needed to reduce and control environmental pollution, it is essential they be designed according to scientific principles. The studies should be cost effective, and the data statistically analyzed so that the maximum amount of information may be extracted.

The purpose of this book is to provide statistical tools for environmental monitoring and research studies. The topics discussed are motivated by the statistical characteristics of environmental data sets that typically occur. In this introductory chapter we discuss these characteristics and give an overview of the principal tasks involved in designing an environmental pollution study. This material sets the stage for Chapter 2, which develops an orientation and understanding of environmental sampling concepts needed before a sampling plan is devised.

1.1 TYPES AND OBJECTIVES OF ENVIRONMENTAL POLLUTION STUDIES

Environmental pollution studies may be divided into the following broad and somewhat overlapping types.

1. *Monitoring.* Data may be collected (a) to monitor or to characterize ambient concentrations in environmental media (air, water, soil, biota) or (b) to monitor concentrations in air and water effluents. The purpose may be to assess the adequacy of controls on the release or containment of pollutants,

to detect long-term trends, unplanned releases, or accidents and their causes, to provide a spatial or temporal summary of average or extreme conditions, to demonstrate or enforce compliance with emission or ambient standards, to establish base-line data for future reference and long-range planning, to indicate whether and to what extent additional information is required, or to assure the public that effluent releases or environmental levels are being adequately controlled.

2. *Research.* Field and laboratory data may be collected (a) to study the transport of pollutants through the environment by means of food chains and aerial pathways to man and (b) to determine and quantitate the cause-and-effect relationships that control the levels and variability of pollution concentrations over time and space.

Many design and statistical analysis problems are common to monitoring and research studies. Environmental data sets also tend to have similar statistical characteristics. These problems and characteristics, discussed in the next section, motivate the topics discussed in this book.

1.2 STATISTICAL DESIGN AND ANALYSIS PROBLEMS

Numerous problems must be faced when applying statistical methods to environmental pollution studies. One problem is how to define the environmental "population" of interest. Unless the population is clearly defined and related to study objectives and field sampling procedures, the collected data may contain very little useful information for the purpose at hand. Chapter 2 gives an approach for conceptualizing and defining populations that leads into the discussion of field sampling (survey) designs in Chapters 3–9. The important role that objectives play in determining sampling designs is discussed in Chapter 3.

Once data are in hand, the data analyst must be aware that many statistical procedures were originally developed for data sets presumed to have been drawn from a population having the symmetric, bell-shaped Gaussian ("normal") distribution. However, environmental data sets are frequently asymmetrical and skewed to the right—that is, with a long tail towards high concentrations, so the validity of classical procedures may be questioned. In this case, nonparametric (distribution-free) statistical procedures are often recommended. These procedures do not require the statistical distribution to be Gaussian. Alternatively, an asymmetrical statistical distribution such as the lognormal may be shown or assumed to apply. Both of these approaches are illustrated in this book. Frequently, a right-skewed distribution can be transformed to be approximately Gaussian by using a logarithmic or square-root transformation. Then the normal-theory procedures can be applied to the transformed data. However, biases can be introduced if results must be expressed in the original scale. Often, other assumptions, such as uncorrelated data and homoscedasticity (constant variance for different populations over time and space), are required by standard statistical analysis procedures. These assumptions also are frequently violated. The problem of correlated data over time and/or space is one of the most serious facing the data analyst. Highly correlated data can seriously affect statistical tests and can give misleading results when estimating the variance of estimated means, computing confidence limits on means, or determining the number of measurements needed to estimate a mean.

Other problems that plague environmental data sets are large measurement errors (both random and systematic, discussed in Chapter 2), data near or below measurement detection limits (Chapter 14), missing and/or suspect data values (Chapter 15), complex trends and patterns in mean concentration levels over time and/or space, complicated cause-and-effect relationships, and the frequent need to measure more than one variable at a time. Berthouex, Hunter, and Pallesen (1981) review these types of problems in the context of wastewater treatment plants. They stress the need for graphical methods to display data, for considering the effect of serial correlation on frequency of sampling, and for conducting designed experiments to study cause-and-effect relationships. Schweitzer and Black (1985) discuss several statistical methods that may be useful for pollution data.

Many routine monitoring programs generate very large data bases. In this situation it is important to develop efficient computer storage, retrieval, and data analysis and graphical software systems so that the data are fully utilized and interpreted. This point is emphasized by Langford (1978). The development of interactive graphics terminals, minicomputers, and personal computers greatly increases the potential for the investigator to view, plot, and statistically analyze data.

In contrast to monitoring programs, some environmental pollution research studies may generate data sets that contain insufficient information to achieve study objectives. Here the challenge is to look carefully at study objectives, the resources available to collect data, and the anticipated variability in the data so that a cost-effective study design can be developed. Whether the study is large or small, it is important to specify the accuracy and precision required of estimated quantities, and the probabilities that can be tolerated of making wrong decisions when using a statistical test. These specifications in conjunction with information on variability can be used to help determine the amount of data needed. These aspects are discussed in Chapters 4–10 and 13.

1.3 OVERVIEW OF THE DESIGN AND ANALYSIS PROCESS

When planning an environmental sampling study, one must plan the major tasks required to conduct a successful study. The following steps give an overview of the process. Schweitzer (1982) gives additional discussion relative to monitoring uncontrolled hazardous waste sites.

1. Clearly define and write down study objectives, including hypotheses to be tested.
2. Define conceptually the time-space population of interest.
3. Collect information on the physical environment, site history, weather patterns, rate and direction of groundwater movement, and so on, needed to develop a sampling plan.
4. Define the types of physical samples to be collected (e.g., 2 L of water or an air filter exposed for 24 h) or field measurements to be made.

5. Develop a quality assurance program pertaining to all aspects of the study, including sample collection, handling, laboratory analysis, data coding and manipulation, statistical analysis, and presenting and reporting results.
6. Examine data from prior studies or conduct pilot or base-line studies to approximate the variability, trends, cycles, and correlations likely to be present in the data.
7. Develop field sampling designs and sample measurement procedures that will yield representative data from the defined population.
8. Determine required statistical data plots, summaries, and statistical analyses, and obtain necessary computer software and personnel for these needs.
9. Conduct the study according to a written protocol that will implement the sampling and quality assurance plans.
10. Summarize, plot, and statistically analyze the data to extract relevant information and to evaluate hypotheses.
11. Assess the uncertainty in estimated quantities such as means, trends, and average maximums.
12. Evaluate whether study objectives have been met, and use the newly acquired information to develop more cost-effective studies in the future.

1.4 SUMMARY

This chapter emphasized the great diversity of environmental monitoring and research studies being conducted, the types of statistical design and analysis problems frequently encountered with pollution data, and the major tasks required to conduct a successful environmental sampling study.

2 Sampling Environmental Populations

Data are easy to collect but difficult to interpret unless they are drawn from a well-defined population of environmental units. The definition of the population is aided by viewing the population in a space-time framework. In addition, sources of variability and error in data from a population should be understood so that a cost-effective sampling plan can be developed. This chapter discusses these concepts to provide a foundation for the discussion of field sampling designs in Chapters 3–9. More specifically, this chapter covers the space-time sampling framework, the population unit, target population, and sampled population, the sources of variability and error in environmental data, and the meaning of accuracy and precision. It concludes with an air pollution example to illustrate these concepts.

2.1 SAMPLING IN SPACE AND TIME

Environmental sampling can be viewed in a structured way by a space-time framework, as illustrated in Figure 2.1. The symbols T_1, T_2, . . . , denote time periods such as hours, days, weeks, seasons, or years. The specific times within the time period T_i when measurements or samples are taken are denoted by t_{i1}, t_{i2}, . . . , and so on. Study sites S_1, S_2, . . . , denote study plots, geographical areas, sections of a city or river basin, and other areas that are part of a larger region. Within study sites specific sampling locations are chosen. The spatial location is determined by east-west, north-south, and vertical coordinates. Measurements or samples (soil, water, air) may be taken at each location and point in time. Alternatively, several samples collected over time or space may be combined and mixed to form a composite sample, and one or more subsamples may be taken for measurement from each sample or composite. Compositing of samples is discussed in Chapter 7.

Figure 2.2 shows four locations being sampled over time. Each location is represented by a box divided into two parts, where each part represents a replicate sample. The volume of each replicate denotes the space being sampled (grab sample of water, a core of soil to specified depth, a volume of air, etc.). This volume is the *support* of the sample, a term used by Journel and Huijbregts (1978) and other writers in the geostatistical literature. For example, suppose a stream is sampled at points in time for dissolved oxygen. At each sampling

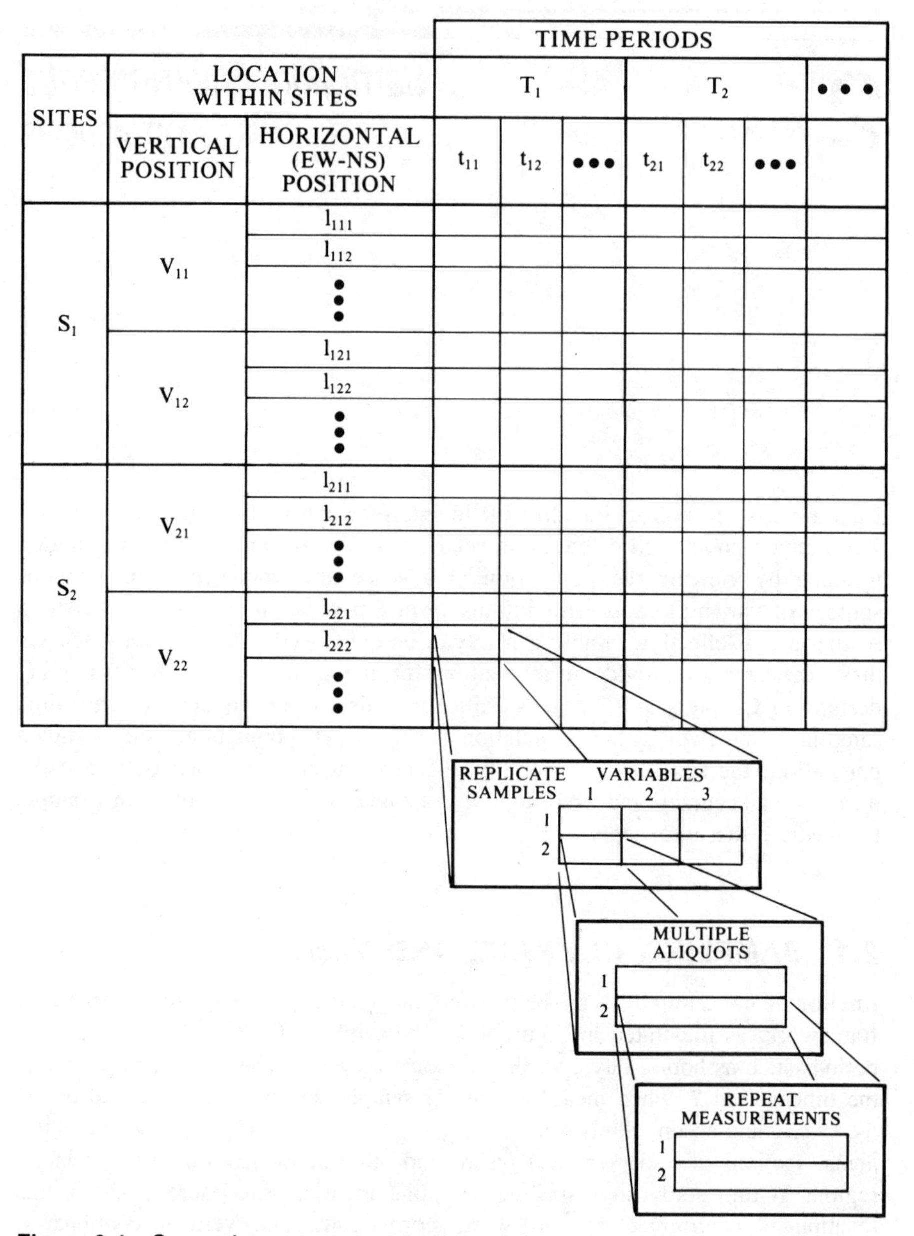

Figure 2.1 Space-time conceptual framework for environmental sampling within a study region (after Green, 1979, Fig. 3.1).

time two equally sized grab samples (replicates) of water are taken at two predetermined depths and two locations along a transect spanning the stream. As another example, air samples are collected at several study sites (S_1, S_2, . . .) located downwind from a pollution source. Several geographical coordinates at each site are selected, and two (replicate) air samplers are placed side by side at each of two heights. Samples are collected continuously at time intervals specified by the study design.

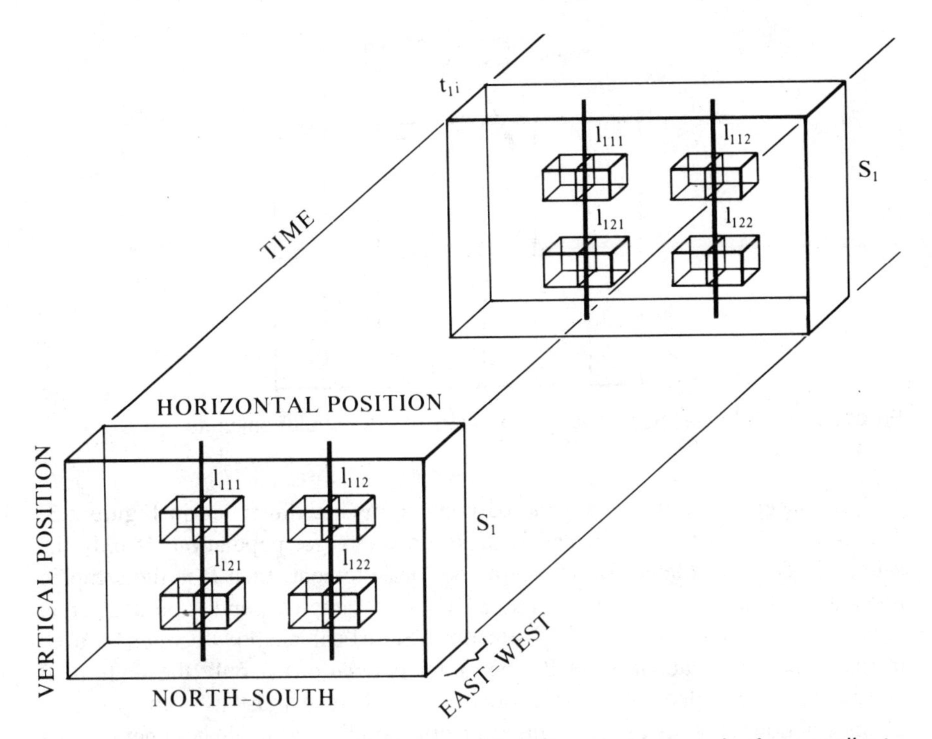

Figure 2.2 Sampling in time and space. Two aliquots from each of two replicate samples at four locations in space at study site 1. Two sampling times are shown.

2.2 TARGET AND SAMPLED POPULATIONS

Designing an environmental sampling program involves defining the environmental units that make up what Cochran, Mosteller, and Tukey (1954) call the *target population* and the *sampled population.* The target population is the set of *N population units* about which inferences will be made. The sampled population is the set of population units directly *available* for measurement. Population units are the *N* objects (environmental units) that make up the target or sampled population. These units can be defined in many ways depending on study objectives, the type of measurement to be made, regulatory requirements, costs, and convenience. Some examples of population units are

Ten-gram aliquots of soil taken from a sample of field soil that undergo specified preparatory procedures, such as drying, grinding, and so on, and that will be analyzed for pollutants according to a specified procedure

Air filters exposed for a specified time interval that undergo specified procedures for chemical analysis

Prescribed areas of ground surface scanned by an in situ radiation detector under specified operating conditions

Aboveground vegetation from 2-*m* × 2-*m* plots that is dried and ashed according to prescribed procedures in preparation for wet chemistry analysis.

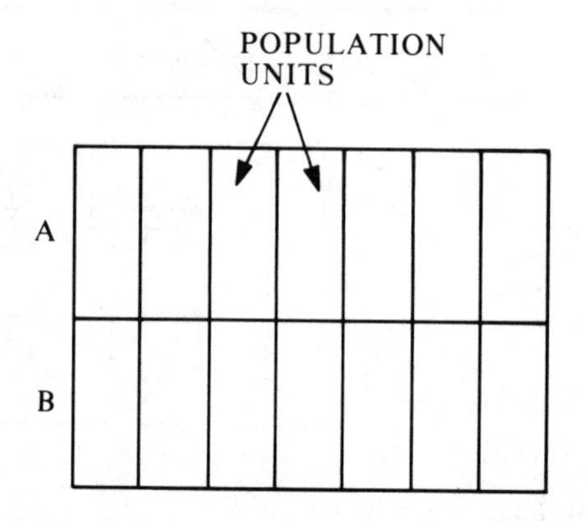

Figure 2.3 A target population comprised of 14 population units.

The concepts of target and sampled populations are illustrated in Figure 2.3. The 14 population units in A and B make up the target population. If only the 7 units in B are available for selection and measurement, then B is the sampled population. If all 14 units are available, then the sampled population and target population are identical, the ideal situation. Statistical methods cannot be used to make inferences about A + B (the target population) if only the units in B are available for selection. In this situation the analyst must use other sources of information, such as expert opinion or prior studies, to evaluate whether data from B have relevance to units in A. The following examples illustrate target and sampled populations.

EXAMPLE 2.1

Information is sought on the average concentration of a radionuclide in the surface 5 cm of soil in a 10-*m* × 10-*m* plot. Surface soil samples will be taken by using a rectangular template of dimensions 10 cm × 10 cm and depth 5 cm. Samples will be dried, ball milled, and sieved. A 10-g aliquot will be removed from each soil sample and radiochemically analyzed according to a written protocol. The target population is the set of all N possible 10-g aliquots (population units) that could be obtained from the 10-*m* × 10-*m* plot according to the sampling, sample preparation, and aliquoting protocols. If all N units are available for selection for radiochemical analyses, then the target and sampled populations are identical.

Note that the sampled and target populations are defined operationally—that is, in terms of collecting, preparing, and analyzing aliquots. Data obtained by using a given operational procedure cannot, in general, be used to make inferences about a target population defined by a different operational procedure. It is difficult to compare data sets from different study sites when the sites use different protocols. Careful consideration should always be given to whether operational procedures result in a target population different from that desired.

EXAMPLE 2.2

A municipal wastewater treatment facility discharges treated water into a nearby river. A three-month study is designed to estimate the average amount of suspended solids and other water-quality variables immediately upriver from a proposed swimming area during June, July, and August of a particular year.

The population unit could be defined to be a volume of water collected in a 1-L grab sample at any location and depth along a sampling line traversing the river a specified distance upstream from the swimming area during the three summer months. The grab samples are collected, handled, and analyzed according to a specified protocol. The target population would then be the set of all N such population units that could conceivably be obtained during the three-month summer period for the specified year.

If the grab samples can only be collected at certain times, depths, and locations along the sampling line, then the sampled and target populations are not identical. Information from prior studies or expert opinion may permit reasonable judgments to be made about the target population on the basis of the collected data.

2.3 REPRESENTATIVE UNITS

The concept of a target population is closely related to that of a representative unit. A *representative unit* is one selected for measurement from the target population in such a way that it, in combination with other representative units, will give an accurate picture of the phenomenon being studied.

Sometimes a list of guidelines for obtaining representative samples will be constructed. Such a list for sampling water and wastewater is the following [from a longer list in Environmental Protection Agency (EPA), 1982].

Collect the sample where water is well mixed—that is, near a Parshall flume or at a point of hydraulic turbulence such as downstream from a hydraulic jump. Certain types of weirs and flumes tend to enhance the settling of solids upstream and accumulate floating solids and oil downstream. Therefore such locations should be avoided as sample sources.

Collect the sample in the center of the channel at 0.4 to 0.6 depth from the bottom where the velocity of flow is average or higher than average and chances of solids settling is minimal.

In a wide channel divide the channel cross section into different vertical sections so that each section is of equal width. Take a representative sample in each vertical section.

When sampling manually with jars, place the mouth of the collecting container below the water surface and facing the flow to avoid an excess of floating material. Keep the hand as far away as possible from the mouth of the jar.

By imposing conditions such as those in the list, one defines the target population. *The crucial point is whether the population so defined is the one needed to achieve study objectives.* For example, the foregoing guidelines may be suitable for estimating the average concentration for well mixed water samples, but be completely inadequate for estimating maximum concentrations where mixing is not complete. The difficulty of obtaining representative samples for determining if air quality standards are being met is illustrated by Ott and Eliassen (1973).

2.4 CHOOSING A SAMPLING PLAN

Once the space-time framework is established and the target and sampled populations are clearly defined, a sampling plan can be chosen for selecting representative units for measurement. The choice of a plan depends on study objectives, patterns of variability in the target population, cost-effectiveness of alternative plans, types of measurements to be made, and convenience. Some plans use a random selection procedure to select units from the target population; others use a systematic approach, such as sampling at equal intervals in time or space beginning at a randomly chosen unit.

In some applications expert knowledge plays an important role, such as when choosing a location for a groundwater monitoring well near a hazardous waste site (HWS) to detect chemicals leaking from the site. The hydrogeologist is needed to define regions where wells will intercept groundwater that has flowed in the vicinity of the HWS. A monitoring well in the wrong location is useless for detecting leaks.

2.5 VARIABILITY AND ERROR IN ENVIRONMENTAL STUDIES

Environmental pollution data are usually highly variable and subject to uncertainties of various types. Sources of this variability and uncertainty are identified in this section. An awareness of these sources allows one to develop more efficient sampling plans and more accurate measurement techniques.

2.5.1 Environmental Variability

Environmental variability is the variation in true pollution levels from one population unit to the next. Some factors that cause this variation are

Distance, direction, and elevation relative to point, area, or mobile pollution sources

Nonuniform distribution of pollution in environmental media due to topography, hydrogeology, meteorology, action of tides, and biological, chemical, and physical redistribution mechanisms

Diversity in species composition, sex, mobility, and preferred habitats of biota

Variation in natural background levels over time and space

Variable source emissions, flow rates, and dispersion parameters over time
Buildup or degradation of pollutants over time.

These factors illustrate the many interacting forces that combine to determine spatial and time patterns of concentrations. When choosing a sampling plan, one must know the concentration patterns likely to be present in the target population. Advance information on these patterns is used to design a plan that will estimate population parameters with greater accuracy and less cost than can otherwise be achieved. An example is to divide a heterogeneous target population into more homogeneous parts or strata and to select samples independently within each part, as discussed in Chapter 5.

2.5.2 Bias, Precision, and Accuracy

Measurement bias is a consistent under- or overestimation of the true values in population units. For example, if the measurement of each unit is consistently 3 ppm too high, then a positive measurement bias of 3 ppm is present. *Random measurement uncertainties* are random (unpredictable) deviations from the true value of a unit. The magnitude of this deviation changes with each repeat measurement on a given unit and may result from the interplay of many factors.

The difference between bias and random measurement uncertainty is illustrated in Figure 2.4. The center of each target in the figure represents the true value for a population unit, and the dots represent several repeat measurements of the unit. Figures 2.4(a) and 2.4(c) indicate bias is present, since the scatter of measurements is not centered on the target.

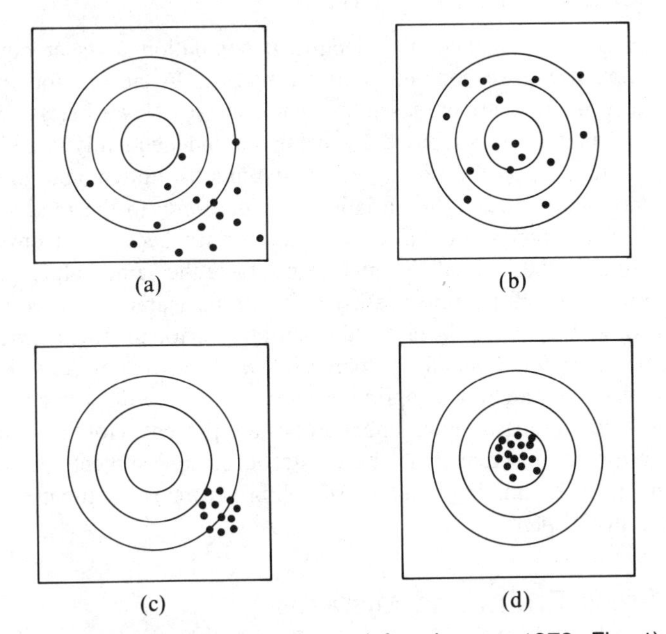

Figure 2.4 Patterns of shots at a target (after Jessen, 1978, Fig. 1). (a): high bias + low precision = low accuracy; (b): low bias + low precision = low accuracy; (c): high bias + high precision = low accuracy; (d): low bias + high precision = high accuracy.

Precision is a measure of the size of the closeness of agreement among individual measurements. Since the dots in Figures 2.4(c) and (d) cluster tightly together, the measurements are said to be precise. *Accuracy* is a measure of the closeness of measurements to the true value. Figure 2.4(d) represents accurate measurements since bias is absent and random measurement uncertainties are relatively small (precision is high).

2.5.3 Statistical Bias

Data are used in algebraic formulas to estimate population parameters of interest. For example, the arithmetic mean $\bar{x} = \Sigma_{i=1}^{n} x_i/n$ is commonly used to estimate the true mean μ of a population of units. If sets of n representative samples are repeatedly drawn from the target population and $\bar{x}$ is computed for each set, all the $\bar{x}$'s could be averaged to estimate μ. If the average of all possible $\bar{x}$'s obtained this way exactly equals μ, then $\bar{x}$ is said to be a statistically *unbiased estimator* of μ. The average of all possible $\bar{x}$'s from the target population is commonly denoted by $E(\bar{x})$ and is called the *expected value* of $\bar{x}$.

If the units selected for measurement are obtained by simple random sampling (discussed in Chapter 4), $\bar{x}$ is indeed an unbiased estimator of μ—that is, $E(\bar{x}) = \mu$. Statistical bias is a discrepancy between the expected value of an estimator and the population parameter being estimated. Some estimators are biased if n is small but become unbiased for n sufficiently large. For examples, see Sections 13.1.2 and 13.3.3. In this book when we say that an estimator is biased, we are referring to statistical bias, not measurement bias.

2.5.4 Random Sampling Errors

Suppose, using data obtained by selecting n population units at random from the target population, we compute $\bar{x}$ to estimate μ. In the previous section we learned that $\bar{x}$ is then an unbiased estimator of μ. Nevertheless, $\bar{x}$ will not exactly equal another $\bar{x}$ computed by using an independently drawn set of n units unless by chance the same set of n units is drawn and there are no measurement uncertainties. The variation in $\bar{x}$ due *only* to the random selection process is called *random sampling error* and arises because of environmental variability—that is, because all N units do not have the same value. The random sampling error in $\bar{x}$ will be zero if all N units are measured and used to compute $\bar{x}$. Of course, if N is infinitely large, the sampling error in $\bar{x}$ will never be zero, although it can be made small by increasing n. Note that if N is finite and if $n = N$ so that the random sampling error is zero, $\bar{x}$ will still not equal μ if bias and/or random measurement uncertainties are present. Note also that random sampling error is different than bias introduced by subjective selection of population units. Estimating random sampling errors for estimators of μ is a major topic in Chapters 4–9.

2.5.5 Gross Errors and Mistakes

Gross errors and mistakes can easily occur during sample collection, laboratory analyses, data reduction, statistical analyses, and modeling. Efforts to carefully follow protocols, double check data transfers, and adequately test computer codes must be made. One must also be careful that statistical computing codes are not applied inappropriately due to unfamiliarity with statistical techniques and their proper use and interpretation. Some types of gross mistakes or unusually deviant data can be identified by statistical tests and data-screening procedures. See Chapter 15 for these methods.

2.6 CASE STUDY

This example is motivated by a similar discussion by Hunter (1977). It illustrates the concepts discussed in this chapter.

A study will be conducted to estimate the average concentration of ozone at noontime (12 noon to 1 P.M.) during June, July, and August at a sampling location over a three-year period (a total of 276 days). The target population consists of 276 population units of 1 h duration at the chosen sampling station that are measured by a particular measurement technique. A single measurement will be made on each of the 276 days. Hence, the sampled population equals the target population (assuming no equipment failure).

Although only a single measurement will be made on each unit, we can conceive of taking $J > 1$ repeat measurements, using the same methods. Suppose the following simple additive model applies:

$$x_{ij} = \mu_i + e_{ij}, \qquad i = 1, 2, \ldots, 276, \quad j = 1, 2, \ldots, J$$

where x_{ij} is the jth measurement of the noontime concentration on the ith day, μ_i is the true value for day (unit) i, and $e_{ij} = x_{ij} - \mu_i$ is the random measurement error.

Figure 2.5 shows hypothetical values of μ_i and x_{i1} for several days. The distribution of possible (conceptual) noontime concentrations is assumed to be normal (Gaussian) with the same spread (variance) for each day. The normal distribution, discussed more fully in Chapter 11, is shown here centered on μ_i, the true concentration for the noontime unit on day i. Hence, the measurement process is unbiased—that is, no measurement bias is present.

In practice, the daily measurement distributions may be asymmetrical, possibly skewed toward high concentrations. Even so, the measurement process is statistically unbiased for a given day i if the true mean of that day's distribution of measurements equals μ_i. The shape of the distribution could also vary from day to day and reflect a changing measurement error variance (precision) over time.

To the far right of Figure 2.5 is shown the hypothetical histogram of the 276 true noontime concentrations μ_i. This histogram arises because of the variability from day to day in true noontime concentrations. In practice, this distribution could take on many shapes, depending on the particular patterns and trends in the true values μ_i over the 276-day period. A hypothetical histogram of the 276 *observed* noontime concentrations x_{i1} is also shown to the far right of Figure 2.5. This latter histogram estimates the histogram of the true values. The true and estimated histograms for this finite sequence will be identical in shape and location only when there is no systematic measurement bias or random measurement uncertainty and no gross errors or mistakes are made in sample collection, measurement, and data reduction. In that special case $\bar{x}$, the arithmetic mean of the 276 observed values, will equal $\mu = \Sigma_{i=1}^{276} \mu_i/276$. In Figure 2.5 the two distributions are offset to illustrate that the random positive and negative

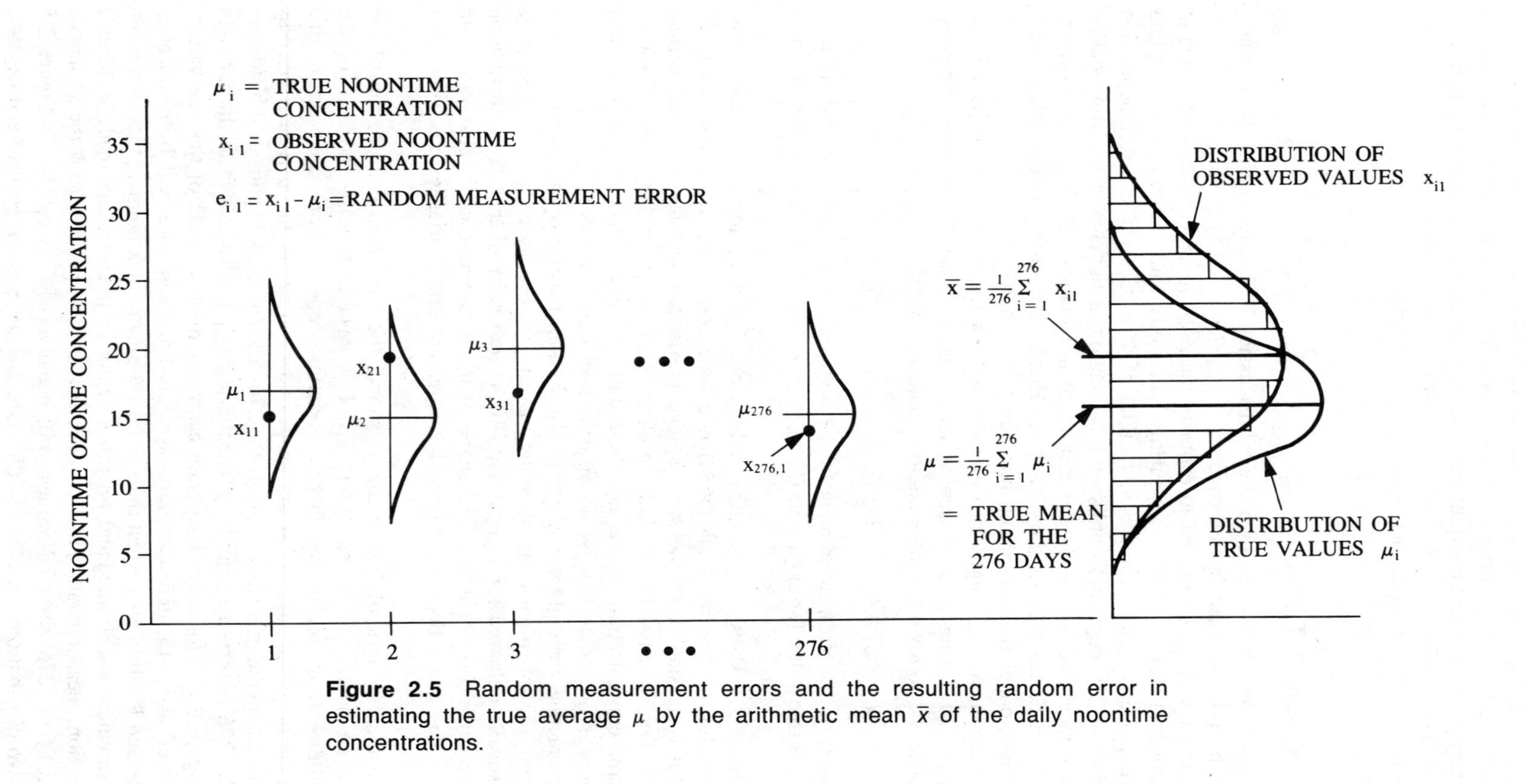

Figure 2.5 Random measurement errors and the resulting random error in estimating the true average μ by the arithmetic mean $\bar{x}$ of the daily noontime concentrations.

measurement uncertainties will not exactly balance each other for any finite sequence of measurements. Also, the histogram of observed data is shown as having a larger spread (variance) than the histogram of the true values. This spread, or the opposite case of smaller spread, can occur whenever random measurement uncertainties are present. If a constant measurement bias were present each day, then the histogram of x_{i1} would be shifted by that amount. A systematic measurement bias that changes over time might also be present. This changing bias would also affect the shape and location of the histogram of data values. Distributions commonly used to model environmental data sets are discussed in Chapters 11–13.

2.7 SUMMARY

This chapter discussed the following points:

A space-time framework should be used to define the environmental population of interest.

Clear definitions of the population unit, target population, and sampled population are needed.

Variation in environmental data is due to variation in true concentration levels between population units, measurement bias, random measurement uncertainties, and gross errors and mistakes. Measurement uncertainties can be controlled by using accurate measuring techniques and careful planning. If the patterns of variation in true levels can be anticipated, more cost-effective sampling plans can be developed.

The error that arises because only a random subset of the population units is selected and measured is called random sampling error. This error can be estimated by statistical methods given in Chapters 4–9.

EXERCISES

2.1 Suppose groundwater monitoring wells will be established in the area around an HWS to quickly detect leaks of chemicals from the HWS. Discuss how the target population and the population unit might be defined. Is it important that the sampled population equal the target population? Why?

2.2 In Exercise 2.1 what role does the hydrogeologist play in choosing locations for wells? Under what conditions would it be acceptable to place wells at random locations down-gradient from the HWS? Are those conditions likely to occur in practice?

2.3 Suppose the true mean of a population of size $N = 1000$ is known to be $\mu = 100$ and that $n = 20$ of the N units are measured. Suppose we decide to estimate μ by computing the arithmetic average of the 5 largest of the 20 measurements obtained. If this average happens to equal 100, does it imply that averaging the highest 5 concentrations in a sample of 20 is an unbiased procedure (estimator) for estimating μ? How does one estimate the bias of an estimator?

ANSWERS

2.1 Discussion.

2.2 The hydrogeologist defines where leaking is most likely to be detected quickly and therefore plays a crucial role in defining the target population. Random locations are acceptable only if the aquifer is known to cover the entire area and contamination is as likely to show up in one spot as in any other spot. No.

2.3 No. Draw a random sample of n measurements and use the estimator (procedure) to compute the estimate from these data. Repeat this, say, m times. Find the arithmetic mean of the m estimates and subtract the true value of the quantity being estimated.

3 Environmental Sampling Design

3.1 INTRODUCTION

Chapter 2 emphasized the need to develop a space-time framework for viewing environmental sampling and for carefully defining the population unit, target population, and sampled population. These actions must precede choosing a sampling plan and the number of units to measure. This chapter identifies criteria for selecting a sampling plan and gives an overview of sampling plans discussed more thoroughly in Chapters 4–9.

The emphasis in Chapters 4–9 is on describing sampling plans and showing how to estimate the mean, the total amount and the sampling errors of these two estimates. Methods are given for determining the number of units that should be measured to estimate the mean and the total with required accuracy within cost constraints. If sampling is done for other objectives (e.g., estimating the temporal or spatial pattern of contamination, finding a hot spot, or detecting trends over time), then the required sampling design (number and location of units) may be quite different from that needed for estimating a mean. Sampling to detect hot spots is discussed in Chapter 10. Sampling for detecting and estimating trends is discussed in Chapters 8, 16, and 17.

3.2 CRITERIA FOR CHOOSING A SAMPLING PLAN

Four criteria for choosing a sampling plan are

1. Objectives of the study
2. Cost-effectiveness of alternative sampling designs
3. Patterns of environmental contamination and variability
4. Practical considerations such as convenience, site assessability and availability, security of sampling equipment, and political considerations.

This book is primarily concerned with the first three criteria. Nonstatistical factors such as listed in item 4 can have a significant impact on the final design, but they are beyond the scope of this book.

3.2.1 Objectives

The importance of objectives in the design of environmental studies was emphasized in Chapter 1. Different objectives require different sampling designs. Van Belle and Hughes (1983) discuss in the context of aquatic sampling the pros and cons of a fixed station network (over time) as opposed to short-term intensive studies. They support the use of fixed sampling stations if the objective is to detect trends in water quality over time, whereas short-term intensive studies are urged for estimating causal relationships, effects of pollution, and mechanisms by which streams recover from pollution. An example of this for river sampling is in Rickert, Hines, and McKenzie (1976), who found that short-term, intensive, synoptic-type aquatic studies were needed to assess cause-and-effect relationships in the Willamette River in Oregon. Loftis, Ward, and Smillie (1983) discuss statistical techniques appropriate for routine, fixed-station monitoring.

Lettenmaier, Conquest, and Hughes (1982) evaluate sampling needs for a metropolitan, routine-stream-quality monitoring network. Six types of monitoring programs were identified by objectives: (a) surveillance, (b) model parameterization, (c) cause and effect, (d) trend detection, (e) water quality control, and (f) base line. They suggest that surveillance and water quality control objectives require relatively high sampling frequencies or continuous monitoring. Model parameterization and cause-and-effect studies require multiple stations per basin, high spatial density, and high sampling frequencies, since data on small time and space scales are needed to estimate transfer coefficients and model parameters. Detecting trends over time requires long sequences of data collected weekly or monthly. Base-line monitoring networks may be similar to those for model parameterization, cause-and-effect, and trend detection but at a lower level of sampling effort. Eberhardt and Gilbert (1980) present a similar set of four classes of sampling discussed in the context of terrestrial sampling.

3.2.2 Cost Effectiveness

A sampling design should be chosen on the basis of cost effectiveness as recommended by National Academy of Sciences (1977). That is, the chosen designs should either achieve a specified level of effectiveness at minimum cost or an acceptable level of effectiveness at specified cost. The magnitude of sampling errors should be evaluated for different sampling designs and levels of effort. For example, suppose effluent wastewater will be sampled to estimate the total amount, I, of a pollutant discharged during a month. We may ask, "What is to be gained in terms of greater accuracy in the estimate of I if the sampling effort is increased by 10%, 50% or 100%?" This kind of question should be addressed before a study begins, so time and monetary resources are effectively used. Approaches to this problem are given in Chapters 4–9 for various designs. Papers that address this issue for environmental studies include Provost (1984), Smith (1984), Skalski and McKenzie (1982), Lettenmaier (1977, 1978), Loftis and Ward (1979, 1980a, 1980b), Montgomery and Hart (1974), Feltz, Sayers, and Nicholson (1971), Gunnerson (1966), Sanders and Adrian (1978), Moore, Dandy, and DeLucia (1976), and Chamberlain et al. (1974).

Nelson and Ward (1981) and Sophocleous, Paschetto, and Olea (1982) discuss the design of groundwater networks. Optimum designs for the spatial sampling of soil are considered by McBratney, Webster, and Burgess (1981), McBratney and Webster (1981), and Berry and Baker (1968). Air network design is considered, for example, by Elsom (1978), Handscombe and Elsom (1982), and Noll and Miller (1977). Rodriguez-Iturbe and Mejia (1974) formulate a methodology for the design of precipitation networks in time and space.

3.2.3 Patterns of Contamination

An important criterion for selecting a sampling plan is the spatial and/or time pattern of concentrations. These patterns are often complex, particularly for air pollutants in urban areas, where topography and meteorology combine to create complex spatial patterns that can change with time. Similar problems exist for aquatic, biotic, soil, and groundwater studies. If the major patterns of contamination are known before the study begins, then more cost-effective sampling plans can be devised. An example is stratified random sampling discussed in Chapter 5.

3.3 METHODS FOR SELECTING SAMPLING LOCATIONS AND TIMES

There are basically four ways of deciding where and when to sample: (1) haphazard sampling, (2) judgment sampling, (3) probability (statistical) sampling in conjunction with prior knowledge of spatial/time variability and other information, and (4) search sampling. Table 3.1 is a guide for when each method is appropriate. This chapter gives an overview of each method but stresses the probability methods because they yield computing formulas for estimating sampling error. Later chapters discuss these probability methods in greater detail.

3.3.1 Haphazard and Judgment Sampling

Haphazard sampling embodies the philosophy of ''any sampling location will do.'' This attitude encourages taking samples at convenient locations or times, which can lead to biased estimates of means and other population characteristics. Haphazard sampling is appropriate if the target population is completely homogeneous in the sense that the variability and average level of the pollutant do not change systematically over the target population. This assumption is highly suspect in most environmental pollution studies.

Judgment sampling means subjective selection of population units by an individual. For example, judgment sampling might be used for target populations where the individual can inspect or see all population units and select those that appear to be ''representative'' of average conditions. However, the individual may select units whose values are systematically too large or too small.

If the individual is sufficiently knowledgeable, judgment sampling can result in accurate estimates of population parameters such as means and totals even if all population units cannot be visually assessed. But it is difficult to measure the accuracy of the estimated parameters estimated. Thus, subjective sampling can be accurate, but the degree of accuracy is difficult to quantitate.

Judgment sampling as used here does *not* refer to using prior information to

Table 3.1 Summary of Sampling Designs and When They Should Be Used for Estimating Means and Totals

Type of Sampling Design	*Conditions When the Sampling Design Is Useful*
Haphazard sampling	A very homogeneous population over time and space is essential if unbiased estimates of population parameters are needed. This method of selection is *not* recommended due to difficulty in verifying this assumption.
Judgment sampling	The target population should be clearly defined, homogeneous, and completely assessable so that sample selection bias is not a problem. Or specific environmental samples are selected for their unique value and interest rather than for making inferences to a wider population.
Probability sampling	
simple random sampling	The simplest random sampling design. Other designs below will frequently give more accurate estimates of means if the population contains trends or patterns of contamination. (Chapter 4)
stratified random sampling	Useful when a heterogeneous population can be broken down into parts that are internally homogeneous. (Chapter 5)
multistage sampling	Needed when measurements are made on subsamples or aliquots of the field sample. (Chapters 6 and 7)
cluster sampling	Useful when population units cluster together (schools of fish, clumps of plants, etc.) and every unit in each randomly selected cluster can be measured.
systematic sampling	Usually the method of choice when estimating trends or patterns of contamination over space. Also useful for estimating the mean when trends and patterns in concentrations are not present or they are known a priori or when strictly random methods are impractical. (Chapter 8)
double sampling	Useful when there is a strong linear relationship between the variable of interest and a less expensive or more easily measured variable. (Chapter 9)
Search sampling	Useful when historical information, site knowledge, or prior samples indicate where the object of the search may be found.

plan the study, define the target population, and choose an efficient probability sampling plan. For example, when sampling environmental media for estimating the degree that humans are exposed to a pollutant via some transport pathway, expert opinion is required to choose the appropriate environmental media, study areas, general time periods of interest, and so forth. As another example, we mention again that air concentrations are highly variable, depending on geographical location, height of the sampler from ground surface, distance from local pollution sources, time of day, day of week, and so on. Here, the use of prior information and expert opinion is clearly needed if the data are to be useful for the intended purpose. Finally, consider sampling groundwater around an HWS. The judgment and knowledge of geochemists and hydrologists is needed to make sure wells are placed in appropriate locations to detect leaks from the waste site.

3.3.2 Probability Sampling

Probability (statistical) *sampling* refers to the use of a specific method of *random* selection. Figure 3.1 shows several probability sampling plans (defined below)

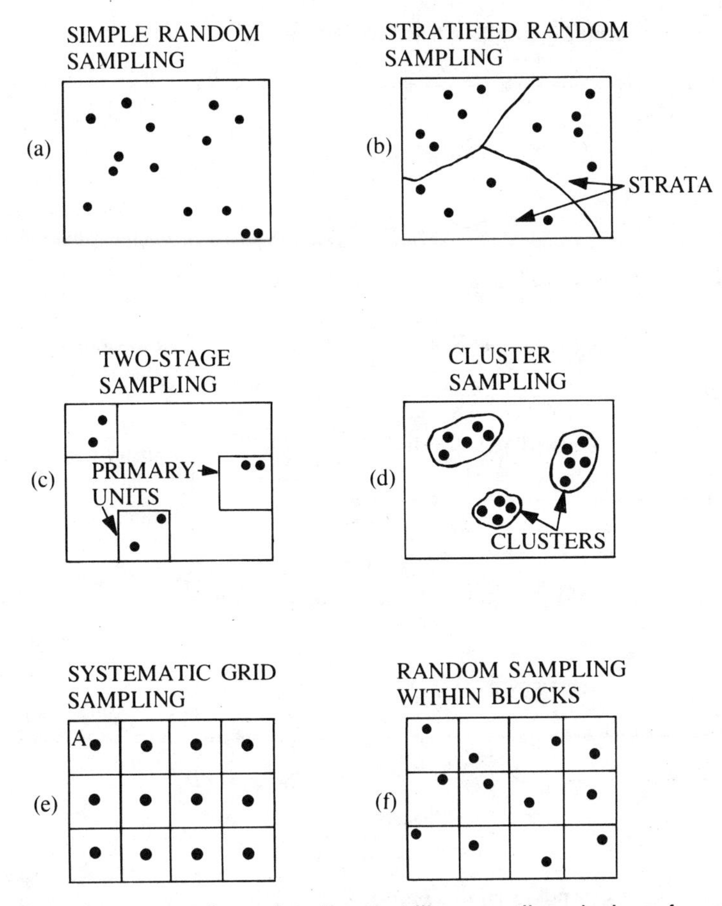

Figure 3.1 Some two-dimensional probability sampling designs for sampling over space.

for spatially distributed variables. These designs can also be applied to sampling along a line in time or space, as illustrated in Figure 3.2. Probability sampling is discussed in some detail in books by Cochran (1977), Jessen (1978), Sukhatme and Sukhatme (1970), Kish (1965), Yates (1981), Deming (1950, 1960), and Hansen, Hurwitz, and Madow (1953). Excellent elementary accounts are given by Williams (1978), Slonim (1957), and Tanur et al. (1972).

The most basic method is *simple random sampling* [Figs. 3.1(a) and 3.2(a); see Chapter 4], where each of the N population units has an equal chance of being one of the n selected for measurement and the selection of one unit does not influence the selection of other units. Simple random sampling is appropriate for estimating means and totals if the population does not contain major trends, cycles, or patterns of contamination. Note that random sampling is not equivalent to picking locations haphazardly.

Stratified random sampling [Figs. 3.1(b) and 3.2(b); see Chapter 5] is a design where the target population is divided into L nonoverlapping parts or subregions called *strata* for the purpose of getting a better estimate of the mean or total for the entire population. Sampling locations are selected from each stratum by simple random sampling. For example, Gilbert et al. (1975) divided nuclear weapons test areas into geographical subregions (strata) and chose sampling locations at random within each stratum. A more precise estimate of

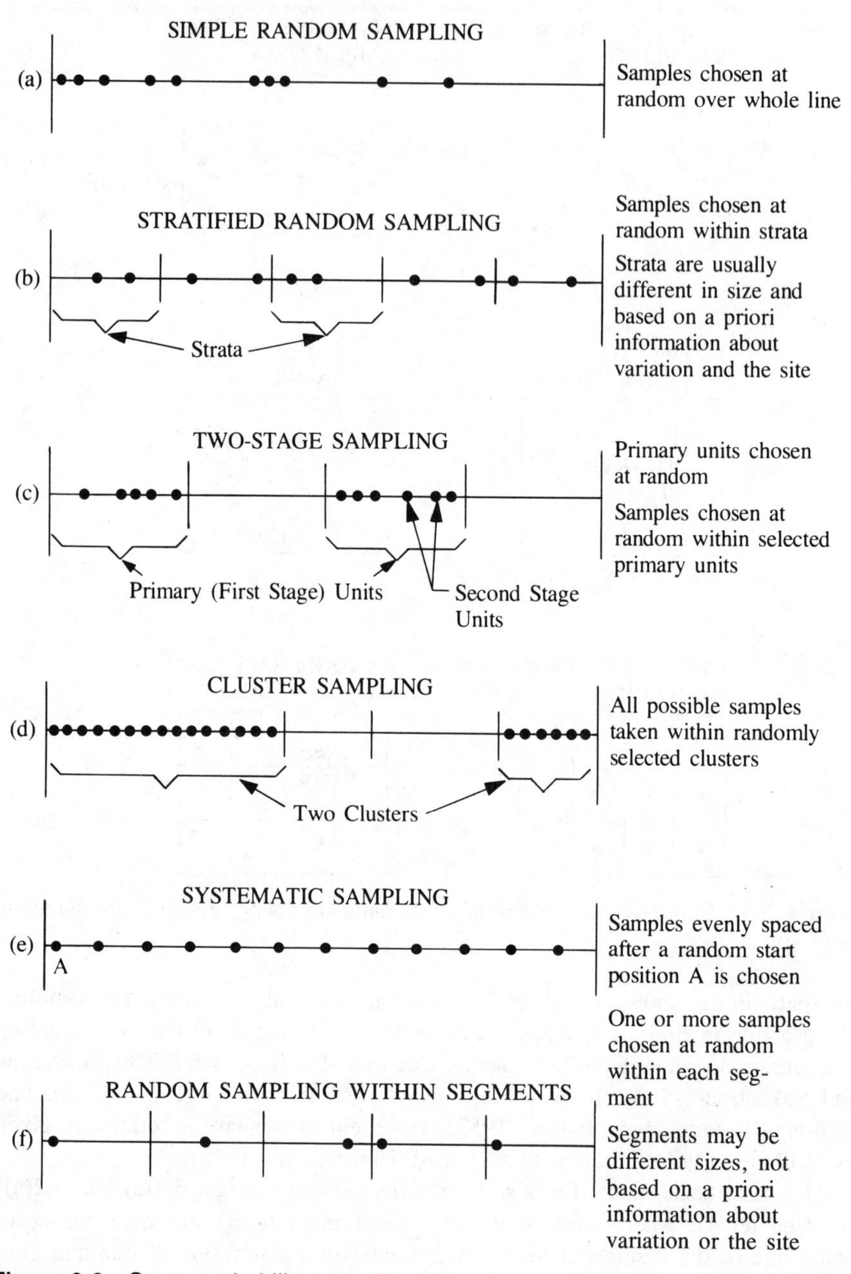

Figure 3.2 Some probability sampling plans for sampling over time or along a transect to estimate a mean or total.

^{239,240}Pu inventory was obtained than if simple random sampling had been used without stratifying the region. If interest is directed at the strata themselves rather than at the site as a whole, then methods by Cochran (1977, pp. 140–144) are useful.

Two-stage sampling [Figs. 3.1(c) and 3.2(c); see Chapters 6 and 7] is where the target population is divided into primary units. Then a set of primary units is selected by using simple random sampling, and each is randomly subsampled.

An example is collecting soil samples (the primary units) at random, then selecting one or more aliquots at random from each soil sample.

Cluster sampling [Figs. 3.1(d) and 3.2(d)] is where clusters of individual units are chosen at random, and all units in the chosen clusters are measured. This method may be needed when it is difficult or impossible to select individual units, but where every unit in the chosen clusters can be measured. For example, it may be difficult to obtain a random sample of plants or organisms (trees, shrubs, fish, etc.) since this would involve numbering the individuals and selecting n at random. Selecting individuals nearest to random points can give biased results, as discussed by Pielou (1977, p. 157). The reader is referred to Cochran (1977) for further discussion of cluster sampling.

Systematic sampling [Figs. 3.1(e) and 3.2(e); see Chapter 8] consists of taking measurements at locations and/or times according to a spatial or temporal pattern: for example, at equidistant intervals along a line or on a grid pattern. Systematic sampling is usually fairly easy to implement under field conditions, and statistical studies indicate it may be preferred over other sampling plans for estimating means, totals, and patterns of contamination. However, it is more difficult to estimate the variance of estimated means and other quantities than if random sampling is used. Also, incorrect conclusions may be drawn if the sampling pattern corresponds to an unsuspected pattern of contamination over space or time.

The sampling plans in Figures 3.1(f) and 3.2(f) combine systematic and random sampling. The site (or line) is divided into blocks of equal size (or segments), and one or more samples are taken at random locations within each block (segment). This procedure allows for a more uniform coverage than simple random sampling does.

Double sampling (Chapter 9) is useful when data using one measurement technique are linearly related to data obtained with less expense or effort using another measurement technique. For example, in situ measurements of radiation by portable detectors may be related linearly to radiochemistry determinations made on the same soil in the laboratory. If the cost of in situ measurements is sufficiently low, double sampling can give more accurate estimates of means and totals than can be obtained by using only wet chemistry data.

3.3.3 Search Sampling

Search sampling is conducted to locate pollution sources or to find "hot spots" of elevated contamination. Examples include the use of gamma radiation surveys from aircraft or portable ground detectors to look for areas of elevated radiation around nuclear installations, taking cores of soil to search for buried waste, and sampling water at strategic locations in a stream network to locate a pollution source (discussed by Sharp, 1970, 1971 and Rajagopal, 1984). The validity of the procedure depends on the accuracy of prior information on where or when to begin the search and on accurate measurements over time or space to guide the search. The most complete search possible is to measure all units in the target population—that is, to conduct a *census*. In practice, this is usually not possible or practical. Chapter 10 discusses methods for determining the optimum spacing between sampling points for finding a target of specified size. Liggett (1984) discusses statistical designs and tests for detecting hot spots by comparing the suspected hot spot area with a background region.

3.4 SUMMARY

This chapter begins by discussing the following criteria for choosing among sampling plans: (1) objectives, (2) cost effectiveness, (3) patterns of environmental contamination, and (4) practical considerations. Then an overview is given of sampling plans that are discussed more thoroughly in Chapters 4–9.

Sampling plans based on a random or systematic selection of population units are highly recommended over haphazard or judgement sampling. Simple random sampling is useful for estimating means and totals if the population is homogeneous—that is, does not have trends or patterns of contamination. Systematic sampling with a random start is preferred to random sampling plans if the objective is to map the variable over time or space. It may also be used to estimate the mean and total if the sampling interval or spacing does not coincide with a recurring pattern of contamination in the population. Stratified random sampling is effective for estimating means and totals if the strata themselves are of interest or if an estimate of the population mean or total over all strata is needed. Two-stage or multistage sampling is frequently necessary because the field sample cannot usually be measured as a whole. Multistage sampling can also be combined with other sampling plans. Cluster sampling is useful for estimating means if it is difficult to randomly select individual units for measurement and if all units in the cluster can be measured. Double sampling can be cost effective for estimating a mean or total if an accurate measurement technique produces data that are linearly related to data obtained using a less accurate and less costly measurement technique. The effectiveness of search sampling depends on the amount of prior information available regarding the location of the object being sought.

EXERCISES

3.1 Discuss the differences between haphazard sampling and simple random sampling. Under what conditions is either method appropriate? Give examples.

3.2 Suppose a chemical processing plant discharges toxic chemicals into a river at unknown periodic intervals. If water samples below the discharge pipe will be collected to estimate the weekly mean concentration of water at that point, does simple random, stratified random, or systematic sampling seem best? Discuss the advantages and disadvantages of these three designs for this situation.

3.3 Suppose the sampling in Exercise 3.2 is for the purpose of estimating the maximum concentration for each week of the year. Is your choice of a sampling plan different than in Exercise 3.2, where the objective was to estimate the weekly mean? Why?

ANSWERS

3.1 Simple random sampling assures that each population unit has the same chance of being selected for measurement. If haphazard sampling is used then the probability of selecting a unit is unknown. Some units may have essentially no chance of being selected if, for example, they are more difficult to sample. Either method is okay if all population units have the same concentration or the same chance of having any particular concentration.

3.2 Systematic sampling will ensure that all periods of time during each week are sampled to the same extent. But contamination could be missed if the plant discharges on a systematic time schedule. Simple random sampling is an unbiased procedure but will require many samples to ensure a good estimate of the weekly average. Stratified random sampling is better than simple random sampling because it ensures that all periods (strata) during the week are sampled.

3.3 To estimate the maximum concentration each week, one must know the patterns of contamination over time. Systematic sampling is effective for this if samples are collected frequently.

4 Simple Random Sampling

Simple random sampling is the simplest sampling plan and is the foundation of other probability sampling plans discussed in Chapters 5 through 9. This chapter

- Shows how to use simple random sampling to estimate the population mean, total amount of pollutant present, and the standard errors of these two estimates
- Gives formulas for determining the number of samples n required to estimate the mean when data are not correlated
- Indicates the effect that measurement uncertainties and correlated data can have on an estimated mean and its variance
- Shows how to determine the number of stations and measurements per station for estimating a regional mean when data are correlated over space and/or over time
- Shows how to estimate the variance of the sample mean when data are spatially and/or temporally correlated.

4.1 BASIC CONCEPTS

Suppose a target population consisting of N population units has been defined. N may be finite, such as the number of hours in a year, or it may be essentially infinite, such as the number of 10-g soil samples in a large field. Simple random sampling may be used to select n of these N units for measurement. If N is finite the first step is to number the units from 1 to N. Then n integers between 1 and N are drawn from a random number table. The use of such a table ensures that every unit not yet selected has an equal chance of being chosen on the next draw. The units labeled with the selected n integers are then measured. They constitute the *sample of size n*. n is also called the *size of the sample*.

Cochran (1977) illustrates how to use a random number table. The table consists of rows and columns of the digits 0, 1, 2, . . . , 9 in random order. First pick a starting place, then follow along horizontally or vertically in the table until the required n numbers are obtained. For example, suppose $N = 50$, $n = 10$, and we decide (perhaps by the flip of a coin) to move horizontally along the rows of the table. Once a starting point is randomly chosen, the first ten distinct two-digit numbers between 01 and 50 along the row identify the population units to be measured. Large random number tables are given by Rand (1955) and by Fisher and Yates (1974). Most statistics textbooks have smaller tables [e.g., Snedecor and Cochran (1980, Table A1)].

4.2 ESTIMATING THE MEAN AND TOTAL AMOUNT

If the target population of N units is heterogeneous—for example, if concentrations cycle with the seasons, are higher in some regions than others, or if there are long-term trends over time and/or space—then the objective of sampling will usually be to describe these features of the population. However, when the population is relatively homogeneous, emphasis is placed on estimating the mean and variance. In this latter situation simple random sampling is an efficient sampling plan. For the more complicated situations, stratified random sampling, systematic sampling, or some combination of the two are required.

We assume the following model for the measurement x_i on the ith unit of the population:

$$x_i = \mu + d_i + e_i = \mu_i + e_i \qquad \textbf{4.1}$$

where: μ = the true mean over all N units in the population,
d_i = the amount by which the true value for the ith unit, μ_i, differs from μ,
$e_i = x_i - \mu_i$ = measurement uncertainty = the amount by which the measured value for the ith unit, x_i, differs from the true value μ_i.

We assume the average of the e_i over the population of N units has zero mean—that is, there are no systematic measurement, sample collection, or handling biases. The true mean concentration is

$$\mu = \frac{1}{N}\sum_{i=1}^{N} \mu_i$$

The total amount (inventory) of pollutant present in the population is $I = N\mu$. The true variance of the N concentrations in the target population is

$$\sigma^2 = \frac{1}{N-1}\sum_{i=1}^{N} (\mu_i - \mu)^2 \qquad \textbf{4.2}$$

Since it is frequently impossible or too costly to measure all N units, and since nonzero e_i are always present, the population parameters μ and σ^2 are unknown. If simple random sampling is used, then statistically unbiased estimates of μ, I, and σ^2 are computed from the data as

$$\bar{x} = \frac{1}{n}\sum_{i=1}^{n} x_i \qquad \textbf{4.3}$$

$\hat{I} = N\bar{x}$, and

$$s^2 = \frac{1}{n-1}\sum_{i=1}^{n} (x_i - \bar{x})^2 \qquad \textbf{4.4}$$

respectively.

A measure of the random sampling error in $\bar{x}$ and $N\bar{x}$ is their variances $\text{Var}(\bar{x})$ and $\text{Var}(N\bar{x})$, which, when simple random sampling is used, are

$$\text{Var}(\bar{x}) = \frac{1}{n}(1 - f)\sigma^2 \qquad \textbf{4.5}$$

and

$$\text{Var}(N\bar{x}) = N^2\,\text{Var}(\bar{x}) = \frac{1}{n}N^2(1 - f)\sigma^2 \qquad \textbf{4.6}$$

respectively, where σ^2 is given by Eq. 4.2 and $f = n/N$ is the sampling fraction—that is, the fraction of the N units in the target population actually measured. The quantity $1 - f$ is called the *finite population correction* (fpc) factor.

Equations 4.5 and 4.6 are appropriate when sampling *without replacement* from a finite population of N units. Sampling without replacement means that once a given population unit has been selected for measurement, it is not replaced into the population for possible reselection and measurement at a later time. If reselection is possible, then we have the option of sampling *with replacement.*

If sampling with replacement is used, then f is set equal to zero in Eqs. 4.5 and 4.6. The equation for $\text{Var}(\bar{x})$ when sampling with replacement is the same as that for sampling without replacement when N is much larger than n, so $f \cong 0$. In most environmental sampling situations it is not feasible to replace a space-time population unit into the population for possible selection at a later time. Hence, the fpc factor should be used unless n is such a small fraction of N that f is essentially zero.

Unbiased estimators of $\text{Var}(\bar{x})$ and $\text{Var}(N\bar{x})$ computed from the n data are (Cochran 1977, p. 26)

$$s^2(\bar{x}) = \frac{1}{n}(1 - f)s^2 \qquad \textbf{4.7}$$

and

$$s^2(N\bar{x}) = N^2 s^2(\bar{x}) = \frac{1}{n}N^2(1 - f)s^2$$

respectively, when the x_i are uncorrelated. The *standard errors* of $\bar{x}$ and $N\bar{x}$ are the square roots of these variances, namely

$$s(\bar{x}) = s\sqrt{\frac{1 - f}{n}}$$

and

$$Ns(\bar{x}) = Ns\sqrt{\frac{1 - f}{n}}$$

respectively.

EXAMPLE 4.1

Suppose it is desired to estimate the mean concentration of ^{222}Ra gas at a location downwind from a uranium mill tailings pile for the month of October for a specified year; concentrations on other months or years are of no interest. Suppose 10 air measurements $x_1, x_2, \ldots, x_{10}$ of 24-h duration will be taken during October of that year, each beginning at 6 A.M. The $n = 10$ sample days will be selected at random from the $N = 31$ potential days by choosing 10 two-digit numbers between 01 and 31 from a random number table. Suppose the 10 ^{222}Ra measurements on the selected days are 1450, 1510, 1480, 1680, 2130, 2110, 1010, 1290, 1150, and 1300 in units of pCi/day.

The true mean for October is $\mu = \Sigma_{i=1}^{31} \mu_i/31$, where μ_i is the true amount (pCi) on day i, and μ is estimated by $\bar{x} = \Sigma_{i=1}^{10} x_i/10 = 1511$ pCi/day. Using Eqs. 4.5 and 4.2, we obtain the true variance

$$\text{Var}(\bar{x}) = \frac{1 - 10/31}{10(30)} \sum_{i=1}^{31} \left(\mu_i - \frac{1}{31}\sum_{i=1}^{31} \mu_i\right)^2$$

$$= \frac{0.67742}{10(30)} \sum_{i=1}^{31} (\mu_i - \mu)^2$$

which is unknown because the μ_i are unknown. But $\text{Var}(\bar{x})$ may be estimated by Eqs. 4.7 and 4.4. We find

$$s^2(\bar{x}) = \frac{0.67742}{10(9)} \sum_{i=1}^{10} (x_i - \bar{x})^2$$

$$= \frac{0.67742}{10(9)} (1{,}249{,}890) = 9407.8$$

Hence, the standard error is $s(\bar{x}) = 97.0$.

In this example the target population was defined to consist of just $N = 31$ unknown daily concentrations for a particular month. Since N is so small, the factor $1 - f$ is needed when computing $s^2(\bar{x})$. However, suppose the ten measurements are considered as representative of concentrations that would arise during October for all years that the pile exists as an entity. Now n may be so large that f is essentially zero. Setting $f = 0$ and using Eq. 4.7 gives $s(\bar{x}) = 118$, which is 21% larger than the $s(\bar{x})$ obtained when $f = 10/31$. Clearly, the conceptualization and resulting size of the target population affects f and, hence, $\text{Var}(\bar{x})$ and the estimate $s(\bar{x})$.

EXAMPLE 4.2

A study will be conducted to estimate the total emissions of carbon monoxide (CO) from a fleet of 500 specially built vehicles, each driven 100 miles. Suppose $n = 5$ vehicles are selected by simple random sampling and the total CO emissions obtained for these vehicles during 100 miles of driving are 501, 1467, 860, 442, and 495 g. (The data are part of a large data set in Table 2 of Lorenzen, 1980.) Since $N = 500$ is the number of vehicles, $n/N = 5/500 = 0.01$. The estimated mean and standard error are $\bar{x} = 753$ and $s(\bar{x}) = s\sqrt{(1 - f)/n} = 432.4\sqrt{0.99/5} = 192$, both in units of g/100 miles. Therefore, the estimated total emissions is $N\bar{x} = 500\ (753) = 376{,}500$ g/100 miles with standard error $Ns(\bar{x}) = 500(192) = 96{,}000$ g/100 miles.

4.3 EFFECT OF MEASUREMENT ERRORS

Measurement uncertainties e_i have an effect on $\bar{x}$ and $s^2(\bar{x})$. Concerning $\bar{x}$, if the uncertainties are random with zero mean (i.e., no systematic tendency to over- or under-estimate the true values of population units), then $\bar{x}$ is still a statistically unbiased estimate of μ. If a constant measurement bias is present, then $\bar{x}$ will also have this same bias. However, estimates of differences over time or space remain unbiased if the measurement bias is the same magnitude for all units. The estimated variance, $s^2(\bar{x})$, is not affected by a constant additive measurement bias for all units since this formula involves computing $\Sigma_{i=1}^{n} (x_i - \bar{x})^2$, where both x_i and $\bar{x}$ contain the same additive bias term. However, $s^2(\bar{x})$ is affected by a measurement bias that changes in magnitude from one unit to another.

For the simple model given by Eq. 4.1, that is, $x_i = \mu_i + e_i$, the usual formula for $s^2(\bar{x})$ (Eq. 4.7) gives an unbiased estimate of $\text{Var}(\bar{x})$ if the uncertainties e_i are uncorrelated and if $f = n/N = 0$. This model is usually assumed in practice. However, if $f = 0$ and the errors have a positive (negative) correlation, $s^2(\bar{x})$ will tend to underestimate (overestimate) $\text{Var}(\bar{x})$, making $\bar{x}$ seem more (less) precise than it really is. If f is nonzero and large random or systematic measurement errors are present, then Eq. 4.7 will not reflect the total uncertainty in $\bar{x}$ because the factor $1 - f$ does not apply to those components of variance. Additional discussion is given by Cochran (1977, pp. 380–384).

4.4 NUMBER OF MEASUREMENTS: INDEPENDENT DATA

This section gives three approaches to determining n. Each approach is concerned with choosing n such that the estimated mean achieves some prespecified accuracy. These methods assume that the data are uncorrelated over time and space, which applies, in general, if the interval or distance measure between samples is sufficiently large. Section 4.5 gives methods for determining n when data are correlated.

4.4.1 Prespecified Variance

Suppose that the true mean μ will be estimated by $\bar{x}$ and that $\text{Var}(\bar{x})$ must be no larger than a prespecified value V. Setting $V = \text{Var}(\bar{x})$ in Eq. 4.5 and solving for n give

$$n = \frac{\sigma^2}{V + \sigma^2/N} \qquad \textbf{4.8}$$

Equation 4.8 reduces to

$$n = \frac{\sigma^2}{V} \qquad \textbf{4.9}$$

if N is very large relative to σ^2. Hence, if the variance σ^2 is known, it is easy to obtain n for any desired V. If σ^2 is not known, it may be estimated by taking an initial set of n_1 measurements and computing s^2 (Eq. 4.4). Cox (1952) showed that enough additional samples should be taken so that the final number of measurements is

$$n = \frac{s^2}{V}\left(1 + \frac{2}{n_1}\right)$$

This approximate result, which assumes $\bar{x}$ is normally distributed, is also given by Cochran (1977, p. 79). (The normal distribution is discussed in the next section and in Chapter 11.)

EXAMPLE 4.3

A study is planned to estimate the mean concentration of a pollutant in stream sediments in a defined area at a given point in time. The sediments have been sampled extensively in past years in this same area, so the variability in the measurements is known to be $\sigma^2 = 100$ (ppm)^2. We assume $\sigma^2/N = 0$, that is, N is very much larger than σ^2. Study objectives require that enough measurements be taken such that $\text{Var}(\bar{x})$ will be no larger than $V = 4$ (ppm)^2. Hence, by Eq. 4.9, approximately $n = \sigma^2/V = 100/4 = 25$ measurements are needed.

Now suppose no prior data are available and that $n_1 = 15$ data are collected and used in Eq. 4.4 to obtain $s^2 = 95$ as an estimate of σ^2. Then the total number of samples to take is

$$n = \frac{95}{4}\left(1 + \frac{2}{15}\right) = 26.9 \cong 27$$

Hence, 27 − 15 = 12 more units must be selected and measured.

4.4.2 Prespecified Margin of Error

Rather than specify V, the variance of $\bar{x}$, we can specify (1) the absolute margin of error d that can be tolerated, and (2) an acceptably small probability α of exceeding that error. The idea is to choose n such that

$$\text{Prob}\left[|\bar{x} - \mu| \geq d\right] \leq \alpha$$

where d and α are prespecified. For example, if $d = 10$ ppm and $\alpha = 0.05$, the value of n must be found such that there is only a $100\alpha = 5\%$ chance that the absolute difference (positive or negative) between the estimate $\bar{x}$ (obtained from the n data to be collected) and the true mean μ is greater than or equal to 10 ppm. If this approach is used, V in Eqs. 4.8 and 4.9 is replaced by $(d/Z_{1-\alpha/2})^2$, where $Z_{1-\alpha/2}$ is the standard normal deviate (discussed in Chapter 11) that cuts off $(100\alpha/2)\%$ of the upper tail of a standard normal distribution. Then Eqs. 4.8 and 4.9 become

$$n = \frac{(Z_{1-\alpha/2}\sigma/d)^2}{1 + (Z_{1-\alpha/2}\sigma/d)^2/N} \qquad \textbf{4.10}$$

and

$$n = (Z_{1-\alpha/2}\sigma/d)^2 \qquad \textbf{4.11}$$

respectively.

The motivation for replacing V by $(d/Z_{1-\alpha/2})^2$ comes from the confidence interval about μ given by Eq. 11.6. That equation gives $\mu - \bar{x} = Z_{1-\alpha/2}\sqrt{V}$, or $(d/Z_{1-\alpha/2})^2 = V$, if we let $d = \mu - \bar{x}$.

Table A1 in the appendix gives values for $Z_{1-\alpha/2}$ for various α. Some selected values from this table for commonly used values of α are

α:	0.20	0.10	0.05	0.01
$Z_{1-\alpha/2}$:	1.2816	1.6449	1.960	2.5758

Equations 4.10 and 4.11 are exact only if $\bar{x}$ is normally distributed, which occurs if the data themselves are normally distributed or if n is sufficiently large.

If the data are approximately normally distributed, but σ^2 is not known with assurance, then the t distribution is used in place of the standard normal distribution. That is, $t_{1-\alpha/2, n-1}$ is used in place of $Z_{1-\alpha/2}$, where $t_{1-\alpha/2, n-1}$ is the value of the t variate that cuts off $(100\alpha/2)\%$ of the upper tail of the t distribution with $n - 1$ degrees of freedom. Hence, Eq. 4.10 becomes

$$n = \frac{(t_{1-\alpha/2, n-1}\,\sigma/d)^2}{1 + (t_{1-\alpha/2, n-1}\,\sigma/d)^2/N}$$

Values of $t_{1-\alpha/2, n-1}$ are given in Table A2.

Since $t_{1-\alpha/2, n-1}$ depends on n, an iterative procedure is used. First, use Eq. 4.10 or 4.11 (depending on the size of N) to determine the initial value of n; call it n_1. ($Z_{1-\alpha/2}$ is used at this first stage since no prior value for n is available for entry into the t tables.) Find $t_{1-\alpha/2, n_1-1}$ in Table A2 and use it in place of $Z_{1-\alpha/2}$ in Eq. 4.10 (or 4.11) to give a value of n; call it n_2. Then $t_{1-\alpha/2, n_2-1}$ is determined and used in Eq. 4.10 (or 4.11) to find n_3, and so on. After a few iterations the value of n will stabilize to the final value.

EXAMPLE 4.4

Suppose we need to estimate the mean concentration of a pollutant in stream sediments, and we can accept a 10% chance (i.e., $\alpha = 0.10$) of getting a set of data for which $d = |\bar{x} - \mu| \geq 20$ μg/L. Suppose a pilot study suggests $\sigma = 50$ μg. If N is very large, Eq. 4.10 gives

$$n_1 = \left[\frac{1.645(50)}{20}\right]^2 = 16.9 \cong 17$$

From Table A2, $t_{0.95, 16} = 1.746$, which when used in place of $Z_{1-\alpha/2}$ in Eq. 4.10 gives

$$n_2 = \left[\frac{1.746(50)}{20}\right]^2 = 19.05 \cong 19$$

Then $t_{0.95, 18} = 1.734$ gives

$$n_3 = \left[\frac{1.734(50)}{20}\right]^2 = 18.8 \cong 19$$

Since $n_3 = n_2$, no further iteration is needed and $n_3 = n = 19$.

4.4.3 Prespecified Relative Error

In some cases a reliable value for σ^2 will not be available for determining n. However, an estimate of the relative standard deviation $\eta = \sigma/\mu$ (the coefficient of variation), may be available. This quantity is usually less variable from one study site or time period to another than σ^2 is. The approach is to specify the *relative* error $d_r = |\bar{x} - \mu|/\mu$ such that

$$\text{Prob}\ [|\bar{x} - \mu| \geq d_r\mu] = \alpha$$

Replacing d with $d_r\mu$ in Eqs. 4.10 and 4.11 gives

$$n = \frac{(Z_{1-\alpha/2}\eta/d_r)^2}{1 + (Z_{1-\alpha/2}\eta/d_r)^2/N}$$

and

$$n = (Z_{1-\alpha/2}\eta/d_r)^2 \qquad \textbf{4.12}$$

respectively, where $\eta = \sigma/\mu$ must be prespecified.

Table 4.1 gives values of n computed by Eq. 4.12 for several values of α, d_r, and η. Large values of n result when η or $1 - \alpha$ is large or when d_r is small. For a given coefficient of variation, one can determine from the table those combinations of α and d_r that result in an n that can be accommodated within funding limitations. Alternatively, if α and d_r are prespecified, then η will determine n. The iterative procedure described in the previous section can also be used. The results in Table 4.1 assume uncorrelated data. If data collected in close proximity in time or space have a positive (negative) correlation, then the values in Table 4.1 are too small (large). Section 4.5 discusses how to determine n when data are correlated.

EXAMPLE 4.5

Consider again Example 4.3 where the goal is to estimate the mean concentration of a pollutant in stream sediments. Suppose we can accept a 10% chance of getting a set of data for which the relative

Table 4.1 Sample Sizes Required for Estimating the True Mean μ

Confidence Level $(1 - \alpha)$	*Relative Error* d_r	*Coefficient of Variation* (η)				
		0.10	0.50	1.00	1.50	2.00
0.80	0.10	2	42	165	370	657
($Z_{0.90} = 1.2816$)	0.25	—	7	27	60	106
	0.50	—	2	7	15	27
	1	—	—	2	4	7
	2	—	—	—	—	2
	5	—	—	—	—	—
0.95	0.10	4	97	385	865	1,537
($Z_{0.975} = 1.96$)	0.25	—	16	62	139	246
	0.50	—	4	16	35	62
	1	—	—	4	9	16
	2	—	—	—	3	4
	5	—	—	—	—	—

error d_r exceeds 20%. Hence, $d_r = 0.2$ and $Z_{1-0.10/2} = 1.645$. Suppose also that $\eta = 0.40$. Then Eq. 4.12 gives

$$n_1 = \left[\frac{1.645(0.40)}{0.20}\right]^2 = 10.8 \cong 11$$

Using the iterative approach, we get $t_{0.95,10} = 1.812$, so

$$n_2 = \left[\frac{1.812(0.40)}{0.20}\right]^2 = 13.1 \cong 13$$

Another iteration gives

$$n_3 = \left[\frac{1.782(0.40)}{0.20}\right]^2 = 12.7 \cong 13$$

Since $n_3 = n_2$, we stop with $n = 13$.

For many environmental pollutants the coefficient of variation will be much larger than the 0.40 used in the preceding example. If $d_r = 0.2$ and $\alpha = 0.10$ but η is increased to 1.0, the reader may verify that the iterative procedure gives $n = 70$, which is about five times larger than when $\eta = 0.40$.

4.4.4 Two-Stage Procedure

To find n using Eqs. 4.9, 4.11, and 4.12, we must have estimates of σ^2 or $\eta = \sigma/\mu$. We can obtain these estimates by one of the following ways:

1. Collect preliminary data from the population to approximate σ^2 or η.
2. Estimate σ^2 or η from data collected from the same population at a prior time or on a population from a similar study site.
3. Use best judgment when reliable data are not available.

Option 1 is effective if there is time to take preliminary measurements at the study site. These measurements should be made under the same environmental conditions and be the same type as in the main study. Option 2 is useful if the variability of measurements obtained at a prior time or at a similar site is representative of current conditions. Option 3 depends entirely on expert opinion without using quantitative information. In most situations quantitative information will be available or can be obtained, so options 1 and 2 can be used.

If very little data are available, an educated guess for the coefficient of variation, η, might be made, and a provisional value of n, say n_1, can be obtained by using Eq. 4.12. Then n_1 or fewer data could be collected and used to compute $\bar{x}$ and $s/\bar{x}$, the latter being an estimate of η. Then n_2 is obtained with $s/\bar{x}$ in place of η in Eq. 4.12. If $n_2 > n_1$, the number of additional data to collect is $n_2 - n_1$. If $n_2 \leq n_1$, then no additional data are needed. The iterative approach using the t distribution could also be used if the data are at least approximately normally distributed.

4.5 NUMBER OF MEASUREMENTS: CORRELATED DATA

4.5.1 Spatial Correlation

The methods for estimating n in previous sections assume measurements are uncorrelated. In practice, a spatial correlation may be present so that part of the information contained in one measurement is also in other measurements taken close by in space. A method by Nelson and Ward (1981) to determine n in this situation is now given.

Suppose n observations are taken at each of n_s stations over the same time period to estimate the regional mean for that period. The regional mean is estimated by

$$\bar{x} = \frac{1}{nn_s} \sum_{i=1}^{n_s} \sum_{j=1}^{n} x_{ij} = \frac{1}{n_s} \sum_{i=1}^{n_s} \bar{x}_i \qquad \textbf{4.13}$$

where x_{ij} is the jth observation at station i and $\bar{x}_i$ is the estimated mean for the ith station. Suppose the variance over time, σ^2, is the same at each station and that the data are correlated over space but not over time. Assume the true station means, μ_i, are equal. Then the variance of the estimated regional mean (from Matalas and Langbein, 1962) is

$$\text{Var}(\bar{x}) = \frac{\sigma^2}{nn_s} [1 + \rho_c(n_s - 1)] \qquad \textbf{4.14}$$

where ρ_c is the average of the $n_s(n_s - 1)/2$ cross-correlations between the n_s stations, where $-1/(n_s - 1) \leq \rho_c \leq 1$.

For this situation the generalization of Eq. 4.11 is

$$nn_s = (Z_{1-\alpha/2}\sigma/d)^2[1 + \rho_c(n_s - 1)] \qquad \textbf{4.15}$$

If n_s is established a priori, then Eq. 4.15 is used to determine n, so the regional mean is estimated with an accuracy and confidence indicated by the prespecified values of d and $Z_{1-\alpha/2}$. If n is prespecified, Eq. 4.15 can be used to determine n_s. If the x_i are quite variable, replace σ^2 in Eq. 4.14 with $\sigma^2 + n\sigma_1^2$ and solve for n, where σ_1^2 is the variance of the station means.

In practice, ρ_c will usually be greater than zero. Theoretically, ρ_c could be negative, which would result in taking fewer measurements than if ρ_c was 0 or positive. If the estimate $\hat{\rho}_c$ is negative, a conservative approach would be to assume ρ_c equals zero or to use the absolute value of $\hat{\rho}_c$. This would guard against taking too few measurements.

EXAMPLE 4.6

Suppose we are interested in estimating the average concentration in groundwater in a defined region, using 3 established wells in the region. Samples have been collected concurrently every month at the 3 wells over a two-year period, yielding 24 measurements per well. Suppose, due to the location of the wells, they tend to give redundant information at each sampling time. However, we suppose there is little or no correlation between measurements made in different months. We wish to determine whether monthly sampling is required to estimate the annual regional mean nitrate concentration with accuracy of at least $d = |\bar{x} - \mu| = 20$ mg/L. Suppose we can accept a 10% chance of getting a data set for which d exceeds 20 mg/L. That is, $\alpha = 0.10$ and $Z_{0.95} = 1.645$. We estimate the pooled σ as follows:

$$\hat{\sigma} = \left[\frac{1}{\sum_{i=1}^{n_s} (n_i - 1)} \sum_{i=1}^{n_s} (n_i - 1) s_i^2 \right]^{1/2} \qquad \textbf{4.16}$$

where $n_s = 3$, s_i^2 is the estimated variance for well i (computed by Eq. 4.4) and $n_i = 24$ for each well. Suppose Eq. 4.16 gives $\hat{\sigma} = 60$ and that $\hat{\rho}_{12} = 0.40$, $\hat{\rho}_{13} = 0.25$, and $\hat{\rho}_{23} = 0.70$, so their average is $\hat{\rho}_c = 0.45$. The estimates $\hat{\rho}_{ii'}$ are calculated by the usual formula, as given by, for example, Snedecor and Cochran (1980, p. 175). For example, for wells 1 and 2,

$$\hat{\rho}_{12} = \frac{\sum_{j=1}^{24} (x_{1j} - \bar{x}_1)(x_{2j} - \bar{x}_2)}{\sqrt{\sum_{j=1}^{24} (x_{1j} - \bar{x}_1)^2 \sum_{j=1}^{24} (x_{2j} - \bar{x}_2)^2}}$$

where $\bar{x}_i = \sum_{j=1}^{24} x_{ij}/24 =$ mean for ith well, $i = 1, 2, 3$. Solving Eq. 4.15 for n gives

$$n = \left(\frac{1.645(60)}{20} \right)^2 \frac{[1 + 0.45(3 - 1)]}{3}$$

$$= \frac{24.35(1.9)}{3} = 15.4 \cong 16$$

Therefore, monthly sampling is not frequent enough to obtain an estimate of the annual regional mean within 20 mg/L of the true value with 90% confidence. If the spatial correlation had not been taken into account, we would have obtained $n = 24.35/3 \cong 8$, a spurious conclusion. The reader may confirm that monthly sampling would be adequate to achieve the desired 20 mg/L accuracy with 80% confidence.

EXAMPLE 4.7

Suppose we redo Example 4.6, keeping everything the same except that $n_s = 15$ wells, instead of 3 wells, are present in the region. Again, we want to determine whether monthly sampling is required to estimate the annual regional mean with accuracy $d = 20$ units and $100(1 - 0.10) = 90\%$ confidence. Solving Eq. 4.15 for n gives

$$n = \frac{24.35[1 + 0.45(15 - 1)]}{15} = 11.85 \cong 12$$

Hence, with 15 wells, monthly sampling is necessary and sufficient.

In Example 4.7 we specified $n_s = 15$ stations were to be sampled and we found $n = 12$ so that $nn_s = (12)(15) = 180$. In Example 4.6 only $n_s = 3$ stations were to be sampled and n was found to be 16 so that $nn_s = 16(3) = 48$ measurements were needed. In both cases we specified $d = 20$ and $\alpha = 0.10$, so why does nn_s differ for the two cases? The answer lies in the positive correlation ρ_c between stations and the a priori specification of n_s. Since the data from different stations are positively correlated, adding more stations does not result in proportionally more information for estimating the regional mean. Examples 4.6 and 4.7 show there is as much information obtained for estimating the regional mean by sampling each of the 3 stations 16 times as there is by sampling each of the 15 stations 12 times. Because of the correlation ρ_c, sampling the additional 12 stations adds little information for estimating the regional mean. In practice, when the number of stations, n_s, is prespecified, we need to estimate the spatial correlation ρ_c and use Eq. 4.15 instead of Eq. 4.11 to determine n.

Equation 4.15 can also be used to determine the number n_s of stations for an a priori specified number n of measurements over time, if we still assume no correlation over time. If Eq. 4.15 is solved for n_s, we obtain

$$n_s = \frac{(Z_{1-\alpha/2}\sigma/d)^2(1 - \rho_c)}{n - (Z_{1-\alpha/2}\sigma/d)^2\rho_c} \qquad \textbf{4.17}$$

where n must be greater than $(Z_{1-\alpha/2}\sigma/d)^2\rho_c$. The use of Eq. 4.17 is illustrated in the following example.

EXAMPLE 4.8

Suppose on the basis of past studies we have the estimates $\hat{\sigma} = 60$ and $\hat{\rho}_c = 0.45$. Further suppose we chose d = 20 mg/L and $\alpha = 0.10$ as in Examples 4.6 and 4.7. Suppose also that we want an annual regional mean and it will be based on $n = 12$ measurements during the year, say one each month. Equation 4.17 gives

$$n_s = \frac{24.35(0.55)}{12 - 24.35(0.45)} = 12.8 \cong 13$$

But suppose that $\hat{\rho}_c = 0.3$ instead of 0.45. Then the reader may verify, using Eq. 4.17, that only $n_s = 4$ stations are needed with monthly sampling. Reducing $\hat{\rho}_c$ still further to zero (no spatial correlation between stations) results in only $n_s = 2$ stations being required. Hence, as $\hat{\rho}_c$ decreases to zero and the number of samples per station remains constant, the number of stations can be reduced and still provide the same amount of information needed to estimate the regional mean with the same prespecified accuracy and confidence.

We have assumed that $\hat{\rho}_c$ is an accurate estimate of the true ρ_c. If this is not true, our estimates of n and n_s may be far from the values that will assure that the a priori accuracy and confidence requirements are met.

4.5.2 Time Correlation

Section 4.5.1 showed how to estimate $\mathrm{Var}(\bar{x})$ and determine the number of measurements to take for estimating $\bar{x}$ when data are correlated over space but not over time. This section shows how to do these things when data are correlated over time but not over space. In Section 4.5.3 both space and time correlation are considered. See Hirtzel and Quon (1979) and Hirtzel, Quon, and Corotis (1982*a*, 1982*b*) for additional information on estimating the mean or maximum of correlated air quality measurements.

Let $x_1, x_2, \ldots, x_n$ denote a sequence of measurements taken at equal intervals along a line in time or space. This section gives a method for estimating the number, n, of these measurements needed at a given station to estimate the true mean μ over a specified time (or space) interval with prespecified accuracy and confidence. Since measurements are collected at equal intervals, we do not require in this section (or Section 4.5.3) that samples be collected by simple random sampling. We do this because the efficient estimation of autocorrelation (see next paragraph) requires equally spaced samples. Collecting samples at equal intervals is an example of systematic sampling, the topic of Chapter 8.

Let ρ_1 be the true correlation of the population of measurements taken 1 lag apart, that is, between measurements x_1 and x_2, x_2 and x_3, x_3 and x_4, and so forth. Similarly, let ρ_2 be the true correlation between measurements 2 lags apart (x_1 and x_3, x_2 and x_4, etc.) and, in general, ρ_l the true correlation between measurements l lags apart, where $-1 \le \rho_l \le 1$. The set $\{\rho_1, \rho_2, \ldots\}$ is called the *autocorrelation function*, and a plot of ρ_l against l for all lags l is the *correlogram*.

We assume that the underlying process being measured does not cycle, does not have long-term trends, and does not make sudden jumps in magnitude or change its autocorrelation function over time. That is, assume that the mean, variance, and serial correlations ρ_l for any segment of the time series do not change as we move along the series—that is, the time series is weakly stationary. If these assumptions are unreasonable in a given situation, then it may not be meaningful to estimate the mean for the entire population. It would be better to divide the population into more homogeneous subpopulations or to take measurements to determine the patterns of variability that are present.

The sample mean $\bar{x} = \Sigma_{i=1}^{n} x_i/n$ is used to estimate the true mean μ. If the foregoing model is appropriate, the variance of $\bar{x}$ is

$$\begin{aligned}\mathrm{Var}(\bar{x}) &= \frac{\sigma^2}{n}\left[1 + \frac{2}{n}\sum_{l=1}^{n-1}(n-l)\rho_l\right] \\ &= \frac{\sigma^2}{n}\left\{1 + \frac{2}{n}\left[(n-1)\rho_1 + (n-2)\rho_2 + \cdots + 2\rho_{n-2} + \rho_{n-1}\right]\right\}\end{aligned} \tag{4.18}$$

where σ^2 is the variance of the population of measurements. (A proof of Eq. 4.18 is given by Loftis and Ward, 1979.) Note that Eq. 4.18 reduces to σ^2/n, the usual formula for $\mathrm{Var}(\bar{x})$ when all ρ_l equal zero.

Now, if all the ρ_l are zero, we know from Eq. 4.11 that the number of samples, n, required to estimate $\bar{x}$ with prespecified accuracy $d = |\bar{x} - \mu|$ and $100(1 - \alpha)\%$ confidence is

$$n = (Z_{1-\alpha/2}\sigma/d)^2$$

which is obtained by solving the equation

$$\left(\frac{d}{Z_{1-\alpha/2}}\right)^2 = \frac{\sigma^2}{n} \qquad \textbf{4.19}$$

for n. When $\text{Var}(\bar{x})$, given by Eq. 4.18, is substituted into Eq. 4.19, we obtain a quadratic equation that can be solved for n by using the quadratic formula. Doing so gives

$$n = \frac{D}{2}\left\{1 + 2\sum_{l=1}^{n-1}\rho_l + \left[\left(1 + 2\sum_{l=1}^{n-1}\rho_l\right)^2 - \frac{8}{D}\sum_{l=1}^{n-1} l\rho_l\right]^{1/2}\right\} \qquad \textbf{4.20}$$

where $D = (Z_{1-\alpha/2}\sigma/d)^2$. The following approximate expression is obtained by ignoring the term in the quadratic equation that has n^2 in the denominator:

$$n = D\left(1 + 2\sum_{l=1}^{n-1}\rho_l\right) \qquad \textbf{4.21}$$

We note that, since $Z_{1-\alpha/2}$ is used in Eqs. 4.20 and 4.21, the measurements x_i are assumed to be normally distributed. Example 4.9 illustrates the computation of Eqs. 4.20 and 4.21.

Equations 4.18, 4.20, and 4.21 require knowledge about the magnitude of the autocorrelation coefficients ρ_l. These coefficients may be estimated from measurements from prior studies obtained at equal intervals (along a line in time or space). If $x_1, x_2, \ldots, x_n$ denotes such a series, Box and Jenkins (1976) indicate that the most satisfactory method of estimating the lth lag autocorrelation, ρ_l, is

$$\hat{\rho}_l = \frac{\sum_{t=1}^{n-l}(x_t - \bar{x})(x_{t+l} - \bar{x})}{\sum_{t=1}^{n}(x_t - \bar{x})^2} \qquad \textbf{4.22}$$

where $\bar{x}$ is the mean of the n data. [Wallis and O'Connell (1972) compare the estimation bias of Eq. 4.22 with that of competing estimators when n is small.] If one or more of the $\hat{\rho}_l$ are negative it is suggested that they be set equal to zero or to their absolute values for the purpose of determining n using Eqs. 4.20 and 4.21 (or Eqs. 4.24, 4.25, and 4.26).

To illustrate the mechanics of calculating Eq. 4.22, consider the time series of five measurements 5, 3, 7, 4, and 10 collected one per month for five consecutive months. We find $\bar{x} = 5.8$. Hence, the estimated autocorrelation between units $l = 2$ units apart is

$$\hat{\rho}_2 = \frac{(5 - 5.8)(7 - 5.8) + (3 - 5.8)(4 - 5.8) + (7 - 5.8)(10 - 5.8)}{(5 - 5.8)^2 + (3 - 5.8)^2 + (7 - 5.8)^2 + (4 - 5.8)^2 + (10 - 5.8)^2}$$
$$= 0.30$$

Estimates $\hat{\rho}_1$, $\hat{\rho}_3$, and $\hat{\rho}_4$ are obtained similarly. In practice, at least $n = 50$ measurements are needed to obtain accurate estimates of the ρ_l for lags $l = 1, 2, \ldots, K$, where K should not exceed $n/4$.

The use of Eqs. 4.20 and 4.21 also requires that we estimate σ^2. The usual estimator s^2 given by Eq. 4.4 may be used for this purpose. This estimator is known to be biased unless the ρ_l are zero (see Scheffé, 1959, p. 338), but the bias will be near zero if the number of data used to compute s^2 is sufficiently large. A suggested alternative estimator of σ^2 that takes into account the correlations ρ_l is given in Exercise 4.4.

EXAMPLE 4.9

Determine the number n of equally spaced daily measurements needed over a 12-month period at a given sampling station to estimate the annual mean at that location with accuracy $d = 20$ units and $100(1 - 0.10) = 90\%$ confidence. Suppose that past data collected daily over several months are normally distributed and give no indication of cycles, trends, or changing variance over time. Suppose these data are used to estimate σ^2 by using Eq. 4.4, which, after taking the square root, gives $s = 60$.

Finally, suppose that Eq. 4.22 is used to estimate ρ_1, ρ_2, and ρ_3, yielding 0.6, 0.4, and 0.2, respectively, and that $\rho_l = 0$ for $l > 3$. Then Eq. 4.21 gives

$$n = 24.354(1 + 2.4) = 82.80 \cong 83$$

Using Eq. 4.20, the exact equation for n, we obtain

$$\begin{aligned} n &= 0.5(24.354)\left\{3.4 + \left[(3.4)^2 - \frac{8(2)}{24.354}\right]^{1/2}\right\} \\ &= 81.6 \cong 82 \end{aligned}$$

The approximation (Eq. 4.21) works well in this case. Taking a measurement every fourth day during the year would result in 91 measurements, more than is required to achieve the desired accuracy and confidence in the annual mean.

4.5.3 Space and Time Correlations

This section considers data collected at multiple stations at equal time intervals that are both spatially and serially correlated. We give expressions for estimating the variance of the estimated regional mean and for determining the number n of data at each station as well as the number n_s of stations.

Let $\bar{x}$ be the estimated regional mean as defined by Eq. 4.13. Furthermore, suppose the correlations over time, ρ_l, $l = 1, 2, \ldots$, are the same at all stations and that the time process is weakly stationary, as assumed in the previous section. Now, suppose we replace σ^2/n in Eq. 4.18 by the variance of the mean of spatially correlated data (Eq. 4.14). Then if the same number, n, of observations is taken at each station, the variance of the regional mean is given by

$$\text{Var}(\bar{x}) = \frac{\sigma^2}{nn_s}\left[1 + \frac{2}{n}\sum_{l=1}^{n-1}(n - l)\rho_l\right][1 + \rho_c(n_s - 1)] \qquad \textbf{4.23}$$

If all the ρ_l equal zero, then Eq. 4.23 reduces to Eq. 4.14. Eq. 4.23 reduces to Eq. 4.18 when $n_s = 1$. If station means, $\bar{x}_i$, are quite variable, use $\sigma^2 + n\sigma_1^2$.

By using data from prior or similar studies to estimate σ^2 and the serial correlations ρ_l, $l = 1, 2, \ldots$, and using these estimates in Eq. 4.23, we can estimate the variance of the regional mean.

Using the general procedure in Section 4.5.2 (following Eq. 4.18), and assuming $\sigma_1^2 = 0$, we obtain the following equation for calculating the number of measurements at each station:

$$n = \frac{D}{2n_s}\left[1 + 2\sum_{l=1}^{n-1}\rho_l + \rho_c(n_s - 1)\left(1 + 2\sum_{l=1}^{n-1}\rho_l\right)\right] + \frac{D}{2}\left\{\frac{1}{n_s^2}\left[1 + 2\sum_{l=1}^{n-1}\rho_l + \rho_c(n_s - 1)\left(1 + 2\sum_{l=1}^{n-1}\rho_l\right)\right]^2 - \frac{8}{n_s D}\left[\sum_{l=1}^{n-1} l\rho_l - \rho_c(n_s - 1)\sum_{l=1}^{n-1} l\rho_l\right]\right\}^{1/2} \quad 4.24$$

where $D = (Z_{1-\alpha/2}\sigma/d)^2$. An approximate expression for n is

$$n \cong \frac{D}{n_s}\left(1 + 2\sum_{l=1}^{n-1}\rho_l\right)[1 + \rho_c(n_s - 1)] \quad 4.25$$

Equations 4.24 and 4.25 reduce to Eqs. 4.20 and 4.21, respectively, if $n_s = 1$. If $\sigma_1^2 > 0$, Eqs. 4.24 and 4.25 are lower bounds on n.

Equation 4.25 may be solved for n_s, giving

$$n_s \cong \frac{D\left(1 + 2\sum_{l=1}^{n-1}\rho_l\right)(1 - \rho_c)}{n - D\left(1 + 2\sum_{l=1}^{n-1}\rho_l\right)\rho_c} \quad 4.26$$

which reduces to Eq. 4.17 when all ρ_l are zero.

EXAMPLE 4.10

Suppose enough equally spaced measurements, n, are to be taken at each of $n_s = 3$ stations to estimate the regional mean $\bar{x}$ with accuracy $d = 20$ units and $100(1 - 0.10) = 90\%$ confidence. Suppose further that data from prior studies have given the estimates $\hat{\sigma} = 60$, $\hat{\rho}_c = 0.45$, $\hat{\rho}_1 = 0.6$, $\hat{\rho}_2 = 0.4$, $\hat{\rho}_3 = 0.2$, and that $\rho_l = 0$ for lags greater than 3. Using Eq. 4.24, we obtain

$$n = \frac{(24.354)}{2(3)}[3.4 + 0.45(2)3.4] + \frac{24.354}{2}\left\{\frac{1}{9}[3.4 + 0.45(2)3.4]^2 - \frac{8}{3(24.354)}[2 - 0.45(2)2]\right\}^{1/2}$$

$$= 52.38 \cong 53$$

The approximate equation for n, Eq. 4.25, gives

$$n = \frac{24.354(3.4)}{3}[1 + 0.45(2)] = 52.44 \cong 53$$

the same result.

Example 4.6 is identical to this example except that no serial correlation was present. In Example 4.6 only 16 measurements at each of three stations were required, much fewer than the 53 required here where positive spatial and temporal correlation is present. This example illustrates again that if data have positive correlation, there is less information per measurement than if the measurements are not correlated. Consequently, more measurements are required to estimate a mean with specified accuracy and confidence.

4.6 ESTIMATING Var($\bar{x}$)

Sections 4.5.1–4.5.3 were primarily concerned with determining n or n_s for a future study. However, once data have been collected that are spatially and/or temporally correlated, an estimate of Var($\bar{x}$) will be needed to evaluate the uncertainty in $\bar{x}$. The quantity Var($\bar{x}$) can be estimated by using Eqs. 4.14, 4.18, or 4.23, whichever is appropriate, if estimates of σ^2, ρ_c, and the serial correlations ρ_l are available. We have already indicated how these estimates can be obtained if enough data have been collected. The estimates of ρ_c and the ρ_l are used when estimating Var($\bar{x}$), even if they are negative. Chapter 11 shows how to use estimates of Var($\bar{x}$) for putting confidence limits about the true mean μ.

Now we give a method for estimating the variance of an estimated regional mean when the observations at a given station are correlated over time, but there is no correlation between stations. This method is simpler than using Eq. 4.18 because it is not necessary to estimate the serial correlations ρ_l.

Suppose n random observations over time have been made at each of n_s monitoring stations. The regional mean may be estimated by using Eq. 4.13 to compute $\bar{x}$. Now, the variance of $\bar{x}$ may be estimated by computing the variance of the station means $\bar{x}_i$:

$$s^2(\bar{x}) = \frac{1}{n_s(n_s - 1)} \sum_{i=1}^{n_s} (\bar{x}_i - \bar{x})^2 \qquad \mathbf{4.27}$$

where $\bar{x}_i = \Sigma_{j=1}^{n} x_{ij}/n$. An advantage of Eq. 4.27 is that temporal correlation does not invalidate its use. Also, the observations at each station may be made systematically—for example, at equal intervals—rather than at random times. The potential benefits and pitfalls of systematic sampling are considered in Chapter 8. Equation 4.27 is the same as Eq. 6.7, which concerns two-stage sampling.

4.7 SUMMARY

This chapter begins by discussing the basic concepts of simple random sampling and estimating the mean and total as well as the variance of these estimates. Then complications that arise when measurement errors are large and correlated are discussed. Methods are given for estimating the number of samples to collect for estimating a mean when the data are uncorrelated or correlated. Also given are equations for estimating $\text{Var}(\bar{x})$ when data are correlated over space or time. These are used in Chapter 11 to compute confidence limits about the true mean.

EXERCISES

4.1 In Example 4.2, N was equal to 500 vehicles. Suppose instead that the target population is $N = 1{,}000{,}000$ vehicles, the total production for a year. Assume the $n = 5$ measurements are a random sample from this larger population, and use the data to compute $s(\bar{x})$. Is this estimate very different from that in Example 4.2? Now, compute $s(\bar{x})$, assuming $N = 50$, and note the amount of change.

4.2 Repeat the iterative computations in Example 4.5 to obtain n, except assume $\eta = 0.20$.

4.3 Doctor and Gilbert (1978) report count-per-minute readings of ^{241}Am in surface soil taken with a handheld detector at 100 adjacent locations along a line transect. The estimated serial correlations for lags 1, 2, . . . , 14 for measurements made 1 ft off the ground surface are (reading left to right)

0.929	0.812	0.646	0.487	0.345	0.250	0.199
0.187	0.184	0.186	0.162	0.122	0.074	0.029

respectively. The sample mean and standard deviation of the $n = 100$ data are $\bar{x} = 18{,}680$ and $s = 4030$. Estimate the standard error of $\bar{x}$, (i) assuming the data are uncorrelated, and (ii), using Eq. 4.18 and the previous correlations, assuming all serial correlations at lags greater than 14 are zero.

4.4 If data are normally distributed and correlated, then it can be shown that the mathematical expected value of s^2 is

$$E(s^2) = \sigma^2\left[1 - \frac{2}{n(n-1)}\sum_{l=1}^{n-1}(n-l)\rho_l\right]$$

This equation suggests using the following estimator for σ^2:

$$\hat{\sigma}^2 = \frac{s^2}{1 - \dfrac{2}{n(n-1)}\displaystyle\sum_{l=1}^{n-1}(n-l)\rho_l}$$

Repeat the computations of part (ii) in Exercise 4.3, except use the preceding estimator for σ^2. Is $s(\bar{x})$ changed very much from that obtained in Exercise 4.3?

ANSWERS

4.1 When $N = 1{,}000{,}000$, $f = n/N = 5/1{,}000{,}000 = 5 \times 10^{-6}$.

$$s(\bar{x}) = 432.4\sqrt{\frac{1 - 5 \times 10^{-6}}{5}} = 193$$

When $N = 50$, $f = 5/50 = 0.10$.

$$s(\bar{x}) = 432.4\sqrt{\frac{0.9}{5}} = 183$$

4.2 $n = [1.645(0.20)/0.20]^2 = 2.706 \cong 3$.

$$t_{0.95,2} = (2.920)^2 = 8.53 \cong 9 = n_2$$

$$t_{0.95,8} = (1.86)^2 = 3.5 \cong 4 = n_3$$

$$t_{0.95,3} = (2.353)^2 = 5.54 \cong 6 = n_4$$

$$t_{0.95,5} = (2.015)^2 = 4.06 \cong 5 = n_5$$

$$t_{0.95,4} = (2.132)^2 = 4.55 \cong 5 = n_6$$

4.3

(i) $s(\bar{x}) = 4030/10 = 403$

(ii) $$s(\bar{x}) = \left[\frac{(4030)^2}{100}\{1 + 0.02[99(0.929) + 98(0.812) + \cdots + 87(0.074) + 86(0.029)]\}\right]^{1/2}$$

$$= 1260$$

4.4

$$\hat{\sigma} = \frac{4030}{\left[1 - \dfrac{2}{100(99)}\,440.5\right]^{1/2}} = 4222.2$$

$$s(\bar{x}) = \frac{4222.2}{10}(9.81)^{1/2} = 1320$$

which is about 5% larger than $s(\bar{x})$ obtained in part (ii) of Exercise 4.3.

5 Stratified Random Sampling

Stratified random sampling is a useful and flexible design for estimating average environmental pollution concentrations and inventories (total amounts). The method makes use of prior information to divide the target population into subgroups that are internally homogeneous. Each subgroup (stratum) is then sampled by simple random sampling to estimate its average or total. These stratum estimates may be the primary focus of the study. However, if the mean and total for the *entire* population are needed, the stratum results may be combined as described in this chapter to give more precise estimates than would be achieved if stratification is not used. Applications of stratified random sampling to environmental studies are found in Jensen (1973), Liebetrau (1979), Environmental Protection Agency (1982), Thornton et al. (1982), Nelson and Ward (1981), Green (1968), Reckhow and Chapra (1983), and Gilbert et al. (1975).

This chapter shows how to estimate the mean and total amount of pollutant in a population composed of or divided into strata, the standard error of the estimated mean and total, and the optimum allocation of samples to the strata under constraints of fixed cost or required precision of estimation. Examples and a case study are given to illustrate these procedures.

5.1. BASIC CONCEPTS

Let N be the total number of population units in the target population. These N units are divided into L nonoverlapping strata such that the variability of the phenomenon *within* strata is less than that over the entire population of N units. Let the number of population units in the L strata be denoted by N_1, N_2, . . . , N_L. The weight of the hth stratum is $W_h = N_h/N$. These weights and the N_h are assumed to be known before any sampling takes place. Once the L strata are defined, a simple random sample of units is drawn from each stratum by the procedures described in Section 4.1. The number of units measured in the hth stratum is denoted by n_h, with $n = \Sigma_{h=1}^{L} n_h$ being the total number of units measured over all L strata. The objective is to first estimate the true mean value and/or total amount (inventory) of the pollutant in each stratum. A weighted average of these stratum means by using the weights W_h is computed to estimate the average and/or inventory over the entire population of N units.

If the stratification has been effective in creating relatively homogeneous strata, then the estimated average and inventory for the entire population of N units should be more accurate than would be obtained if a simple random sample had been collected from the N units without first stratifying the population.

EXAMPLE 5.1

We want to estimate the average amount of a contaminant ingested by a specified population that eats a particular food. The population consists of N men, women, and children; men and women eat about the same amount of this food each day, but children eat less. Two groups (strata) are defined: stratum 1 consists of N_1 adult men and women, and stratum 2 consists of N_2 children, and $N = N_1 + N_2$. A simple random sample of individuals is drawn independently from both strata: n_1 in stratum 1 and n_2 in stratum 2. These $n_1 + n_2 = n$ individuals are interviewed, and the data are used to estimate the average amount of food ingested per day per individual for each stratum. Then the average μ over the population of N individuals is estimated by computing the weighted average of the stratum means. The weights for strata 1 and 2 are N_1/N and N_2/N, respectively.

5.2 ESTIMATING THE MEAN

This section gives equations for estimating the stratum means μ_h, the overall mean μ, and the standard errors of the estimated means. These equations and those used in Section 5.3 through 5.6 are from Cochran (1977). The mean μ of the population of N units is

$$\mu = \frac{1}{N} \sum_{h=1}^{L} N_h \mu_h = \sum_{h=1}^{L} W_h \mu_h \qquad 5.1$$

Hence, μ is the weighted average of the true stratum means μ_h, where

$$\mu_h = \frac{1}{N_h} \sum_{i=1}^{N_h} \mu_{hi}$$

and where μ_{hi} is the true value for the ith unit in the hth stratum.

The stratum mean μ_h is estimated by randomly selecting n_h units from stratum h and computing

$$\bar{x}_h = \frac{1}{n_h} \sum_{i=1}^{n_h} x_{hi} \qquad 5.2$$

An unbiased estimator of μ is

$$\bar{x}_{st} = \sum_{h=1}^{L} W_h \bar{x}_h \qquad 5.3$$

Note that $\bar{x}_{st}$ is a weighted mean, the weights W_h being the relative sizes of the strata. If $N_h/N = n_h/n$ in all strata—that is, if the proportion of samples collected in the hth stratum equals the proportion of the N units in stratum h—then $\bar{x}_{st}$ reduces to

$$\bar{x}_{st} = \sum_{h=1}^{L} \frac{n_h}{n} \bar{x}_h = \frac{1}{n} \sum_{h=1}^{L} \sum_{i=1}^{n_h} x_{hi}$$

which is the simple arithmetic mean of the n data collected over all L strata. This type of sample allocation is called *proportional allocation*. It is simple to use but will not be optimal in the sense of giving the most accurate estimate of μ if sampling costs or data variability differ for the various strata.

Since only a portion of the population units in each stratum have been measured, $\bar{x}_{st}$ has a variance due to the random sampling procedure, of

$$\mathrm{Var}(\bar{x}_{st}) = \sum_{h=1}^{L} W_h^2 \frac{\sigma_h^2}{n_h} (1 - f_h)$$

where σ_h^2 is the (unknown) variance over all N_h units in stratum h, and $f_h = n_h/N_h$. If proportional allocation is used, this expression reduces to

$$\mathrm{Var}(\bar{x}_{st}) = \frac{1 - f}{n} \sum_{h=1}^{L} W_h \sigma_h^2$$

where $f = n/N$. The square root of $\mathrm{Var}(\bar{x}_{st})$ is the true standard error of $\bar{x}_{st}$.

If simple random sampling is used in each stratum, a statistically unbiased estimate of $\mathrm{Var}(\bar{x}_{st})$ is obtained by computing

$$s^2(\bar{x}_{st}) = \frac{1}{N^2} \sum_{h=1}^{L} N_h^2 \left(1 - \frac{n_h}{N_h}\right) \frac{s_h^2}{n_h} \qquad 5.4$$

where

$$s_h^2 = \frac{1}{n_h - 1} \sum_{i=1}^{n_h} (x_{hi} - \bar{x}_h)^2$$

Equation 5.4 reduces to

$$s^2(\bar{x}_{st}) = \sum_{h=1}^{L} \frac{W_h^2 s_h^2}{n_h} \qquad 5.5$$

if N is very large.

In summary, $\bar{x}_{st}$ (Eq. 5.3) estimates the population mean μ of the N population units. The uncertainty in $\bar{x}_{st}$ due to sampling only a portion of the units is estimated by computing $s^2(\bar{x}_{st})$ with Eq. 5.4 or Eq. 5.5. An estimate of the standard error of $\bar{x}_{st}$ is given by $s(\bar{x}_{st})$. The stratum means are estimated by Eq. 5.2. The estimated variance of $\bar{x}_h$ is $(1 - f_h)s_h^2/n_h$. If random uncertainties due to sample collection, handling, and measurement are present, then Eq. 5.4 is still an unbiased estimator of $\mathrm{Var}(\bar{x}_{st})$ if $f_h = 0$ for all strata.

5.3 ESTIMATING THE TOTAL AMOUNT

The total amount (inventory) over all strata is

$$I = \sum_{h=1}^{L} N_h \mu_h = N\mu$$

where μ is defined by Eq. 5.1, and $N_h \mu_h$ is the inventory for stratum h. The quantity I is estimated by computing

$$\hat{I} = \sum_{h=1}^{L} N_h \bar{x}_h = N\bar{x}_{st} \qquad 5.6$$

where $N_h \bar{x}_h$ is the estimated total amount in stratum h.

The variance of $\hat{I}$ is

$$\text{Var}(\hat{I}) = \sum_{h=1}^{L} N_h^2 \left(1 - \frac{n_h}{N_h}\right) \frac{\sigma_h^2}{n_h}$$

An estimate of this variance is obtained by computing Eq. 5.4 and multiplying by N^2. That is,

$$s^2(\hat{I}) = \sum_{h=1}^{L} N_h^2 \left(1 - \frac{n_h}{N_h}\right) \frac{s_h^2}{n_h} \qquad 5.7$$

EXAMPLE 5.2

This example is patterned after a similar one in Reckhow and Chapra (1983). Suppose we want to estimate the average concentration and total amount of phosphorus in the water of a pond. Data from similar ponds suggest that phosphorus concentrations may be higher at depth than near the surface of the water. Hence, the pond is divided into 3 depth strata; a surface layer, a bottom layer, and the intermediate zone. Within each stratum 1-L water samples are taken at random locations with respect to depth and horizontal position. The phosphorus concentration of a 100-mL aliquot from each 1-L sample will be measured. The target population is the total number N of 100-mL water samples in the pond. The number of such aliquots in stratum h is N_h. Suppose the funds available for sampling will allow for $n = 30$ samples and that proportional allocation to the 3 strata is used. Table 5.1 gives the values of N_h for our hypothetical pond along with the resulting values of n_h. Hypothetical data and the resulting stratum means, totals, and variances are also given. Using Eqs. 5.3, 5.4, 5.6, and 5.7, we obtain $\bar{x}_{st} = 3.45$ μg/100 mL, $s(\bar{x}_{st}) = 0.146$ μg/100 mL, $\hat{I} = 33.3 \times 10^6$ μg, and $s(\hat{I}) = 1.41 \times 10^6$ μg, respectively.

5.4 ARBITRARY SELECTION OF STRATA

When sampling in one or two dimensions, such as along a line or over a land area, it may be desirable to distribute sampling points more uniformly than can be achieved by simple random sampling. One way to do it is to use systematic sampling discussed in Chapter 8. Another approach that still uses random sampling is to divide the area into blocks of equal or unequal sizes and choose one or more locations at random within each block. The case of one point per equal-sized block is illustrated in Figures 3.1(f) and 3.2(f). Also, strata might be defined on the basis of topography, political boundaries, roads, rivers, and so on. This design might be considered to be stratified random sampling with

Table 5.1 Data and Computations for Stratified Random Sampling to Estimate Average Phosphorus Concentration (μg/100 mL) and Total Amount (μg) in Pond Water

Strata	N_h	W_h	n_h	*Measurements* (μg/100 mL)	$\bar{x}_h$	$N_h\bar{x}_h$ (μg)	s_h^2	$W_h\bar{x}_h$ (μg/100 mL)	$W_h^2 s_h^2/n_h$
1	3,940,000	0.409	12	2.1 1.2 1.5 2.5 3.0 1.4 1.1 1.9 1.7 3.1 2.3 2.1	1.99	7.8×10^6	0.4299	0.814	0.005993
2	3,200,000	0.332	10	4.0 3.6 4.1 2.9 4.8 3.3 3.7 3.0 4.4 4.5	3.83	12.3×10^6	0.4134	1.27	0.004557
3	2,500,000	0.259	8	6.0 5.8 3.4 5.1 3.9 5.3 5.7 6.9	5.26	13.2×10^6	1.294	1.36	0.01085
Sum	9,640,000	1.00	30			$\hat{I} = 33.3 \times 10^6$		$\bar{x}_{st} = 3.45$	$s^2(\bar{x}_{st}) = 0.0214$

one or more observations per stratum, except that the stratum (block) boundaries are arbitrarily determined rather than being based on a priori information on spatial or time variability.

If equal-sized blocks are used, the main concern will probably be to estimate μ, the mean of the entire population. If two or more observations per block are obtained, the usual stratified random sampling estimate $\bar{x}_{st}$ (Eq. 5.3) may be used to estimate the population mean μ. Equation 5.4 is used to estimate $\mathrm{Var}(\bar{x}_{st})$. However, since the strata (blocks) have been arbitrarily defined, there is no assurance that this approach will result in a more precise estimate of μ than if the simple mean $\bar{x} = \Sigma_{i=1}^{n} x_i/n$ of all the data is used. If there is only one observation per block as in Figure. 3.1(f), one could still compute $\bar{x}_{st}$, but $\mathrm{Var}(\bar{x}_{st})$ cannot be estimated using Eq. 5.4 since two or more measurements per block are required. Some approaches for estimating $\mathrm{Var}(\bar{x}_{st})$ for this situation are given by Cochran (1977, p. 138).

5.5 ALLOCATION OF SAMPLES TO STRATA

An important aspect of the design of a stratified random sampling plan is to decide how many samples n_h to collect within the hth stratum. If the main objective is to estimate the overall population mean μ or population total Nμ, Cochran (1977, pp. 96–99) describes a procedure for choosing the n_h that will either minimize $\mathrm{Var}(\bar{x}_{st})$ for fixed cost of sampling or minimize cost C for a fixed prespecified $\mathrm{Var}(\bar{x}_{st})$ that must not be exceeded. He uses the cost (in dollars) function

$$\text{cost} = C = c_0 + \sum_{h=1}^{L} c_h n_h \qquad \mathbf{5.8}$$

where c_h is the cost per population unit in the hth stratum and c_0 is the fixed overhead cost. With this cost function the optimum number of samples in the hth stratum is then

$$n_h = n \frac{W_h \sigma_h / \sqrt{c_h}}{\sum_{h=1}^{L} (W_h \sigma_h / \sqrt{c_h})} \qquad \mathbf{5.9}$$

where σ_h is the true population standard deviation for the hth stratum and n is the total number of samples collected in all L strata. In practice, σ_h is replaced with an estimate, s_h, obtained from data from prior studies. Methods for determining n are given in Section 5.6.

Equation 5.9 says to take more observations in a stratum if the stratum is larger (i.e., if W_h is large), more variable internally, or cheaper to sample. If the cost per population unit is the same for all strata, Eq. 5.9 reduces to

$$n_h = \frac{n W_h \sigma_h}{\sum_{h=1}^{L} W_h \sigma_h} \qquad \mathbf{5.10}$$

Equation 5.10 is frequently called Neyman allocation.

A simple alternative to Eqs. 5.9 or 5.10 is to use proportional allocation:

$$n_h = nW_h = \frac{nN_h}{N}$$

For example, if a given stratum is 30% of the entire population, then 30% of the samples are allocated to that stratum. With proportional allocation there is no need to know the stratum standard deviations σ_h to find n_h. However, a more accurate estimate of μ will usually result if a reasonably accurate estimate of σ_h is available so that Eqs. 5.9 or 5.10 can be used.

5.6 NUMBER OF SAMPLES

The methods for estimating n_h in Section 5.5 require knowing n, the total number of population units that will be measured in the L strata. This section shows how to estimate n.

5.6.1 Prespecified Fixed Cost

If total cost C (Eq. 5.8) for the study is fixed a priori, the optimum n is estimated by

$$n = \frac{(C - c_0) \sum_{h=1}^{L} (W_h s_h / \sqrt{c_h})}{\sum_{h=1}^{L} (W_h s_h \sqrt{c_h})} \qquad \textbf{5.11}$$

where s_h is obtained from prior studies and c_0 represents overhead costs. Hence, $C - c_0$ is the total dollars available for collecting and measuring samples, not including overhead expenses.

5.6.2 Prespecified Variance

If $\text{Var}(\bar{x}_{st})$ is a prespecified value V, then n is obtained by computing

$$n = \frac{\left(\sum_{h=1}^{L} W_h s_h \sqrt{c_h}\right) \sum_{h=1}^{L} W_h s_h / \sqrt{c_h}}{V + \frac{1}{N} \sum_{i=1}^{L} W_h s_h^2}$$

Another approach for determining n when V is prespecified is to also prespecify the stratum sampling fractions $w_h = n_h/n$ (not to be confused with the stratum weights $W_h = N_h/N$). Costs or other constraints might dictate the sampling fractions w_h. These fractions may not necessarily be optimum in the sense of minimizing cost for fixed precision or minimizing precision for fixed cost. Nevertheless, if both V and the w_h are prespecified, n is computed as

$$n = \frac{\sum_{h=1}^{L} W_h^2 s_h^2 / w_h}{V + \frac{1}{N} \sum_{h=1}^{L} W_h s_h^2} \qquad \textbf{5.12}$$

Since the sampling fractions n_h/n have been prespecified, the n_h can be determined using n calculated from Eq. 5.12. If proportional allocation is used so that $w_h = n_h/n = W_h$, then Eq. 5.12 reduces to

$$n = \frac{\sum_{h=1}^{L} W_h s_h^2}{V + \frac{1}{N} \sum_{h=1}^{L} W_h s_h^2}$$

5.6.3 Prespecified Margin of Error

Rather than prespecify V, we may specify a margin of error $d = |\bar{x}_{st} - \mu|$ that can be tolerated and a small probability α of exceeding that error. This general approach was also used in Section 4.4.2 for simple random sampling. If proportional allocation is used, and if $\bar{x}_{st}$ is approximately normally distributed, then the optimum value for n may be estimated by computing (see Cochran, 1977, p. 105)

$$n = \frac{Z_{1-\alpha/2}^2 \sum_{h=1}^{L} W_h s_h^2/d^2}{1 + Z_{1-\alpha/2}^2 \sum_{h=1}^{L} W_h s_h^2/d^2 N} \qquad \textbf{5.13}$$

where $Z_{1-\alpha/2}$ is the standard normal deviate defined in Section 4.4.2 and tabulated in Table A1.

EXAMPLE 5.3

Let us return to Example 5.2, where interest centered on estimating the average concentration of phosphorus in water of a pond. We use the stratum weights (W_h) and estimated variances (s_h^2) in Table 5.1 to determine n such that

$$\text{Prob}[|\bar{x}_{st} - \mu| \geq 0.2\ \mu\text{g}/100\ \text{mL}] = \alpha$$

where $\bar{x}_{st}$ will be estimated from the n measurements. From Table 5.1 we have

$$\sum_{h=1}^{3} W_h s_h^2 = 0.409(0.4299) + 0.332(0.4134) + 0.259(1.294)$$

$$= 0.6482$$

Suppose we specify $\alpha = 0.10$. Then $Z_{1-\alpha/2} = 1.645$ and Eq. 5.13 gives

$$n = \frac{(1.645)^2\ 0.6482/(0.2)^2}{1 + 0} = 43.8 \cong 44$$

If sampling and analysis costs are equal in all strata, we may use Eq. 5.10 to determine the optimum allocation of these 44 samples to the 3 strata. Using s_h from Table 5.1 in place of σ_h in Eq. 5.10, we obtain

$$n_1 = \frac{44(0.409)(0.6557)}{0.7764} = 15.2 \cong 15$$

$$n_2 = \frac{44(0.332)(0.6430)}{0.7764} = 12.1 \cong 12$$

$$n_3 = \frac{44(0.259)(1.138)}{0.7764} = 16.7 \cong 17$$

5.7 CASE STUDY

Hakonson and Bostick (1976) report ^{238}Pu and other radionuclide average concentrations in sediments from three canyons in Los Alamos, N. M. that receive liquid waste containing small amounts of these radionuclides. Three sediment core samples were taken at −100, 0, 20, 40, 80, 160, 320, 640, 1280, 2560, 5120, and 10,240 m from the point of liquid waste discharge. The core sampling tool was a section of 2.4-cm-diameter plastic pipe driven to a maximum depth of 30 cm. Sampling depths varied from location to location as well as between the three cores at each location. Only the ^{238}Pu data in one of the canyons (Mortandad) are considered here (data in Table 5.2). Suppose the objective is to estimate the total amount of ^{238}Pu in the top 2.5 cm of sediment in the stream channel from the point of discharge out to 10,000 m. One approach is to use stratified random sampling. Suppose six strata are defined as in Table 5.3 and that the data in each stratum may be considered as representative for the stratum. Following are a series of three questions and answers concerning the design of the study.

Question 1. Suppose cost considerations limit the number of sediment samples to $n = 200$. What is the optimum allocation of these 200 samples to the 6 strata? Assume that sampling and radiochemical analysis costs are equal for all strata and that all cores will be taken to a depth of 12.5 cm.

Since n is fixed and costs are equal, Eq. 5.10 is used to determine $n_1, n_2, \ldots, n_6$. First determine the stratum weight W_h, which, since sampling depth is the same for all strata, is the proportion of the total surface area of the stream channel that lies in stratum h. The width of the channel varies from less than 1 m at distances less than 2000 m, to 2–3 m thereafter. For illustration purposes we assume that the width is 0.8 m from the discharge point out to 2000 m, and 2.5 m beyond 2000 m. The resulting values of W_h are given in Table 5.3 along with values of s_h approximated using the values of s in Table 5.2. The stratum sampling fractions n_h/n computed by using Eq. 5.10 are given in the last column in Table 5.3. We see that the greatest proportion of the samples (78%) should be taken in stratum 5. Although stratum 6 is larger than stratum 5, its s_h is very small. Hence, only about 1% of the samples are allocated to stratum 6. Since the total n is 200, the allocation in Table 5.3 gives $n_1 = 2$, $n_2 = 4$, $n_3 = 20$, $n_4 = 16$, $n_5 = 156$, $n_6 = 2$.

Question 2. Hakonson and Bostick (1976) indicate that sediment depth increased with distance from the waste discharge point. Suppose that in strata 1 through 5 sediment cores will be taken to 12.5 cm,

Table 5.2 Average ^{238}Pu (pCi/g dry) Concentrations in Sediments from Mortandad Canyon in Los Alamos, New Mexico

Meters Downstream	*Average Concentration*[a]	*Coefficient of Variation*	*s*
0	160	0.40	64
20	110	0.94	103.4
40	65	0.91	59.2
80	39	0.62	24.2
160	86	0.87	74.8
320	20	0.93	18.6
640	17	0.06	1.0
1,280	6.8	0.42	2.86
2,560	8.6	1.10	9.46
5,120	0.05	1.40	0.07
10,280	0.01	1.20	0.01

Source: After Hakonson and Bostick, 1976, Table 1.
[a]Three observations at each location.

but in stratum 6 the depth will be increased to 25 cm. What are the stratum weights W_h? What is the optimum allocation for estimating total ^{238}Pu inventory by this approach?

Since sampling depths are not constant for all strata, W_h is now the proportion of sediment *volume* (rather than surface area) in stratum h. We find that $W_1 = 0.0002$, $W_2 = 0.0005$, $W_3 = 0.0040$, $W_4 = 0.014$, $W_5 = 0.248$, $W_6 = 0.733$. As expected, the sixth stratum gets more "weight," since the sampling depth was increased in that stratum. The reader may determine the optimum allocation for estimating the total inventory.

Question 3. Define the population units for the core sampling plans as in Question 1, and estimate the stratum and total inventory of ^{238}Pu.

The population unit is a sediment core the size of the coring tool, a 2.4-cm-diameter cylinder to a depth of 12.5 cm (volume = 56.55 cm^3). Hence, N_h in Eq. 5.6 is the number of these unit volumes in stratum h, and $\bar{x}_h$ is the average amount of ^{238}Pu (in, say, pCi) in such units.

The area of the sediment core is $\pi r^2 = 4.52$ cm^2, since $r = 1.2$ cm. Using the strata in Table 5.3, we obtain the number of 4.52-cm^2 unit areas in the six strata by multiplying the stratum areas by (10,000 cm^2)/m^2 × (1/4.52 cm^2) = 2212.4/m^2. We obtain $N_1 = 17{,}699$, $N_2 = 35{,}398$, $N_3 = 300{,}886$, $N_4 = 1{,}061{,}952$, $N_5 = 18{,}716{,}904$, $N_6 = 27{,}655{,}000$. The required number n_h of core samples (given in the answer to Question 2) is collected at random locations within the strata and radiochemistry determinations x_{hi} are made and expressed as pCi per core. The mean $\bar{x}_h$ and standard deviations are computed, and Eqs. 5.6 and 5.7 are used to compute $N\bar{x}_{st}$ and $s^2(N\bar{x}_{st})$, the estimated inventory and its estimated variance, respectively.

5.8 SUMMARY

Pollutant concentrations in environmental media are frequently highly variable over space and time. This variability means that estimates of average concen-

Table 5.3 Determining the Allocation of Samples to Strata

Strata	*Surface Area* (m^2)	*Meters Downstream*	*Stratum Weights* W_h	*Standard Deviation* s_h (pCi/g)	$W_h s_h$	$n_h/n = w_h s_h / \Sigma_{h=1}^{6} W_h s_h$
1	0.8 × 10 = 8	0–10	0.0004	64	0.0256	0.0073
2	0.8 × 20 = 16	10–30	0.0007	103	0.0721	0.0205
3	0.8 × 170 = 136	30–200	0.0063	57	0.3591	0.1021
4	0.8 × 600 = 480	200–800	0.0222	13	0.2886	0.0821
5	(0.8 × 1200) +(2.5 × 3000) = 8,460	800–5,000	0.3917	7	2.7419	0.7798
6	2.5 × 5000 = 12,500	5,000–10,000	0.5787	0.05	0.0289	0.0082
	21,600		$\Sigma_{h=1}^{6} W_h = 1.0$		$\Sigma_{h=1}^{6} W_h s_h = 3.5162$	1.00

trations and total amounts will also be highly variable. If the main patterns of spatial-time variability are known at least approximately, the population may be broken down into strata that are internally less variable. Then stratified random sampling can be used to obtain more accurate estimates of means and totals for the population. Methods for estimating means and totals are given and illustrated using examples and a case study. Also, the allocation of samples to strata is illustrated when the goal is to estimate the overall population mean.

EXERCISES

5.1 Gilbert (1977) reports the results of sampling soil at a nuclear weapons test area by stratified random sampling to estimate the total amount of $^{239,240}Pu$ in surface soil. Use the following information from Gilbert (1977) to estimate the plutonium inventory in surface soil and its standard error (S.E.).

Strata	*Size of the Stratum* N_h *(m²)*	n_h	Mean $\bar{x}_h$ *(μCi/m²)*	Variance s_h^2
1	351,000	18	4.1	30.42
2	82,300	12	73	10,800
3	26,200	13	270	127,413
4	11,000	20	260	84,500

5.2 Use the data in Exercise 5.1 to determine the optimum number of population units to measure in each of the 4 strata. Also find the total number of units to sample, assuming cost is fixed. Assume the total money available for sampling and analysis, excluding overhead expenses, is $50,000 and that cost per population unit is c = $500 in all strata.

5.3 Repeat Exercise 5.2 but let the costs per population unit in the 4 strata be c_1 = $800, c_2 = $600, c_3 = $600, c_4 = $300. Does the allocation change very much from that in Exercise 5.2?

ANSWERS

5.1 Using Eq. 5.6, we obtain $\hat{I}$ = 17,380,000 μCi = 17.4 Ci. Using Eq. 5.7 gives

$$\text{S.E.} = s(\hat{I}) = (1.354 \times 10^{13})^{1/2} \ \mu\text{Ci} = 3.7 \text{ Ci}$$

5.2 Using Eq. 5.11, we have

$$n = \frac{50{,}000 \sum_{h=1}^{4} W_h s_h}{\left(500 \sum_{h=1}^{4} W_h s_h\right)} = 100$$

Using Eq. 5.10 with s_h in place of σ_h, we then have

$$n_1 = \frac{100\ W_1 s_1}{\sum_{h=1}^{4} W_h s_h} = \frac{100(4.1146)}{48.96} = 8.4 \cong 8$$

Similarly, $n_2 = 37$, $n_3 = 41$, $n_4 = 14$.

5.3 Using Eq. 5.11, we have

$$n = \frac{50{,}000(0.14547 + 0.74212 + 0.81139 + 0.39237)}{116.38 + 445.27 + 486.83 + 117.71}$$

$$= 89.7 \cong 90$$

Using Eq. 5.9 gives

$$n_1 = \frac{90(0.14547)}{2.0914} = 6.3 \cong 6$$

Similarly, $n_2 = 32$, $n_3 = 35$ and $n_4 = 17$.

6 Two-Stage Sampling

When environmental samples are collected, the entire sample mass may not be measured. For example, even though 500 g of soil are collected at a given point in time and space, a subsample (aliquot) of only 50 g may be actually measured. This procedure is called *two-stage sampling* or *subsampling*. The first stage consists of collecting the environmental sample. The second stage occurs when one or more aliquots are selected from each environmental sample for measurement. Subsampling introduces additional uncertainty into estimates of means and totals because the entire sample mass is not measured. This additional sampling variability may be large, particularly if the pollutant is in particulate form and very small subsamples are used.

This chapter gives methods to estimate the mean and total when subsampling is used, to estimate the standard errors of these estimates, and to determine the optimum number of samples to collect and subsamples to measure. The two cases of equal and unequal numbers of population units in each sample are considered in Sections 6.2 and 6.3, respectively.

6.1 BASIC CONCEPTS

When two-stage sampling is applicable, the target population is divisible into N primary (first stage) units, and the ith primary unit is divisible into M_i subunits (second-stage units). Some of the N primary units are chosen at random and within the ith chosen primary unit m_i subunits are selected for measurement in their entirety. The subunit corresponds to the population unit discussed in previous chapters. There is no requirement that the M_i be equal for all primary units, but the formulas for estimating the mean, total, and their standard errors are simpler in that case (see Section 6.2). The two situations of constant and variable M_i are illustrated in Figure 6.1.

Two examples of two-stage sampling are the following: (1) N soil samples of specified dimensions can be potentially collected from a study site. The ith sample can be divided into M_i subsamples (aliquots). A set of n samples is selected at random. From the ith selected sample, m_i subsamples are measured for a pollutant. (2) A month has $N = 31$ days (primary units). Each day may be divided into $M = 6$ periods of 4-h duration. A sample of $n = 5$ days is chosen at random and air samples are collected over $m = 3$ randomly chosen 4-h periods on each of these days.

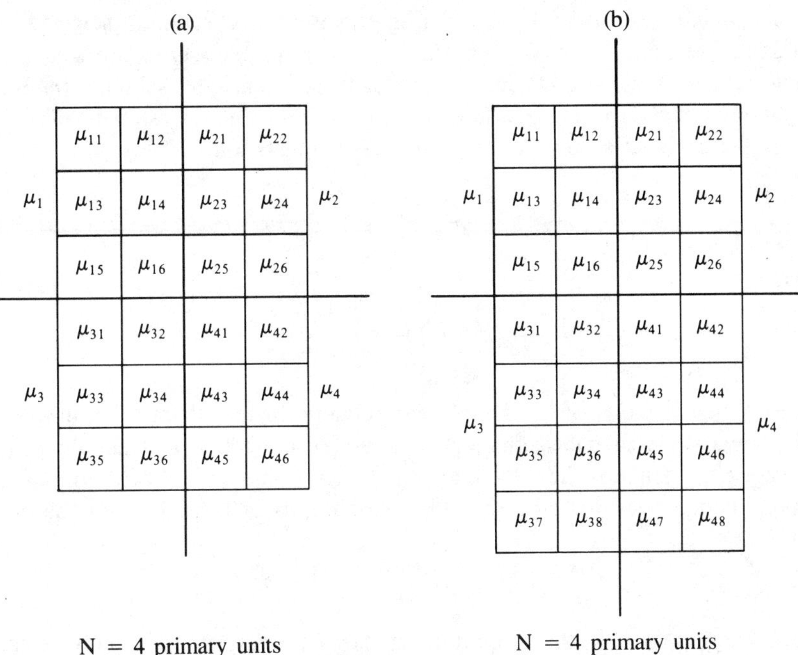

Figure 6.1 Two-stage sampling for cases of (a) equal and (b) unequal numbers *M* of subunits within primary units.

In two-stage sampling the subunits chosen for measurement are measured in their entirety. Two-stage sampling is easily extended to three stages, where the subunit is itself subsampled. Then the third-stage unit becomes the population unit that is measured. Three-stage sampling is discussed in Chapter 7 along with compositing.

The next section focuses on the case where all primary units contain the same number, M, of subunits. The case of unequal numbers of subunits per primary unit is considered in Section 6.3.

6.2 PRIMARY UNITS OF EQUAL SIZE

6.2.1 Estimating the Mean

Suppose that the target population contains N primary units and each primary unit contains M subunits—that is, $M_1 = M_2 = \cdots = M_N = M$. Then NM is the total number of potential units available for measurement from the population. The true mean for these NM units is

$$\mu = \frac{1}{NM} \sum_{i=1}^{N} \sum_{j=1}^{M} \mu_{ij} \tag{6.1}$$

where μ_{ij} is the true amount of pollutant present in the jth subunit of the ith primary unit. The true mean for the ith primary units is $\mu_i = \Sigma_{j=1}^{M} \mu_{ij}/M$. The true mean μ is estimated by measuring representative subunits selected from representative primary units. Suppose n of the N primary units are selected by simple random sampling. Then for each of these n units, m subunits ($m \leq M$) are randomly selected for measurement. Then an unbiased estimate of μ (Eq. 6.1) is the arithmetic mean of the nm measurements x_{ij}:

$$\bar{x} = \frac{1}{nm} \sum_{i=1}^{n} \sum_{j=1}^{m} x_{ij} = \frac{1}{n} \sum_{i=1}^{n} \bar{x}_i \qquad \textbf{6.2}$$

where

$$\bar{x}_i = \frac{1}{m} \sum_{j=1}^{m} x_{ij}$$

is the estimated mean of the m selected subunits in the ith primary unit.

Let us assume that measurement and sample collection and handling errors are negligibly small. Then the uncertainty in $\bar{x}$ due to only nm of the NM subunits being measured (the sampling variance) is estimated by computing

$$s^2(\bar{x}) = (1 - f_N) \frac{s_1^2}{n} + f_N(1 - f_M) \frac{s_2^2}{nm} \qquad \textbf{6.3}$$

where: $f_N = n/N$ = the proportion of the N primary units actually selected,
$f_M = m/M$ = the proportion of the M subunits in each primary unit actually measured,

$$s_1^2 = \frac{1}{n-1} \sum_{i=1}^{n} (\bar{x}_i - \bar{x})^2 \qquad \textbf{6.4}$$

= estimated variance among the N primary unit means,

$$s_2^2 = \frac{1}{n(m-1)} \sum_{i=1}^{n} \sum_{j=1}^{m} (x_{ij} - \bar{x}_i)^2 \qquad \textbf{6.5}$$

= estimated variance among M subunits within primary units.

Equation 6.3 takes on different forms as the sampling fractions f_N and f_M are 1 or less than 1. Four cases can be identified

Case 1. $f_N = 1$, $f_M = 1$, that is, *all* subunits in *all* primary units are measured. Then Eq. 6.3 is zero, since all population units have been measured and by assumption there are no measurement or sample collection and handling errors.

Case 2. $f_N < 1$, $f_M = 1$, that is, all M subunits within each unit of a random set of n primary units are measured. Then Eq. 6.3 becomes

$$s^2(\bar{x}) = (1 - f_N) \frac{s_1^2}{n} \quad \text{for } n \geq 2 \qquad \textbf{6.6}$$

There is no need to compute s_2^2 in this case.

Case 3. $f_N = 1$, $f_M < 1$, that is, a random set of m subunits is measured within all N primary units. Then

$$s^2(\bar{x}) = (1 - f_M) \frac{s_2^2}{nm}$$

Case 4. $f_N \cong 0$, $f_M \leq 1$, that is, one or more subunits are measured in a random set of n primary units, where N is much larger than n. Then Eq. 6.3 becomes

$$s^2(\bar{x}) = \frac{s_1^2}{n} \qquad \textbf{6.7}$$

This case occurs frequently in practice. As in Case 2 there is no need to compute s_2^2, which is a real advantage if subunits are selected by systematic sampling, since in that case s_2^2 may be biased, as discussed in Chapter 8. Note that if several subsamples from each primary unit are composited and a single measurement made, then Eq. 6.7 can still be used. Chapter 7 discusses compositing.

Two-stage sampling can also be viewed as a one-way analysis of variance (AOV), random effects model, as described, for example, by Snedecor and Cochran (1980, pp. 238–242). The estimate of $\text{Var}(\bar{x})$ is obtained by dividing the "between primary units" mean square (from the analysis of variance table) by nm, the total number of observations. This approach assumes f_N is zero.

EXAMPLE 6.1

Suppose that hourly average oxidant air concentrations are routinely recorded at numerous sampling locations in an urban area. Interest centers on the maximum hourly average for each day. A study is planned to estimate the true average, μ, of these daily maximums for a specific set of $N = 30$ stations during the months of July and August for 1973. Since there are $M = 62$ days in July and August, the target population consists of $NM = (30)(62) = 1860$ population units. Suppose that budget constraints restrict the number of operating stations to $n = 5$. Then μ is estimated by computing $\bar{x}$ by using Eq. 6.2 with $n = 5$ and $m = 62$, since data are available for all 62 days at all 5 operating stations. The variance of $\bar{x}$ is estimated by computing $s^2(\bar{x})$ by using Eq. 6.6 (since $f_M = 62/62 = 1$) with $f_N = 5/30 = 0.1667$.

We illustrate, using the maximum hourly oxidant data at 5 stations in the San Francisco area for 1973 listed by Grivet (1980, pp. 65–68). The means for the 5 stations are $\bar{x}_1 = 4.95$, $\bar{x}_2 = 3.71$, $\bar{x}_3 = 6.95$, $\bar{x}_4 = 9.26$, and $\bar{x}_5 = 5.64$. (Units are parts per hundred million, pphm.) Therefore, using Eqs. 6.2 and 6.6, we find $\bar{x} = 6.1$ pphm and $s^2(\bar{x}) = 0.748\ (\text{pphm})^2$, or $s(\bar{x}) = 0.86$ pphm.

If large measurement or other errors are in the data, then Eq. 6.6 will tend to be too small, since the factor f_N does not apply to those components of variance. The one-way AOV approach is preferred to Eq. 6.6 in that case.

6.2.2 Estimating the Total Amount

From Eq. 6.1 we see that the total amount of pollutant present in the NM subunits of the target population is

$$I = \sum_{i=1}^{N} \sum_{j=1}^{M} \mu_{ij}$$

An unbiased estimate of I is

$$\hat{I} = NM\bar{x} \tag{6.8}$$

where $\bar{x}$ is given by Eq. 6.2. The sampling variance of $\hat{I}$ is estimated by computing

$$s^2(\hat{I}) = (NM)^2 s^2(\bar{x}) \tag{6.9}$$

where $s^2(\bar{x})$ is given by Eq. 6.3.

EXAMPLE 6.2

Suppose the liquid effluent from a chemical manufacturing plant will be sampled to estimate the average concentration, μ, and total amount, I, of a toxic chemical discharged continuously to a river over the 80-h operating period of the plant during a given week. Measurements will be made on 100-mL aliquots of effluent. Hence, the target population consists of the total number of 100-mL aliquots discharged during that week. To estimate μ and I, an 8-L sample of effluent will be collected at n randomly selected times during the week. The choice of n would depend, in practice, on cost considerations and the variability in concentrations over time. Suppose, for illustration's sake, that $n = 5$. Two 100-mL aliquots are withdrawn from each of the 5 samples and chemically analyzed. The data are as follows:

	Concentrations (μg/100 mL)		
Sample	x_{1i}	x_{2i}	$\bar{x}_i$
1	0.013	0.010	0.0115
2	0.011	0.014	0.0125
3	0.043	0.036	0.0395
4	0.027	0.034	0.0305
5	0.033	0.025	0.0290

Equation 6.2 is used to compute $\bar{x}$ if the effluent flow rate is constant over time. Otherwise, a weighted mean is required. This latter case is considered in Chapter 7 (Section 7.3). We assume here that the flow rate is constant. Using Eq. 6.2 we find $\bar{x} = 0.0246$ μg/100 mL, our estimate of μ.

Now, to estimate I, Var($\bar{x}$), and Var(I), we need to determine N and M. Suppose the constant flow rate is 2000 L/h. Then

$$N = \frac{(80 \text{ h})(2000 \text{ L/h})}{(8 \text{ L/sample})} = 20{,}000 \text{ samples of size 8 L}$$

Also, the number M of 100-mL subsamples in each 8-L sample is $M = 8000$ mL/100 mL $= 80$. Therefore, by Eq. 6.8

$$\begin{aligned}\hat{I} &= 20{,}000(80)\bar{x} \\ &= (1{,}600{,}000 \text{ subsamples})(0.0246\ \mu\text{g/subsample}) \\ &= 39{,}360\ \mu\text{g}\end{aligned}$$

Since $n/N = 5/20{,}000 = 0.000250$, we set $n/N = 0$ and use Eq. 6.7 to compute $s^2(\bar{x}) = 0.00002971$, or $s(\bar{x}) = 0.00545$ μg/100 mL. Using Eq. 6.9, we obtain $s(\hat{I}) = 20{,}000(80)0.00545$ μg $= 8720$ μg.

To summarize, if the flow rate is a constant 2000 L/h, the estimated mean and inventory for the 80-h period is 0.0246 μg/100 mL and 39,360 μg, respectively. The standard errors of these estimates are 0.00545 μg/100 mL and 8720 μg, respectively.

6.2.3 Number of Primary Units and Subsamples

When planning two-stage sampling, it is necessary to decide on n and m. If the primary objective is to estimate μ or the inventory I with maximum precision for fixed total cost, the approach described here may be used. This method is also applicable if the goal is to minimize costs, given some prespecified value for Var($\bar{x}$) that must not be exceeded. Budget constraints must be considered, since they limit the number of samples that can be collected. If travel costs between primary units are of little importance, then the cost function $C = c_1 n + c_2 nm$ is useful. Using this function and assuming M is large, one can show (see, e.g., Snedecor and Cochran, 1967, p. 532) that the optimum value for m can be estimated from the equation

$$\hat{m}_{\text{opt}} = \left(\frac{c_1/c_2}{s_1^2/s_2^2}\right)^{1/2} \qquad \textbf{6.10}$$

where s_1^2 and s_2^2 are computed by using Eqs. 6.4 and 6.5, respectively, with data from prior or pilot studies. If M is small, Eq. 6.10 becomes

$$\hat{m}_{\text{opt}} = \left(\frac{c_1/c_2}{s_1^2/s_2^2 - 1/M}\right)^{1/2}$$

If the denominator is negative, m_{opt} should be made as large as practical. Once m is chosen, n may be determined from the cost function $C = c_1 n + c_2 nm$.

Note from Eq. 6.3 that if f_N is close to zero, then the second term in that equation is also near zero, so $s^2(\bar{x})$ is not reduced very much by using a large m. If f_N is not near zero and s_1^2 and s_2^2 are reliable and appropriate estimates from prior studies, then the effect on $s^2(\bar{x})$ (Eq. 6.3) of varying n and m can be determined. This effect is illustrated in the following example.

EXAMPLE 6.3

Suppose the oxidant air concentration data from Example 6.1 will be used to decide the number of stations to operate the following year (1974) for the purpose of estimating μ for July and August. If plans are to operate samplers on each day of the 2-month period, then $m = M = 62$, and there is no need to estimate m by Eq. 6.10. Suppose the total money available for sampling and analysis in 1974 is $C = \$100{,}000$. Furthermore, suppose $c_1 = \$2000$ and $c_2 = \$100$, where c_1 is the cost associated with operating each station and c_2 is the cost of collecting and measuring an air sample. Then $100{,}000 = 2000n + 100(62)n$, or $n = 12.2$, which is rounded to 12. Hence, the budget will permit about 12 stations to operate daily during July and August.

Now suppose there is no requirement that the air samplers operate each and every day. What would be the optimum values for n and m if the costs are as given in the previous paragraph? From the data listed by Grivet (1980) we calculate $s_1^2 = 4.49$ and $s_2^2 = 9.55$ by Eqs. 6.4 and 6.5. Therefore, using Eq. 6.10 gives

$$\hat{m}_{\text{opt}} = \left(\frac{2000/100}{4.49/9.55}\right)^{1/2} = 6.5$$

Using $m = 7$ in the cost equation, we obtain $100{,}000 = 2000n + 100(7)n$, or $n = 37$. Hence, the data and cost considerations suggest the most accurate estimate of μ for July and August of 1974 would be obtained by selecting 37 representative stations and taking oxidant measurements on 7 representative days during the July-August period.

The assumption has been made that the 1973 data are representative of conditions in 1974. We also note that maximum oxidant concentrations may be different on weekdays than on weekends. Therefore, careful definition of the target population is required before selecting days on which data will be taken.

To see the effect on $s^2(\bar{x})$ of changing n and m, we can compute Eq. 6.3 for various combinations of n and m. Assume there are $N = 100$ potential sampling stations. Then Eq. 6.3 becomes

$$s^2(\bar{x}) = \left(1 - \frac{n}{100}\right)\frac{4.49}{n} + \frac{n}{100}\left(1 - \frac{m}{62}\right)\frac{9.55}{nm}$$

If the optimum $n = 37$ and $m = 7$ are used, then $s(\bar{x}) = 0.298$. If n is decreased to 20 and m is increased to 30, then $s(\bar{x})$ increases to 0.426. If n is increased to 45 and m is decreased to 2, the $s(\bar{x}) = 0.318$. If the cost equation is $100{,}000 = 2000n + 100nm$, all these combinations of n and m would cost approximately \$100,000, but the standard errors $s(\bar{x})$ vary from 0.30 to 0.43.

6.3 PRIMARY UNITS OF UNEQUAL SIZE

Often it is not convenient or possible to have the same number of subunits M in each primary unit. For example, suppose an area of land is contaminated with a pollutant. Information is sought on the average concentration of the pollutant and on the total amount present within a prescribed contaminated area. A series of soil samples (primary units) are collected, and one or more aliquots (subunits) of soil are taken from each soil sample for chemical analysis. If the sizes (volumes) of the soil samples differ, the number of potential aliquots and the sample weight will vary from sample to sample. Even if the soil sample volumes or weights are identical, the number of aliquots may differ due to different proportions of stones and other debris in the samples.

This section gives methods for estimating average pollutant concentrations and total amounts in environmental media when primary units have different numbers of subunits. Our discussion will be in terms of the soil example given in the preceding example, but the procedures have wide applicability. The methods presented here are from Cochran (1977, p. 305) and are for the case when all primary units have the same chance of being selected. Cochran (1977, pp. 308–310) considers estimators that are appropriate when this is not the case.

6.3.1 Estimating the Mean

We want to estimate the true mean μ over all $\Sigma_{i=1}^{N} M_i$ subunits. Let x_{ij} be the amount of pollutant in the jth subunit of the ith primary unit. Then

$$\mu = \frac{\sum_{i=1}^{N} \sum_{j=1}^{M_i} \mu_{ij}}{\sum_{i=1}^{N} M_i} = \frac{\sum_{i=1}^{N} M_i \mu_i}{\sum_{i=1}^{N} M_i} \qquad \textbf{6.11}$$

where the numerator is the total amount in the $\Sigma_{i=1}^{N} M_i$ subunits, and

$$\mu_i = \frac{1}{M_i} \sum_{j=1}^{M_i} \mu_{ij}$$

is the true mean per subunit in the ith primary unit. Equation 6.11 is a weighted mean of the true primary unit means μ_i, the weight $W_i = M_i/\Sigma_{i=1}^{N} M_i$ being the proportion of all subunits in the ith primary unit. Hence, if the ith unit constitutes, say, 80% of the target population, the weight for the ith true mean μ_i is 0.80.

We may estimate μ by first selecting n of the N primary units by simple random sampling. Then m_i of the M_i subunits in the ith selected primary unit are randomly selected and measured. An estimate of μ may then be computed as

$$\bar{x} = \frac{\sum_{i=1}^{n} M_i \bar{x}_i}{\sum_{i=1}^{n} M_i} \qquad \textbf{6.12}$$

where

$$\bar{x} = \frac{1}{m_i} \sum_{j=1}^{m_i} x_{ij}$$

is an estimate of μ_i.

Var($\bar{x}$) may be estimated by computing

$$s^2(\bar{x}) = \frac{1 - (n/N)}{\bar{M}^2 n(n-1)} \sum_{i=1}^{n} M_i^2 (\bar{x}_i - \bar{x})^2 + \frac{1}{nN\bar{M}^2} \left[\sum_{i=1}^{n} M_i^2 \frac{(1 - m_i/M_i)\, s_{2i}^2}{m_i} \right] \qquad \textbf{6.13}$$

where

$$s_{2i}^2 = \frac{1}{m_i - 1} \sum_{j=1}^{m_i} (x_{ij} - \bar{x}_i)^2$$

the mean $\bar{x}$ is calculated using Eq. 6.12, and $\bar{M} = \Sigma_{i=1}^{n} M_i/n$. Note that the second term of Eq. 6.13 is negligibly small if $nN\bar{M}^2$ is sufficiently large.

EXAMPLE 6.4

Suppose we want to estimate the true average concentration of $^{239,240}Pu$ in surface soil over a 100-m^2 area at a nuclear weapons test area on the Nevada Test Site. Suppose $n = 5$ soil samples of area 10 cm $\times$ 10 cm to a depth of 5 cm are collected at random locations within the 100-m^2 plot. All Pu measurements are made on 10-g aliquots of soil that pass through a fine mesh screen. The weight in grams of fine soil differs for the 5 samples because some samples contain more stones (that are discarded) than other samples. Two aliquots of fine soil are removed from 3 of the 5 samples and 3 aliquots from the remaining 2 samples. The Pu concentrations (nCi/10 g) are given in Table 6.1 along with M_1 through M_5 (number of possible 10-g aliquots in the 5 soil samples). For illustration, assume no Pu is present in soil not passing through the fine mesh screen. Using Eq. 6.12 gives $\bar{x} = 26{,}989/170.5 = 158$ nCi/10 g.

The variance of this estimate is approximated by Eq. 6.13 with $n/N = 0$, since N is very large. Also, the second term of Eq. 6.13 is essentially zero since N is so large. As pointed out by Cochran (1977, p. 305), the numerator of the first term of Eq. 6.13 may be computed as

$$\sum_{i=1}^{n} (M_i\bar{x}_i)^2 - 2\bar{x} \sum_{i=1}^{n} (M_i\bar{x}_i)\, M_i + \bar{x}^2 \sum_{i=1}^{n} M_i^2$$

Hence

$$s^2(\bar{x}) = \frac{229{,}269{,}141 - 320{,}092{,}421 + 149{,}323{,}414}{23{,}256}$$

$$= \frac{58.500 \times 10^6}{23{,}256} = 2515\ (\text{nCi}/10\ \text{g})^2$$

so that $s(\bar{x}) = 50$ nCi/10 g.

In summary, based on the available data, the true concentration over the 100-m^2 area is estimated to be 16 nCi/g. The estimated standard error of this estimate is 5 nCi/g.

Table 6.1 $^{239,240}Pu$ Concentrations in Soil (nCi/10 g-aliquot)

Soil Sample Number	*Weight of Sample (g)*[a]	*Number of 10-g aliquots* (M_i)	*Number of Aliquots Selected* (m_i)	*$^{239,240}Pu$ (nCi/10-g aliquot)*[b] x_{i1}	x_{i2}	x_{i3}	$\bar{x}_i$	*nCi/Soil Sample* $M_i\bar{x}_i$
1	435	43.5	2	203	226		214.5	9,331
2	259	25.9	2	65	52		58.5	1,515
3	363	36.3	2	227	396		311.5	11,307
4	316	31.6	3	44	77	68	63	1,991
5	332	33.2	3	57	68	132	85.7	2,845
	Sum = 1,705	Sum = 170.5 $\bar{M}$ = 34.1						Sum = 26,989

[a] Hypothetical weight of soil that passes through the fine mesh screen.
[b] Unpublished data obtained by the Nevada Applied Ecology Group, U.S. Department of Energy, Las Vegas. Used by permission of the Nevada Applied Ecology Group.

6.3.2 Estimating the Total Amount

The total amount of pollutant present in the $\Sigma_{i=1}^{N} M_i$ subunits of the target population is

$$I = \sum_{i=1}^{N} \sum_{j=1}^{M_i} \mu_{ij} = \mu \sum_{i=1}^{N} M_i \qquad \textbf{6.14}$$

An unbiased estimate of I is obtained by estimating μ by $\bar{x}$ (Eq. 6.12) and by estimating μ by $\Sigma_{i=1}^{N} M_i$ by $N\bar{M}$. Making these substitutions in Eq. 6.14, we obtain

$$\hat{I} = \frac{N}{n} \sum_{i=1}^{n} M_i \bar{x}_i \qquad \textbf{6.15}$$

The Var($\hat{I}$) may be approximated by computing

$$s^2(\hat{I}) = (N\bar{M})^2 \, s^2(\bar{x}) \qquad \textbf{6.16}$$

where $s^2(\bar{x})$ is computed using Eq. 6.13.

EXAMPLE 6.5

Continuing with Example 6.4, suppose there is a need to estimate the total amount, I, of $^{239,240}Pu$ in the fine surface soil over the 100-m^2 area. The potential number, N, of surface soil samples in the 100-m^2 area must be known before I can be computed. Since each soil sample has a surface area of 100 cm^2, it follows that N = 10,000. Hence, using Eq. 6.15 and the data in Table 6.1, we have

$$\hat{I} = (10{,}000)\left(\frac{26{,}989}{5}\right)$$

$$= (10{,}000 \text{ soil samples}) (5398 \text{ nCi/soil sample})$$

$$= 53.98 \times 10^6 \text{ nCi} = 0.054 \text{ Ci}$$

The variance of $\hat{I}$ due to sampling only a portion of the 10-g aliquots in the 100 m^2 area is estimated by Eq. 6.16:

$$s^2(\hat{I}) = [(10{,}000)(34.1)]^2 \, 2515 = 292.4 \times 10^{12} \text{ (nCi)}^2$$

or $s(\hat{I})$ = 17.1 × 10^6 nCi = 0.017 Ci. In summary we estimate that 0.054 Ci of $^{239,240}Pu$ is present in the fine surface soil of the defined 100-m^2 area. The standard error of this estimate is approximately 0.017 Ci.

To use the methods in this chapter, we assume that all subunits are exactly the same size (e.g., soil aliquots all of size 10 g), but in practice this assumption may not be valid. Even so, the methods given here may be used if M_i is redefined to be the weight (in grams, say) of the ith primary unit, and x_{ij} is the amount *per gram* in the jth subunit of the ith primary unit. If this is done for the Pu soil data in Table 6.1, then

$$\bar{x} = [435(21.45) + \cdots + 332(8.57)] \text{ nCi/1705 g}$$

$$= 26989.39 \text{ nCi/1705 g}$$

$$= 15.8 \text{ nCi/g}$$

and

$$\hat{I} = (10{,}000)\left(\frac{26989.39}{5}\right)$$

$$= (10{,}000 \text{ soil samples})(5398 \text{ nCi/soil sample})$$

$$= 53.98 \times 10^6 \text{ nCi}$$

which are the same results as we obtained before. The reader may verify that $s^2(\bar{x})$ and $s^2(\hat{I})$ are also unchanged.

6.3.3 Number of Primary Units and Subsamples

Section 6.2.3 showed how to determine the optimum n and m when each primary unit contains the same number M of subunits. The same method may be used when M varies between primary units if sampling costs and within-unit variability are the same for each unit and if the N values of M_i are roughly equal. These conditions will result in equal m for all units, with m determined using Eq. 6.10. A special case is to take $m = 1$. However, we have seen from Example 6.3 that this choice may be far from optimum.

Ideally, m_i should be greater for units with greater variability between subunits. However, in practice, it is impractical to estimate this variability for each unit. An alternative approach is to choose m_i proportional to M_i. This implies that the within-unit variability is approximately equal for all units.

Once a method for determining the m_i has been chosen, the number of primary units n may be found by a suitable cost equation; for example,

$$C = c_1 n + c_2 n\bar{m} \qquad \textbf{6.17}$$

where $\bar{m} = \Sigma_{i=1}^{n} m_i/n$ is the average number of measured subunits per unit. Now $\bar{m}$ cannot be determined exactly until n is known. However, $\bar{m}$ can be estimated by choosing n' primary units at random, determining m for each of these units, and computing $\Sigma_{i=1}^{n'} m_i/n'$. For example, if proportional allocation is used, $\bar{m}$ is estimated by computing $\Sigma_{i=1}^{n'} kM_i/n'$, where $k = m_i/M_i$ for all i. Then solving Eq. 6.17 for n gives

$$n = \frac{C}{c_1 + c_2 k \sum_{i=1}^{n'} M_i/n'} \qquad \textbf{6.18}$$

If this procedure results in $n > n'$, then $n - n'$ additional units are randomly selected, all n units then being subsampled proportional to their size M_i. If $n < n'$, then $n' - n$ units (selected at random) are not used. Hence, n' should not be taken too large so as to avoid unnecessary sampling.

Different values of k will result in different n and m. If the sampling program will be repeated many times in the future, it may be cost effective to conduct sampling experiments using several values of k. That is, select units and subunits accordingly for each k, measure the subunits, and see which k results in the smallest $s^2(\bar{x})$ computed by Eq. 6.13. A choice of k to use in the future based on cost and precision in $\bar{x}$ can then be made.

EXAMPLE 6.6

Suppose a total of \$10,000 is available for sample collection and analysis and that $c_1 = \$200$ and $c_2 = \$100$. Furthermore, suppose proportional allocation will be used, and $n' = 5$ primary units from the target population are chosen at random, the M_i for these units being $M_1 = 50$, $M_2 = M_3 = 100$, $M_4 = M_5 = 200$, so $\Sigma_{i=1}^{5} M_i/5 = 130$. If $k = 0.02$ is used, then $m_1 = 1$, $m_2 = m_3 = 2$, and $m_4 = m_5 = 4$. Also, Eq. 6.18 gives

$$n = \frac{10{,}000}{200 + 100(0.02)130} \cong 22$$

Therefore, $n - n' = 22 - 5 = 17$ additional primary units are selected. If the 17 new units tend to have more than the average of 130 subunits obtained from the preliminary 5 units, then total costs will exceed $C = \$10{,}000$. Then some of the 22 primary units may not be used.

If $k = 0.04$ had been used instead of $k = 0.02$, then $m_1 = 2$, $m_2 = m_3 = 4$, $m_4 = m_5 = 8$, and $n = 14$. If a sampling study is conducted for these values of n and m_i, the resulting value for $s^2(\bar{x})$ may be compared with the value of $s^2(\bar{x})$ obtained by using the n and m_i given before for $k = 0.02$. Then the k giving the smallest $s^2(\bar{x})$ would be preferred, unless differences in $s^2(\bar{x})$ are small.

6.4 SUMMARY

It is common practice to measure only a portion (subsample) of environmental samples. This chapter showed how to estimate the uncertainty of estimated means and total amounts that arise from variability between samples and between subsamples within samples. Methods are also given for choosing the number of samples and subsamples that will minimize the variance of estimated means and totals for a prespecified cost.

EXERCISES

6.1 Consider Example 6.2. Can you think of situations when it might be better to sample at equidistant points in time during the 80-h period rather than at randomly chosen times? Consider factors such as cost, convenience, and accuracy of the estimated mean and total amount.

6.2 Suppose we want to estimate the average value, μ, for specific conductance in groundwater for a defined region at a given point in time. There are $n = 5$ wells more or less uniformly spaced over the region. A sample from each well is taken on the same day and $m = 4$ aliquots from each sample are measured, giving the following data:

	Aliquot			
Well	1	2	3	4
1	3,150	3,150	3,145	3,150
2	1,400	1,380	1,390	1,380
3	2,610	2,610	2,610	2,610
4	3,720	3,720	3,719	3,710
5	2,100	2,120	2,110	2,080

Estimate μ and the standard error of that estimate. Assume that $f_N = 0$ and that each groundwater sample has the same volume—that is, M is the same for all wells.

6.3 Suppose the cost of collecting a sample is $c_1 = \$500$, the cost of measuring an aliquot is $c_2 = \$100$, and the cost function $C = c_1 n + c_2 nm$ applies. Estimate the optimum number of aliquots to measure for specific conductance, using the data in Exercise 6.2. How many wells should be sampled if $C = \$2{,}000$ is available for sampling and analyses?

6.4 Suppose the data in Exercise 6.2 are replaced by

	Aliquot			
Well	1	2	3	4
1	3110	3180	3180	3190
2	3400	3400	3400	3400
3	3090	3100	3080	3040
4	3750	3750	3700	3750
5	3100	3100	3140	3100

Compute the mean, standard error of the mean, and the optimum m and n if $c_1 = \$500$, $c_2 = \$100$, and $C = \$2000$. Is m_{opt} larger than in Exercise 6.3?

ANSWERS

6.1 Discussion.

6.2 We use Eq. 6.2 and obtain $\bar{x} = 51{,}864/20 = 2593$. Since $f_N = 0$, we use Eq. 6.7 to obtain $s^2(\bar{x})$. The means for the 5 wells are 3149, 1388, 2610, 3717, and 2102. Hence, $s^2(\bar{x}) = 163{,}345$ or $s(\bar{x}) = 404$.

6.3 Use Eq. 6.10. We have $m = 4$, $n = 5$, and we obtain $s_1^2 = 816{,}726.4$ and $s_2^2 = 1239.5/15 = 82.63$ using Eqs. 6.4 and 6.5. Therefore, Eq. 6.10 gives

$$m_{\text{opt}} = \left(\frac{500/100}{816{,}726.4/82.63}\right)^{1/2} = (5/9884)^{1/2} = 0.02$$

that is, $m_{\text{opt}} = 1$. To determine n, we have $2000 = 500n + 100n(1)$, or $n = 2000/600 = 3.3$ or 4 wells.

6.4 By Eq. 6.2, $\bar{x} = 3298$. By Eq. 6.7, $s^2(\bar{x}) = 15{,}272$, or $s(\bar{x}) = 124$. Now $s_1^2 = 76{,}360$, $s_2^2 = 9250/15 = 616.67$. Therefore,

$$m_{\text{opt}} = \left[\frac{5}{76{,}360/616.67}\right]^{1/2} = 0.2$$

That is, $m_{\text{opt}} = 1$, the same as in Exercise 6.3. Also, $n = 4$ as in Example 6.3.

7 Compositing and Three-Stage Sampling

One way to estimate an average is to collect individual grab samples, measure each, and mathematically compute their average value. An alternative approach is to collect several grab samples and thoroughly mix them together into a composite sample. Then either the entire composite is measured, or one or more random subsamples from the composite are withdrawn and measured. If the mixing process is thorough, a physical averaging process takes place, so the subsamples represent the average concentration of the original grab samples.

Compositing is useful if the cost of analyzing individual grab samples for contaminants is high, if the mixing process is thorough, and if information on the variability or extreme concentrations for grab samples is not needed. Also, the total amount of pollutant present in the composite is equal to or greater than any single grab sample making up the composite. Hence, if the entire composite or large subsamples are analyzed, the pollutant may be more easily detected.

Adequacy of the mixing process must be assessed before compositing is done. Ideally, equal proportions of each grab sample should be in every possible subsample that could be withdrawn for measurement from the composite. These proportions may vary widely between subsamples if the mixing process is inadequate. This variation in turn will tend to increase the variance between subsample measurements. Note that even if the proportions in each subsample are the same, the mixing process may not remove the heterogeneity that originally existed *within* each grab sample.

This chapter discusses compositing in the setting of subsampling, since composite samples are usually subsampled rather than measured in their entirety. Examples show how to estimate averages, totals, and variances of these estimates when the composite samples are formed from a random partitioning of field samples. Methods for deciding on the number of composites and subsamples within composites are also given.

Compositing is frequently done in the monitoring of rivers and wastewater discharges. Guidelines when compositing should be preferred to grab sampling in such situations are given by Schaeffer and co-workers (1980, 1983), Schaeffer and Janardan (1978), Marsalek (1975), Montgomery and Hart (1974), and Rabosky and Koraido (1973). Brumelle, Nemetz, and Casey (1984), Elder, Thompson, and Myers (1980), and Rohde (1976) discuss statistical aspects of compositing.

7.1 BASIC CONCEPTS

When compositing is done, the target population is still defined in terms of population units, as discussed in previous chapters. But now these units are thought of as grouped into composite samples from which randomly selected units (subsamples) are withdrawn for measurement.

Figure 7.1 illustrates one type of compositing design. The irregularly shaped area represents the target population of population units for which an estiamte of μ is needed, say an area of land contaminated with a pollutant. A set of ng soil (grab) samples are collected randomly or systematically over the area. (Chapter 8 discusses systematic sampling.) These samples are randomly grouped into n composite samples, each composed of g soil samples. Then m subsamples are withdrawn from each composite, and s repeat measurements made on each subsample. The illustration in Figure 7.1 uses $n = 4$ and $g = m = s = 2$. The population unit is defined to be the subsample, since each subsample selected is measured in its entirety. The target population is the set of all possible subsamples to which inferences will be made.

The compositing design just described could be embedded in a more complicated design that could be used to take account of changing concentrations over space or time. For example, the target population of Figure 7.1 might correspond to one of many small local areas within a larger region, which might be one of several strata in a stratified sampling plan. This situation is discussed in the next section.

In practice, it is common to set $m = s = 1$, that is, to make just one measurement on one subsample from each composite. This practice may be adequate for estimating a mean or total, but it does not provide information on subsampling and repeat-measurement variability. Such information in conjunction with estimates of variability between composite samples can be used to guide the choice of n, m, and s for future sampling and compositing efforts. This approach was illustrated in Section 6.2.3 for two-stage sampling and is considered in Section 7.2.4 for three-stage sampling.

7.2 EQUAL-SIZED UNITS

Suppose the goal is to estimate the mean, μ, and total amount, I, for a defined target population of units. These population units can be thought of as being grouped into *batches*, where one or more composite samples are formed from a random or systematic selection of units from each batch. For example the target population in Figure 7.1 corresponds to a batch. A batch might be a study plot of dimension 10 m × 10 m from which soil samples are collected and used to form one or more composites. Several such batches (plots) might be located randomly or systematically over a large area. Information on the mean or total for each batch as well as for the larger area can be obtained from this design. A batch might also be the total flow of water from an effluent pipe during a 24-h period. Water samples could be collected at random times during this time span and pooled to form one or more composite samples. Each new 24-h period would correspond to a new batch.

Let B denote the number of batches in the population. Also, let N_i be the number of composites that could possibly be formed in the ith batch, and let

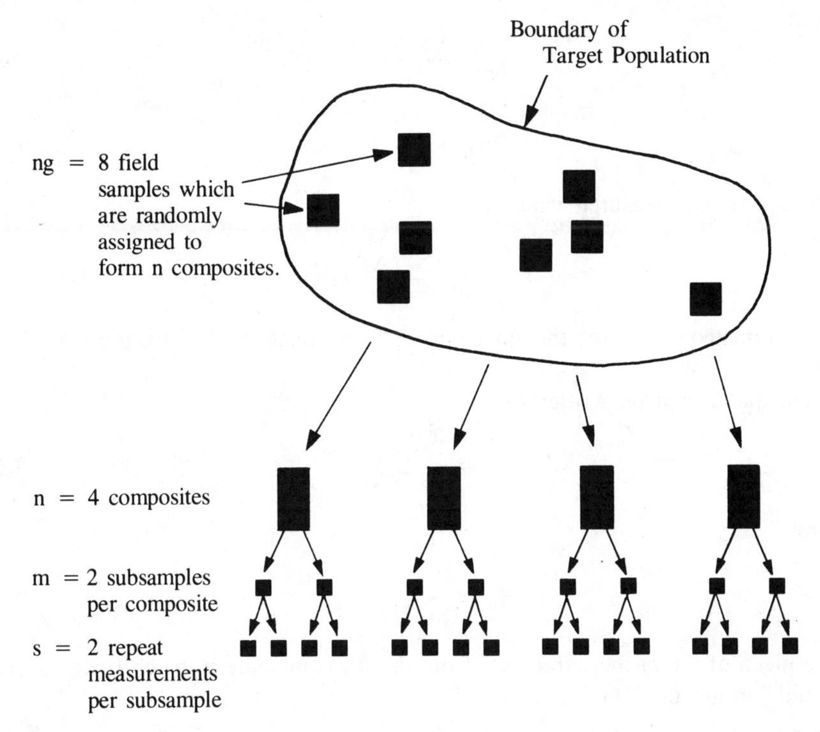

Figure 7.1 Formation of composite samples by random grouping of field samples.

M_{ij} be the number of possible subsamples in the jth composite in the ith batch. In this section we assume that all batches are equal in size, that is, $N_1 = N_2 = \cdots = N_B = N$, and that all composites have the same number of subsamples, that is, $M_{ij} = M$ for all i and j. The general case of unequal N_i and M_{ij} is considered in Section 7.3.

7.2.1 Estimating the Mean and Total Amount

For equal N_i and equal M_{ij} the parameters μ and I are

$$\mu = \frac{1}{BNM} \sum_{i=1}^{B} \sum_{j=1}^{N} \sum_{k=1}^{M} \mu_{ijk}$$

and

$$I = BNM\mu \qquad \textbf{7.1}$$

respectively, where μ_{ijk} is the true value for the kth subsample (population unit) of the jth composite of the ith batch.

These parameters may be estimated by selecting b of the B ($b \leq B$) batches at random, from each of which ng field samples are selected and randomly grouped to form n composite samples each containing g field samples. Then m subsamples from each composite are withdrawn and measured s times. Henceforth we assume $s = 1$, the usual case in practice. Then μ is estimated by computing

$$\bar{x} = \frac{1}{bnm} \sum_{i=1}^{b} \sum_{j=1}^{n} \sum_{k=1}^{m} x_{ijk} \tag{7.2}$$

$$= \frac{1}{b} \sum_{i=1}^{b} \bar{x}_i \tag{7.3}$$

where x_{ijk} is the measured value and

$$\bar{x}_i = \frac{1}{nm} \sum_{j=1}^{n} \sum_{k=1}^{m} x_{ijk} \tag{7.4}$$

is the estimated mean for the ith batch. The estimate of the inventory is $\hat{I} = BNM\bar{x}$.

Note that $\bar{x}_i$ can be written as

$$\bar{x}_i = \frac{1}{n} \sum_{j=1}^{n} \bar{x}_{ij} \tag{7.5}$$

where

$$\bar{x}_{ij} = \frac{1}{m} \sum_{k=1}^{m} x_{ijk} \tag{7.6}$$

is the mean of the m measurements from the jth composite in batch i. Equations 7.5 and 7.6 are used in Equation 7.7.

7.2.2 Variance of Batch Means

Consider the case where n/N is very near zero and an estimate is required of $\mathrm{Var}(\bar{x}_i)$, the variance of the ith batch mean. Elder, Thompson, and Myers (1980) showed that when compositing is used as discussed here, an unbiased estimate of this variance may be obtained by calculating a one-way analysis of variance (AOV) and dividing the between composites mean square by the number of measurements nm for the ith batch. This procedure is equivalent to computing the estimate of $\mathrm{Var}(\bar{x}_i)$ as follows:

$$s^2(\bar{x}_i) = \frac{1}{n(n-1)} \sum_{j=1}^{n} (\bar{x}_{ij} - \bar{x}_i)^2 \tag{7.7}$$

where $\bar{x}_i$ and $\bar{x}_{ij}$ are computed by Eqs. 7.5 and 7.6, respectively.

If n/N is not near zero, then $\mathrm{Var}(\bar{x}_i)$ may be estimated by Eq. 6.3, the usual formula for two-stage sampling, that is, by computing

$$s^2(\bar{x}_i) = \frac{(1 - n/N)}{n(n-1)} \sum_{j=1}^{n} (\bar{x}_{ij} - \bar{x}_i)^2 + \frac{(n/N)(1 - m/M)}{n^2 m(m-1)} \sum_{j=1}^{n} \sum_{k=1}^{m} (x_{ijk} - \bar{x}_{ij})^2 \tag{7.8}$$

In practice, g, n, and m may vary from batch to batch. If so, then n and m in Eqs. 7.2–7.8 are replaced by n_i and m_i, respectively, where n_i is the number of composites in batch i, and m_i is the number of subsamples withdrawn from each composite in batch i.

7.2.3 Variance of the Mean and Total Amount

If b/B is very near zero, an unbiased estimate of $\mathrm{Var}(\bar{x})$ ($\bar{x}$ as computed by Eqs. 7.2 or 7.3) may be obtained by computing a nested AOV (for methodology see, e.g., Snedecor and Cochran, 1980, pp. 248–250) and dividing the between batches mean square by the total number of measurements, bnm. This approach (when $b/B = n/N = 0$) is identical to computing

$$s^2(\bar{x}) = \frac{1}{b(b-1)} \sum_{i=1}^{b} (\bar{x}_i - \bar{x})^2 \qquad \mathbf{7.9}$$

For some applications, b/B or n/N may not be near zero. Cochran (1977, p. 287) then shows that $\mathrm{Var}(\bar{x})$ may be estimated by

$$s^2(\bar{x}) = \left(1 - \frac{b}{B}\right)\frac{s_1^2}{b} + \frac{b}{B}\left(1 - \frac{n}{N}\right)\frac{s_2^2}{bn} + \frac{b}{B}\left(\frac{n}{N}\right)\left(1 - \frac{m}{M}\right)\frac{s_3^2}{bnm} \qquad \mathbf{7.10}$$

where

$$s_1^2 = \frac{1}{b-1} \sum_{i=1}^{b} (\bar{x}_i - \bar{x})^2 \qquad \mathbf{7.11}$$

$$s_2^2 = \frac{1}{b(n-1)} \sum_{i=1}^{b} \sum_{j=1}^{n} (\bar{x}_{ij} - \bar{x}_i)^2 \qquad \mathbf{7.12}$$

and

$$s_3^2 = \frac{1}{bn(m-1)} \sum_{i=1}^{b} \sum_{j=1}^{n} \sum_{k=1}^{m} (x_{ijk} - \bar{x}_{ij})^2 \qquad \mathbf{7.13}$$

Note that Eq. 7.9 is a special case of Eq. 7.10 when $b/B = 0$. The quantity $\mathrm{Var}(\hat{I})$ is estimated as $(BNM)^2 s^2(\bar{x})$, where $s^2(\bar{x})$ is computed by Eqs. 7.9 or 7.10, whichever is appropriate.

If b, n, or m are large, estimating s_1^2, s_2^2, and s_3^2 from Eqs. 7.11, 7.12, and 7.13 can be tedious if done by hand. Of course, these equations can be programmed on a digital computer. Alternatively, the variances can be easily obtained using the computed mean squares from a nested AOV which can also be programmed. The appropriate AOV is illustrated by Snedecor and Cochran (1980, p. 248). Table 7.1 shows that $s_1^2 = (\mathrm{MS})_B/nm$, $s_2^2 = (\mathrm{MS})_C/m$ and s_3^2

Table 7.1 Nested Analysis of Variance Showing the Relationships between the Mean Squares and s_1^2, s_2^2, and s_3^2 [The method for computing the mean squares is given by Snedecor and Cochran (1980, p. 248).]

Source of Variation	*Mean Squares*
Batches	$(\mathrm{MS})_B = nms_1^2$
Composites within batches	$(\mathrm{MS})_C = ms_2^2$
Measurements within composites	$(\mathrm{MS})_E = s_3^2$

= $(MS)_E$. This AOV method cannot be used unless m and n are the same for all composites and batches, respectively.

EXAMPLE 7.1

This example extends Example 6.2. The liquid effluent discharged by a manufacturing plant contains a toxic chemical. Samples of the effluent are to be collected to estimate the true concentration ($\mu g/L$) and the total amount (μg) of this chemical in the total effluent discharged over a given 80-h work week (two 8-h shifts per day). Measurements will be made on 100-mL aliquots of effluent. The population unit is a 100-mL aliquot, and the target population consists of the total number of 100-mL aliquots discharged into the river during the designated 80-h period. We shall assume that the effluent flow rate is a constant 2000 L/h over the 80-h. A constant flow rate will allow us to use the unweighted mean (Eq. 7.2) to estimate μ. Unequal flow rates are considered in Section 7.3.

Suppose one effluent 1-L water sample is collected at a random time during each hour throughout the 80-h week. The $ng = 16$ samples for a day are randomly grouped into $n = 2$ composite samples each consisting of $g = 8$ one-liter samples. Then $m = 2$ subsamples of size 100-mL are withdrawn from each thoroughly mixed composite and measured for the chemical of interest. This sampling and compositing design is illustrated in Figure 7.2.

Suppose the subsample data are as given in Table 7.2. Using Eqs. 7.2 or 7.3, $\bar{x} = 0.0483\ \mu g/100$ mL is an estimate of μ. Using Eq. 7.10, which requires that B, N, and M be determined, we estimate $\mathrm{Var}(\bar{x})$. Since the goal is to estimate μ for the 5 days of a given week, we have $B = 5$. Hence, $b/B = 1$, since subsamples were measured on each of the 5 days. Each composite sample contains 8000 mL, so there are $M = 8000$ mL/100 mL $= 80$ subsamples in each composite. Hence, $m/M = 0.025$, since $m = 2$. The number N of potential composites per day depends on the total flow. Multiplying the flow rate by 16 h gives

Table 7.2 Hypothetical Concentrations for a Chemical in Effluent Water (for Example 7.1)

Day	*Composite*	*Measurements* x_{ijk} (μg/100 mL)		*Composite Mean* $\bar{x}_{ij}$	*Day Mean* $\bar{x}_i$	$s(\bar{x}_i)$
1	1	0.021	0.025	0.023	0.027	0.00400
	2	0.032	0.030	0.031		
2	1	0.045	0.048	0.0465	0.04275	0.00375
	2	0.038	0.040	0.039		
3	1	0.035	0.036	0.0355	0.03425	0.00125
	2	0.031	0.035	0.033		
4	1	0.065	0.066	0.0655	0.06475	0.00075
	2	0.068	0.060	0.064		
5	1	0.070	0.071	0.0705	0.07275	0.00225
	2	0.075	0.075	0.075		

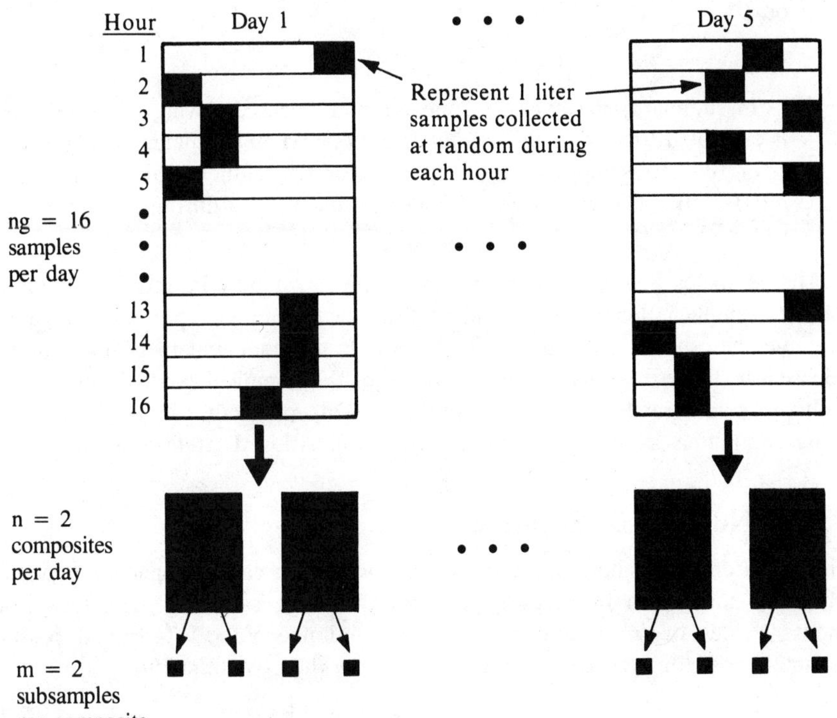

Figure 7.2 An illustration for Example 7.1. Sampling the flow from an effluent pipe at a random time each hour. Flow is assumed constant over time.

$$N = (2000 \text{ L/h}) (16 \text{ h/day}) (1 \text{ composite/8 L}) = 4000$$

composites per day. Hence, $n/N = 2/4000 = 0.0005$. Using $b/B = 1$, $m/M = 0.025$, and $n/N = 0$ in Eq. 7.10 gives

$$s^2(\bar{x}) = \frac{s_2^2}{bn} = \frac{0.0000149}{10} = 0.00000149$$

or

$$s(\bar{x}) = 0.00122\ \mu\text{g}/100 \text{ mL}$$

The inventory is estimated as

$$\begin{aligned}\hat{I} &= BNM\bar{x} \\ &= (5 \text{ days})(4000 \text{ composites/day}) \\ &\qquad (80 \text{ subsamples/composite})(0.0483\ \mu\text{g/subsample}) \\ &= 77{,}300\ \mu\text{g for the week}\end{aligned}$$

Then $\text{Var}(\hat{I})$ is estimated as

$$\begin{aligned}s^2(\hat{I}) &= (BNM)^2 s^2(\bar{x}) \\ &= (5 \times 4000 \times 80)^2\ (0.00000149) \\ &= 3.81 \times 10^6\ (\mu\text{g})^2\end{aligned}$$

or

$$s(\hat{I}) = 1950\ \mu g$$

In summary, the average concentration released during the week is estimated to be $\bar{x} = 0.0483\ \mu g/100$ mL with an estimated standard error (due to sampling only a portion of the effluent) of 0.00122 $\mu g/100$ mL. The estimated total amount discharged during the week is $\hat{I} = 77{,}300\ \mu g$ with a standard error of 1950 μg.

The estimates $\bar{x}$, $s(\bar{x})$, $\hat{I}$, and $s(\hat{I})$ in Example 7.1 may be misleading unless the 1-L sample collected at a random time each hour is representative of the true average concentration during that hour. If the chemical to be measured is released at discrete points in time, a pulse of the chemical could be missed by taking a single random sample each hour. This problem can be avoided by using continuous sampling rather than by compositing discrete grab samples.

7.2.4 Number of Samples

The selection of the optimum values of b, n, and m can be made in a manner similar to that given in Section 6.2.3 for two-stage sampling. The idea is to choose values of b, n, and m that will minimize $\text{Var}(\bar{x})$ for fixed cost or minimize cost for prespecified $\text{Var}(\bar{x})$. Assume the cost function

$$C = c_1 b + c_2 bn + c_3 bnm \qquad \textbf{7.14}$$

is applicable, where c_1 is the average cost per batch associated with selecting batches, c_2 is the average cost per composite of forming bn composites, and c_3 is the cost per subunit of selecting and measuring subunits. Then the optimum values of m and n may be approximated by computing

$$m = \left(\frac{c_2/c_3}{s_2^2/s_3^2}\right)^{1/2} \qquad \textbf{7.15}$$

and

$$n = \left(\frac{c_1/c_2}{s_1^2/s_2^2}\right)^{1/2} \qquad \textbf{7.16}$$

where s_1^2, s_2^2, and s_3^2 are computed using Eqs. 7.11, 7.12, and 7.13, respectively. Equations 7.15 and 7.16 are derived, for example, by Snedecor and Cochran (1967, p. 533). (We note that Eqs. 7.15 and 7.16 are derived assuming M and N are large. When M and N are not large, the formulas for m and n are such that negative estimates can occur. Sukhatme and Sukhatme (1970, pp. 280–282) give appropriate procedures for this case.)

The data for computing s_1^2, s_2^2, and s_3^2 should be obtained from prior studies at the same site. These studies should have used the same compositing design and mixing procedure as will be used for the new study. If not, the estimates s_1^2, s_2^2, and s_3^2 may not apply to the new study. Once m and n are determined, the cost equation (Eq. 7.14) may be solved to obtain b.

Once best values for b, n, and m have been determined, we must still choose g, the number of field samples that should be collected and mixed to form each composite. Elder, Thompson, and Myers (1980, p. 185) discuss the problems involved in choosing g. The main complication is that mixing a larger number

g of field samples to form each composite will not in all cases reduce $\mathrm{Var}(\bar{x})$. If by increasing g the mixing process can no longer achieve a homogeneous composite, then $\mathrm{Var}(\bar{x})$ could actually increase. Practical considerations such as ease of sampling, costs, and capacity of the mixing apparatus will restrict g to some feasible range of values $[g_{\min}, g_{\max}]$. Elder, Thompson, and Myers (1980) suggest that a pilot study using $g_{\min}$ and another study using $g_{\max}$ could be performed to see if $\mathrm{Var}(\bar{x})$ changes significantly. This experimental approach may be cost effective for a long-term study where it is important to maintain $\mathrm{Var}(\bar{x})$ at a low level.

EXAMPLE 7.2

Consider Example 7.1, where the average concentration and total amount of a toxic chemical discharged in an effluent stream was estimated for a given week. We shall use those data (Table 7.2) to estimate the optimum values of m, n, and b for estimating the mean and the total for a future week. Assume that the ratios s_1^2/s_2^2 and s_2^2/s_3^2 do not change appreciably over time and that the cost function given by Eq. 7.14 is appropriate.

Using the data in Table 7.2 and Eqs. 7.11, 7.12, and 7.13, we find that $s_1^2 = 0.000388$, $s_2^2 = 0.0000149$, and $s_3^2 = 0.0000058$. Hence, $s_1^2/s_2^2 = 26$ and $s_2^2/s_3^2 = 2.6$. Suppose the costs for a week's sampling and analyses effort are $C = \$2000$, and that $c_1 = \$50$, $c_2 = \$50$, and $c_3 = \$100$; then $c_1/c_2 = 1$ and $c_2/c_3 = 0.5$. Using Eqs. 7.16 and 7.17, we have

$$m = \left(\frac{0.5}{2.6}\right)^{1/2} = 0.4$$

$$n = \left(\frac{1}{26}\right)^{1/2} = 0.2$$

Using $m = 1$ and $n = 1$ in Eq. 7.14, we find that $b = 10$. Hence, based on the available data and the cost function, the allocation that will minimize $\mathrm{Var}(\bar{x})$ is estimated to be $b = 10$, $n = 1$, and $m = 1$. However, if $n = m = 1$, it will not be possible to compute s_2^2 or s_3^2 (Eqs. 7.12 and 7.13). If s_2^2 and s_3^2 are needed for planning future allocations or for other purposes, then n and m must be greater than 1.

The effect on $s^2(\bar{x})$ of varying b, n, and m can be evaluated by computing Eq. 7.10 for $B = 5$, $N = 4000$, $M = 80$, and the values of s_1^2, s_2^2, and s_3^2 given in the example. See Exercise 7.3.

7.3 UNEQUAL-SIZED UNITS

Section 7.2 required that (a) each batch have the same number N of potential composite samples, and (b) each composite have the same number M of potential subsamples. In this section a method discussed by Sukhatme and Sukhatme (1970, pp. 302–304) is given for estimating μ and I when these requirements cannot be met.

7.3.1 Estimating the Mean and Total Amount

Recall that N_i is the number of possible composite samples in the ith batch, and that M_{ij} is the number of possible subsamples in the jth composite of the ith batch. Our goal is to estimate μ, which is now defined to be

$$\mu = \frac{\sum_{i=1}^{B} \sum_{j=1}^{N_i} \sum_{k=1}^{M_{ij}} \mu_{ijk}}{\sum_{i=1}^{B} \sum_{j=1}^{N_i} M_{ij}} \tag{7.17}$$

$$= \frac{\text{total amount present in the target population}}{\text{total number of subsamples in the target population}}$$

$$= \frac{1}{B} \sum_{i=1}^{B} w_i \mu_i$$

where

$$w_i = \frac{\sum_{j=1}^{N_i} M_{ij}}{\frac{1}{B} \sum_{i=1}^{B} \sum_{j=1}^{N_i} M_{ij}}$$

$$= \frac{\text{number of possible subsamples in } i\text{th batch}}{\text{average number of possible subsamples per batch}} \tag{7.18}$$

and

$$\mu_i = \frac{\sum_{j=1}^{N_i} \sum_{k=1}^{M_{ij}} \mu_{ijk}}{\sum_{j=1}^{N_i} M_{ij}}$$

$$= \text{average amount per subsample in the } i\text{th batch} \tag{7.19}$$

It is clear from Eq. 7.17 that the inventory is just

$$I = \mu \sum_{i=1}^{B} \sum_{j=1}^{N_i} M_{ij}$$

The parameters μ and I are estimated by selecting b of the B batches at grouped into n_i composite samples, perhaps of different sizes. Each composite is thoroughly mixed, and m_{ij} subsamples are selected randomly from the jth composite in the ith batch. Each selected subsample is then measured in its entirety for the constituent of interest, yielding the data x_{ijk}.

The x_{ijk} may be used as follows to estimate μ:

$$\bar{x} = \frac{1}{b} \sum_{i=1}^{b} w_i \bar{x}_i \tag{7.20}$$

where

$$\bar{x}_i = \frac{1}{n_i} \sum_{j=1}^{n_i} v_{ij}\bar{x}_{ij}$$

$$= \text{estimated mean for the } i\text{th batch},$$

where w_i was defined in Eq. 7.18,

$$\bar{x}_{ij} = \frac{1}{m_{ij}} \sum_{k=1}^{m_{ij}} x_{ijk}$$

$$= \text{mean of the } m_{ij} \text{ measured subsamples in the } j\text{th composite of the } i\text{th batch}$$

and

$$v_{ij} = \frac{M_{ij}}{\sum_{j=1}^{N_i} M_{ij}/N_i}$$

$$= \frac{\text{number of possible subsamples in the } j\text{th composite of batch } i}{\text{average number of possible subsamples per composite in the } i\text{th batch}} \qquad \textbf{7.21}$$

Equation 7.20 reduces to Eq. 7.3 when $N_i = N$, $M_{ij} = M$, $n_i = n$, and $m_{ij} = m$ for all batches i and composites j.

An estimate of the inventory I is

$$\hat{I} = \bar{x} \sum_{i=1}^{B} \sum_{j=1}^{N_i} M_{ij} \qquad \textbf{7.22}$$

where $\bar{x}$ is computed using Eq. 7.20.

7.3.2 Variance of the Mean and the Total Amount

An estimate of $\mathrm{Var}(\bar{x})$ may be calculated from the x_{ijk} data as follows, if all M_{ij} are known:

$$s^2(\bar{x}) = \left(1 - \frac{b}{B}\right)\frac{s_1^2}{b} + \frac{b/B}{b^2} \sum_{i=1}^{b} w_i^2 \left(1 - \frac{n_i}{N_i}\right)\frac{s_{i2}^2}{n_i} + \frac{b/B}{b^2} \sum_{i=1}^{b} \frac{w_i^2 (n_i/N_i)}{n_i^2} \sum_{j=1}^{n_i} v_{ij}^2 \left(1 - \frac{m_{ij}}{M_{ij}}\right)\frac{s_{ij3}^2}{m_{ij}} \qquad \textbf{7.23}$$

where

$$s_1^2 = \frac{1}{b-1} \sum_{i=1}^{b} (w_i\bar{x}_i - \bar{x})^2$$

$$s_{i2}^2 = \frac{1}{n_i - 1} \sum_{j=1}^{n_i} (v_{ij}\bar{x}_{ij} - \bar{x}_i)^2$$

and

$$s_{ij3}^2 = \frac{1}{m_{ij} - 1} \sum_{k=1}^{m_{ij}} (x_{ijk} - \bar{x}_{ij})^2$$

Equation 7.23 reduces to Eq. 7.10 when $N_i = N$, $M_{ij} = M$, $n_i = n$, and $m_{ij} = m$ for all batches i and composites j.

The quantity $\mathrm{Var}(\hat{I})$ is estimated by computing

$$s^2(\hat{I}) = s^2(\bar{x}) \left(\sum_{i=1}^{B} \sum_{j=1}^{N_i} M_{ij} \right)^2 \qquad 7.24$$

where $s^2(\bar{x})$ is computed by Eq. 7.23.

As noted, Eqs. 7.23 and 7.24 can be used if all M_{ij} are known. In general, however, the M_{ij} will be unknown for composites not among the n_i actually formed unless all composites in the target population are the same size. If some of the M_{ij} are unknown, then v_{ij}, w_i, $\bar{x}$, $\hat{I}$, and $s^2(\hat{I})$ cannot be computed, since $\sum_{j=1}^{N_i} M_{ij}$ is required. However, we may estimate v_{ij} and w_i as follows, using only knowledge of N_i and the M_{ij} for the composites actually formed:

$$\hat{v}_{ij} = \frac{M_{ij}}{\frac{1}{n_i} \sum_{j=1}^{n_i} M_{ij}} \qquad 7.25$$

$$\hat{w}_i = \frac{N_i \sum_{j=1}^{n_i} M_{ij}/n_i}{\frac{1}{b} \sum_{i=1}^{b} \left(N_i \sum_{j=1}^{n_i} M_{ij}/n_i \right)} \qquad 7.26$$

Now, we estimate μ by computing

$$\bar{x} = \frac{1}{b} \sum_{i=1}^{b} \hat{w}_i \bar{x}_i \qquad 7.27$$

where

$$\bar{x}_i = \frac{1}{n_i} \sum_{j=1}^{n_i} \hat{v}_{ij} \bar{x}_{ij}$$

Then we estimate $\mathrm{Var}(\bar{x})$ as before, using $\hat{v}_{ij}$ and $\hat{w}_i$ in Eq. 7.23, and the inventory I by computing

$$\hat{I} = \frac{\bar{x}B}{b} \left(\sum_{i=1}^{b} N_i \sum_{j=1}^{n_i} M_{ij}/n_i \right) \qquad 7.28$$

where $\bar{x}$ is computed from Eq. 7.27. The quantity $\mathrm{Var}(\hat{I})$ is estimated by

$$s^2(\hat{I}) = s^2(\bar{x}) \left[\frac{B}{b} \sum_{i=1}^{b} N_i \sum_{j=1}^{n_i} M_{ij}/n_i \right] \qquad 7.29$$

where $s^2(\bar{x})$ is computed from Eq. 7.23 using $\hat{v}_{ij}$ and $\hat{w}_{ij}$ in place of v_{ij} and w_{ij}.

EXAMPLE 7.3

Consider again Example 7.1, where a chemical plant discharges liquid effluent containing a toxic chemical into a river. The effluent must be sampled to estimate the average concentration μ and total amount I of the chemical discharged during a given 5-day work week. The effluent flow rate and the concentration (μg/L) of the toxic chemical vary considerably during each day.

The target population is defined as in Example 7.1—that is, as the total number of 100-mL aliquots of effluent discharged to the river during a particular 80-h work week (5 days, 16 h per day). Suppose cost considerations will restrict the collection of samples to only $b = 3$ of the $B = 5$ days, the 3 days being selected at random. Suppose these days are Monday, Wednesday, and Thursday, denoted henceforth as day 1, day 2, and day 3, respectively. Composite samples will be formed manually by first collecting one 4-L sample at a random time during each hour of each selected day. Then the 16 samples collected on the ith day will be used to form n_i composites, where n_i is chosen to be proportional to the total flow for that day. Suppose that cost constraints limit the total number of composites that can be formed to $n_1 + n_2 + n_3 = 9$. All composites will be constructed to be 8 L in size, and $m = 2$ subsamples each of size 100 mL will be withdrawn from each composite.

Table 7.3 lists hypothetical flow data for days 1, 2, and 3. Since flow varies over time, the amount of the 4-L sample collected during hour h that goes into forming a composite is VP_h, where V is the desired volume of the composite sample ($V = 8$ in this example) and P_h is the proportion of that day's flow that occurs during hour h. For example, using Table 7.3, we have that on day 1, 0.182 L of the 4-L sample collected during the first hour is combined with 0.217 L from the second hour, 0.317 L from the third hour, and so on, to form the 8-L composite sample. The same procedure is used to form each composite on day 1, and similarly for the other days. The total flow for the 3 selected days is 131,955 L, of which 29%, 49%, and 22% occurred on days 1, 2 and 3, respectively. For the number of composites per day to be proportional to daily flow, we multiply these percents by 9, the allowed total number of composites. We get $n_1 = 3$, $n_2 = 4$, and $n_3 = 2$.

The manner in which the composites are formed in this example is not in accordance with random grouping of $n_i g$ samples into n_i composites, as described following Eq. 7.19. In this example the several composites formed on a given day have a positive correlation because all of them are formed from the same set of effluent samples. The result is that $s^2(\bar{\bar{x}})$ given by Eq. 7.23 may be too small, that is, it may underestimate $\text{Var}(\bar{\bar{x}})$. To obtain a more accurate estimate of $\text{Var}(\bar{\bar{x}})$, we need an estimate of the variability due to the mixing process as well as that due to the variation in mixing proportions. The latter variance may be estimated by using the flow proportion P_h, but the mixing variance will usually be unknown in practice. An exception is when the constituent of interest is dissolved in water, so a well-mixed composite of water samples implies a mixing variance of zero. In this case Schaeffer and Janarden (1978) and Schaeffer, Kerster, and Janardan (1980) give a method for estimating $\text{Var}(\bar{\bar{x}})$ from composite samples all obtained for the same flow conditions. In this example we shall use Eq. 7.23 to estimate $\text{Var}(\bar{\bar{x}})$, with the understanding that it may be somewhat too small.

Table 7.4 gives daily total flow data for each day of the week. The number N_i of possible composites on day i is obtained by dividing these flows by 8 L. The total number of possible subsamples

Table 7.3 Flow (in Liters) from an Effluent Pipe for Each Hour of the Three Days Selected at Random from a Five-Day Work Week (for Example 7.3) (hypothetical data)

	Monday			Wednesday			Thursday		
Hour	*Flow*	P_h[a]	VP_h[b]	*Flow*	P_h	VP_h	*Flow*	P_h	VP_h
1	867	0.0228	0.182	1,570	0.0244	0.195	932	0.0314	0.251
2	1,027	0.0271	0.217	1,864	0.0290	0.232	987	0.0333	0.266
3	1,502	0.0396	0.317	2,288	0.0355	0.284	1,057	0.0357	0.286
4	1,952	0.0514	0.411	3,160	0.0491	0.393	1,271	0.0429	0.343
5	2,121	0.0559	0.447	4,558	0.0708	0.566	1,593	0.0537	0.430
6	2,567	0.0676	0.541	6,080	0.0945	0.756	2,076	0.0700	0.560
7	3,893	0.1026	0.821	6,440	0.1001	0.800	2,552	0.0861	0.689
8	3,750	0.0988	0.790	5,988	0.0930	0.745	2,743	0.0925	0.740
9	3,339	0.0880	0.704	6,052	0.0940	0.752	3,072	0.1036	0.829
10	3,054	0.0805	0.644	5,460	0.0848	0.678	3,250	0.1096	0.877
11	2,987	0.0787	0.630	5,058	0.0786	0.629	3,074	0.1037	0.830
12	3,117	0.0821	0.657	4,396	0.0683	0.546	2,891	0.0975	0.780
13	2,700	0.0711	0.569	3,686	0.0573	0.458	1,598	0.0539	0.431
14	2,079	0.0548	0.438	3,400	0.0528	0.422	1,270	0.0428	0.342
15	1,501	0.0395	0.316	3,300	0.0513	0.410	900	0.0304	0.243
16	1,498	0.0395	0.316	1,060	0.0165	0.132	375	0.0127	0.102
Totals	37,954	1.000	8	64,360	1.000	8	29,641	1.000	8

[a] P_h = proportion of the day's flow that occurs during hour h.
[b] V = 8 L = desired volume of the composite for this example. VP_h = amount of the 4 L sample collected during hour h that goes into forming each 8 L composite sample for the day.

Table 7.4 Data for Computing the w_i and v_{ij}

Day	*Flow* (L)	*Number of Possible Composites*[a] N_i	*Number of Possible Subsamples*[b] $\sum_{j=1}^{N_i} M_{ij}$	w_i[c]	v_{ij}[d]
Monday	37,954	4,744	379,520	0.925	1
Tuesday	34,500	4,312	345,000	0.841	1
Wednesday	64,360	8,045	643,600	1.569	1
Thursday	29,641	3,705	296,400	0.722	1
Friday	38,700	4,837	387,000	0.943	1
	Total	25,643	2,051,520		

[a]Obtained by dividing total daily flow by the size of the composite (8 L in this example).
[b]Equal to $80N_i$ in this example.
[c]Computed from Eq. 7.18.
[d]Computed from Eq. 7.21.

on day i is obtained as $\Sigma_{j=1}^{N_i} M_{ij}$, which reduces to $80N_i$ because each composite contains $M_{ij} = 80$ subsamples. This information allows us to compute the w_i and v_{ij} using Eqs. 7.18 and 7.21. These results are given in Table 7.4. We note that total flows are required to compute w_i and v_{ij}, even for those days when composites are not formed.

Table 7.5 gives hypothetical concentrations for the two subsamples selected from each composite. Using Eqs. 7.20 and 7.22, we estimate the true mean μ and inventory I to be $\bar{x} = 0.053$ μg/100 mL and $\hat{I} = 109{,}000$ μg for the 5-day period. Equations 7.23 and 7.24 give $s^2(\bar{x}) = 0.00021744$ or $s(\bar{x}) = 0.015$ μg/100 mL and $s(\hat{I}) = 0.0147(2{,}051{,}520) = 30{,}157$, or 30,000 μg.

The third term of $s^2(\bar{x})$ in Eq. 7.23 is essentially zero because $n_i/N_i \cong 0$ for all 3 batches. Also, the second term contributes a negligible amount because the variances s_{i2}^2 of the composite means within each batch are so small relative to the variance $s_1^2 = 0.00162$ between the batch means.

Now suppose the daily flows were recorded only on days when composites were formed. In that case Eq. 7.27 is used to compute $\bar{x}$, and the v_{ij} and w_i must be estimated by Eqs. 7.25 and 7.26. Using $M_{ij} = 80$ and the N_i data for Monday, Wednesday, and Thursday in Table 7.4, we find $v_{ij} = 1$ (as before) and $\hat{w}_1 = 0.863$, $\hat{w}_2 = 1.463$, and $\hat{w}_3 = 0.6740$. Hence, from Eq. 7.27, $\bar{x}$ is equal to 0.0496, as compared to 0.0531, obtained when the w_i did not have to be estimated. Using Eqs. 7.28 and 7.29, we find $\hat{I} = 0.0495(5)(439840) = 108{,}860$, or 109,000 μg and $s(\hat{I}) = (0.0138)(5)(439840) = 30{,}349$, or 30,000 μg, the same values obtained previously.

7.4 SUMMARY

This chapter gives formulas for estimating the mean and the total amount of environmental contaminants when measurements are made on composite samples

Table 7.5 Estimating the Average Concentration and Inventory of a Chemical by Analyzing Subsamples

Day	n_i	*Composite*	n_i/N_i	*Measurements* x_{ijk} (μg/100 mL)		s_{ij3}^2	$\bar{x}_{ij}$	v_{ij}	$v_{ij}\bar{x}_{ij}$	s_{i2}^2	$\bar{x}_i$	w_i	$w_i\bar{x}_i$
Monday	3	1	0.000632	0.025	0.036	6.05×10^{-5}	0.0305	1	0.0305	2.33×10^{-6}	0.03083	0.925	0.0285
		2		0.031	0.028	4.50×10^{-6}	0.0295	1	0.0295				
		3		0.035	0.030	1.25×10^{-5}	0.0325	1	0.0325				
Tuesday	No composites formed			—	—	—	—	—	—	—	—	—	—
Wednesday	4	1	0.000497	0.065	0.063	2.0×10^{-6}	0.0640	1	0.0640	9.83×10^{-6}	0.0635	1.569	0.0996
		2		0.059	0.061	2.0×10^{-6}	0.0600	1	0.0600				
		3		0.057	0.078	2.2×10^{-4}	0.0675	1	0.0675				
		4		0.061	0.064	4.5×10^{-6}	0.0625	1	0.0625				
Thursday	2	1	0.000540	0.043	0.051	3.2×10^{-5}	0.0470	1	0.0470	2.81×10^{-5}	0.04325	0.722	0.0312
		2		0.038	0.041	4.5×10^{-6}	0.0395	1	0.0395				
Friday	No composites formed			—	—	—	—	—	—	—	—	—	—

formed from a random grouping of field samples. These methods involve two- and three-stage subsampling since composite samples are frequently subsampled rather than measured in their entirety. Formulas for estimating the variance of estimated means are also provided, as are methods for choosing the optimum number of composites and subsamples within composites.

EXERCISES

7.1 Suppose in Example 7.1 that the target population is redefined to be a period of 52 weeks rather than 1 week and that the data in Table 7.2 were obtained for 5 days chosen at random over the 365 days of the year. Estimate μ and I and compute the sample standard errors $s(\bar{x})$ and $s(\hat{I})$ of those estimates. Are $s(\bar{x})$ and $s(\hat{I})$ larger than obtained in Example 7.1? If so, why?

7.2 In Example 7.1 suppose the flow rate is 100,000 L/h and the target population is as described in Exercise 7.1. Compute $\bar{x}$, $s(\bar{x})$, $\hat{I}$, and $s(\hat{I})$. Do these estimates differ from those in Example 7.1 or Exercise 7.1? Why or why not?

7.3 For Example 7.2 use Equation 7.11 to evaluate the effect on $s(\bar{x})$ of varying the number of days (b), composites (n), and subsamples (m) when $B = 5$, $N = 4000$, $M = 80$, $s_1^2 = 0.000388$, $s_2^2 = 0.0000149$, and $s_3^2 = 0.0000058$. These 3 sample variances were computed by using the data in Table 7.2 in Equations 7.11, 7.12, and 7.13.

ANSWERS

7.1 $\bar{x} = 0.0483$ μg/100 mL as in Example 7.1. But now $b/B = 5/365 = 0.013699$. Equation 7.10 gives

$$s(\bar{x}) = \left[(1 - 0.013699)\frac{s_1^2}{5} + 0.013699(0.9995)\frac{s_2^2}{10} + 0.013699(0.0005)(0.975)\frac{s_3^2}{20}\right]^{1/2}$$

$$= (7.64 \times 10^{-5} + 2.04 \times 10^{-8} + 1.94 \times 10^{-12})^{1/2}$$

$$= 0.00874$$

which is 7 times larger than in Example 7.1 since b/B is smaller—that is, we are sampling only 5 of the 365 days in the year.

$$\hat{I} = (365 \times 4000 \times 80)0.0483 = 5{,}641{,}440 \text{ or } 5{,}640{,}000\ \mu\text{g/year}$$

$$s(I) = (365 \times 4000 \times 80)0.00874 = 1{,}020{,}832 \text{ or } 1{,}020{,}000\ \mu\text{g}$$

7.2 $\bar{x} = 0.0483$ μg/100 mL as in Example 7.1 and Exercise 7.1.

$$N = (100{,}000 \text{ L/h})(16 \text{ h})(1 \text{ composite/8 L})$$

$$= 200{,}000 \text{ composites per day.}$$

Therefore, $n/N = 2/100{,}000 = 0.00002 \cong 0$. Hence, $s(\bar{x})$ is the same as obtained in Exercise 7.1.

$$\hat{I} = (365 \times 100{,}000 \times 80)0.0483 = 141{,}036{,}000 \text{ or } 191{,}000{,}000\ \mu\text{g/year}$$

and

$$s(\hat{I}) = (365 \times 100{,}000 \times 80)0.00874 = 25{,}520{,}800 \text{ or } 25{,}500{,}000\ \mu\text{g}$$

which are larger than in Exercise 7.1, since the number of potential composites, N, is larger.

7.3 $$s(\bar{x}) = \left[\left(1 - \frac{b}{5}\right)\frac{0.000387575}{b} + \left(\frac{b}{5}\right)\left(1 - \frac{n}{4000}\right)\frac{0.0000149}{bn} + \left(\frac{b}{5}\right)\left(\frac{n}{4000}\right)\left(1 - \frac{m}{80}\right)\frac{0.0000058}{bnm}\right]^{1/2}$$

Days b	*Composites* n	*Subsamples* m	$s(\bar{x})$
5	2	2	0.00122 (as in Example 7.1)
5	10	2	0.000545
5	10	4	0.000545
3	2	2	0.00729
3	10	2	0.00721
3	10	4	0.00721

Taking more than 2 subsamples has no effect on $s(\bar{x})$. Increasing n from 2 to 10 cuts $s(\bar{x})$ in half if $b = 5$, but has almost no effect when $b = 3$.

8 Systematic Sampling

Chapters 4–7 discuss sampling designs for estimating a mean or a total that usually use a random selection procedure for choosing all population units to be measured. However, it is frequently desirable to choose the units systematically. This chapter discusses systematic sampling, a method where, in general, only one of the units is randomly selected. This randomly selected unit establishes the starting place of a systematic pattern of sampling that is repeated throughout the population of N units.

A simple example is choosing a point at random along a transect, then sampling at equidistant intervals thereafter. When sampling a geographical area, a common systematic design is to sample on a square grid. The location of one grid point, say that in the upper left corner, is determined at random within a local region. Then the locations of the remaining grid points are completely determined by the fixed spacing between grid lines.

Systematic sampling is usually easier to implement under field conditions than are simple random or stratified random sampling plans. Moreover, it provides for a uniform coverage of the target population that, in many cases, will yield more accurate estimates of mean concentrations. However, if the process being measured follows unsuspected periodicities over time and/or space, systematic sampling can give misleading and biased estimates of the population mean and total. Another problem with systematic sampling is the difficulty of obtaining an accurate estimate of the sampling error of the estimated mean and total unless the population is in random order.

In practice, a monitoring program is seldom conducted solely to estimate means. Usually, there is a need to estimate long-term trends, to define seasonal or other cycles, or to forecast pollution concentrations. For these objectives the use of equal-spaced observation points is recommended. An important design question is how frequently to sample, a topic discussed in this chapter.

This chapter

- Shows how to select a systematic sample
- Shows the dangers of using systematic sampling for estimating the population mean when periodicities are present
- Compares the accuracy of simple random sampling, stratified random sampling, and systematic sampling for estimating means when the population is in random order or has trends, periodicities, or correlations

Illustrates methods for estimating μ and $\mathrm{Var}(\bar{x})$ when systematic sampling is used and the population is not in random order

Briefly discusses methods for estimating spatial distributions of pollutants and provides references to the geostatistical literature in this area

A review of the statistical literature on systematic sampling is given by Iachan (1982).

8.1 SAMPLING ALONG A LINE

8.1.1. Selecting Systematic Samples

Consider sampling over time at an air monitoring station to estimate the annual average μ at that location. Suppose air measurements are made on air filters exposed for 24-h periods so that the target population consists of $N = 365$ population units (days) in the calendar year. A systematic sampling of these N units can be set up to obtain an unbiased estimate of μ.

First, an interval k (period between collection times) is chosen: for example, $k = 5$ days. Then a number between 1 and k inclusively is selected at random, say 3. A 24-h air sample is then collected on January 3 and every fifth day thereafter—that is, on January 3, 8, 13, 18, 23, and so on, as illustrated in Figure 8.1. The total number of observations for the year is then $n = N/k = 365/5 = 73$. Note that k is the number of possible systematic samples of size n that can be collected. These samples are enumerated in Table 8.1 for $N = 365$ and $k = 5$. In practice, usually only one of the k systematic samples of size n is collected. The mean $\bar{x}$ of these n values is an unbiased estimate of the true mean μ of the N population units. Then $\mathrm{Var}(\bar{x})$ may be approximated by one of the methods in Section 8.6.

In the preceding example the interval k between sample collection points or times was selected first, and then the number n of observations was determined as $n = N/k$. With this approach it is possible to get a fractional value for n (e.g., $n = 36.5$) if $N = 365$ and $k = 10$. A fractional value means some of the k systematic samples have a larger number of samples n than others. For the $N = 365$, $k = 10$ case, five systematic samples have $n = 37$ and five have $n = 36$.

To avoid this complication, one can first select n and then determine k as $k = N/n$ using the following procedure by Lahiri (1951) (discussed by Murthy, 1967, p. 139) that provides a statistically unbiased estimate of the mean. First decide how many units n are to be selected. Then let k be the integer nearest to N/n. Regard the N units as ordered around a circle. Choose a random number between 1 and N and select every kth unit thereafter going around the circle until n units are obtained. The following example is given in Figure 8.2. Suppose a systematic sample with $n = 4$ observations is desired from a target population of $N = 15$ units. Since $N/n = 3.75$, we set $k = 4$. Suppose the random number selected between 1 and 15 is 7. Then the $n = 4$ units in the systematic sample are 7, 11, 15, and 4.

When sampling along a transect in space, one need not actually specify N and k. Rather, the distance d between sampling points or times is specified. Then a random starting point P and the total length of line are specified. The

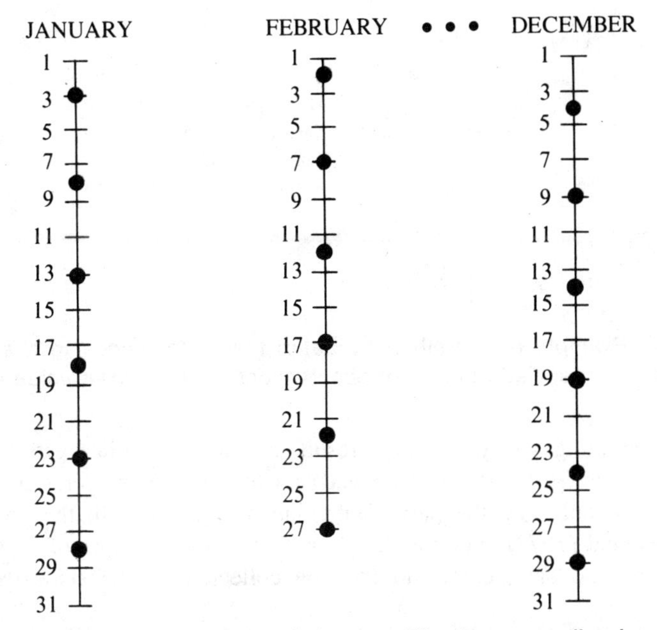

Figure 8.1 A systematic sample of size $n = 73$ with a sampling interval of $k = 5$ days and a random start position on January 3. $N = 365$ days.

sampling locations are at the points P, $P + d$, $P + 2d$, and so on. When the end of the transect is reached, one goes to the beginning of the transect and progresses until the point P is reached. An example is given in Figure 8.3 for a transect of 2500 m and a distance d of 500 m.

8.1.2 Cyclical Processes

Akland (1972) used suspended particulate air data to compare systematic sampling with a stratified random sampling procedure. He found that a biweekly ($k = 14$) or every third day ($k = 3$) systematic plan generally gave more accurate estimates of the annual mean than did the stratified random procedure. However,

Table 8.1 The k Possible Systematic Samples, Each Containing n Observations, When $N = 365$ and $k = 5$. Table Entries Are the Day of the Year When Observations Are Taken

Observation Numbers	*Possible Systematic Samples* 1	2	3[a]	4	5
1	1	2	3	4	5
2	6	7	8	9	10
3	11	12	13	14	15
⋮	⋮	⋮	⋮	⋮	⋮
72	356	357	358	359	360
$n = 73$	361	362	363	364	365

[a] Illustrated in Figure 8.1.

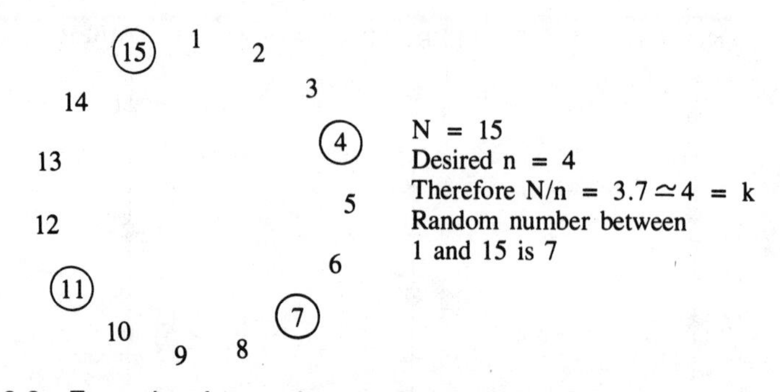

Figure 8.2 Example of a method by Lahiri (1951) for choosing a systematic sample with prespecified number of observations n and known value of N.

systematic sampling will not always result in a more accurate estimate of the mean. The accuracy of estimates depends on the particular cycles (if any) that exist in the variable and the particular value of k chosen. In the previous air example, Akland (1972) cautions that $k = 7$ or some multiple of 7 should not be used, since all samples would then be collected on the same day of the week.

Figure 8.4 illustrates the importance of knowing the particular pattern likely to be present before choosing k (this example is from Cochran, 1977, p. 218). Shown is a pure sine wave fluctuating about a constant mean concentration with a period of 30 days. If k is chosen equal to the period, as illustrated by the sampling points A, or an even multiple of the period, then every datum will have the same value. In that case there is no more information in the n measurements than for a single observation chosen at random. Furthermore, the estimate of μ is biased unless the points A happen to fall exactly on the mean line. However, suppose k is chosen to be an odd multiple of the half-period, for example, $k = 15$ as illustrated by the points B in Figure 8.4. Then every datum in the systematic sample will be the same distance above or below the true mean. Hence their average, $\bar{x}$, will equal μ, and the sampling variance of $\bar{x}$ is zero. For values of k between the extremes illustrated by points A and B, systematic sampling will give various degrees of accuracy in estimating μ.

Figure 8.4 illustrates the danger of not knowing about the presence of periodic variation before using systematic sampling to estimate a mean or total. When estimating a mean, the general strategy in choosing k is to ensure that all portions of the cyclical curve are represented. For example, if the cycle is weekly, all days of the week should be equally represented.

Power spectrum analysis may be used to study the periodic features of a

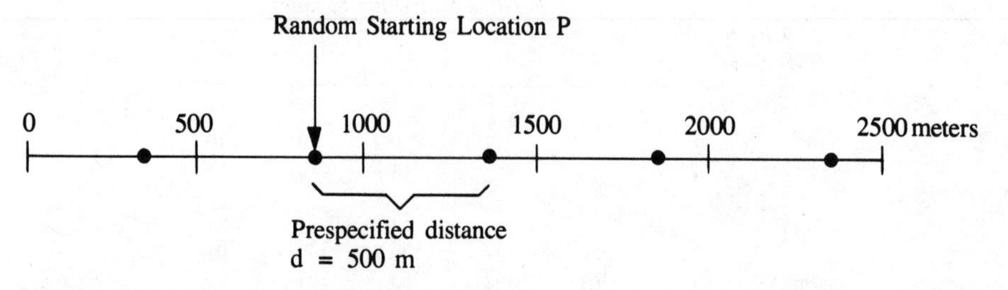

Figure 8.3 Choosing sampling locations for a systematic sample along a transect.

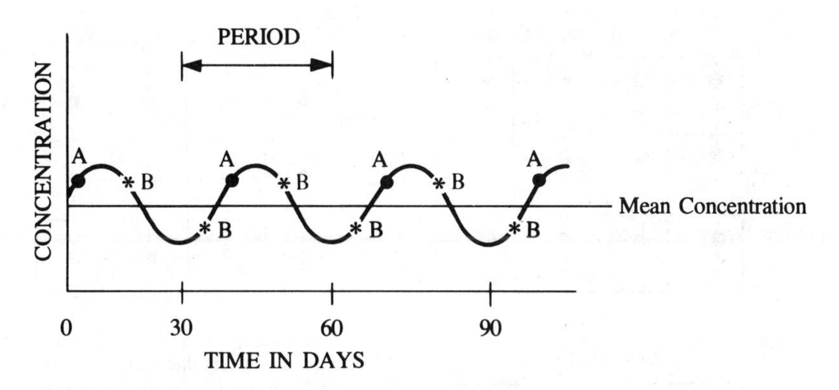

Figure 8.4 Pure sine wave periodic variation.

long series of data over time or space. The method, as applied to determining the frequency of sampling water or wastewater, is illustrated in Environmental Protection Agency (1982), Gunnerson (1966, 1968), and Sparr and Schaezler (1974). Sanders and Adrian (1978) base the sampling frequency in river quality monitoring on the magnitude of half the confidence interval of the random component of an annual mean. Loftis and Ward (1980b) consider the effects of seasonality and temporal correlation on the frequency of sampling needed for estimating a geometric mean. Ward et al. (1979) demonstrate a procedure for using statistics to evaluate a regulatory water-quality monitoring network's sampling frequencies.

If time or resources do not permit studying a suspected periodic structure before sampling begins, it may be preferable to estimate the mean by stratified random sampling with numerous strata. Systematic sampling rather than random sampling could be used within each stratum, with the starting point independently determined for each stratum.

8.2 SAMPLING OVER SPACE

The simplest systematic designs for sampling an area are the aligned and central aligned square grids illustrated in Figure 8.5(a) and (b), respectively. To determine the population units to be sampled for the aligned grid, first choose the distance between grid lines, which is equivalent to fixing n. Then two random coordinate numbers are drawn to fix the location of the point A in Figure 8.5(a). The remaining grid points are then fixed by the prespecified grid spacing. Deliberately placing A at the center of that square gives Figure 8.5(b).

To guard against bias in the estimated mean due to unsuspected periodicities over space, one can use the unaligned grid pattern in Figure 8.5(c). Berry and Baker (1968) recommend this design because it combines the useful aspects of random, stratified, and systematic sampling methods. Cochran (1977, p. 228) cites studies by Quenouille (1949) and Das (1950) that suggest this design is superior to both the square grid and stratified random sampling at least for some simple spatial correlation functions.

The sampling locations for the unaligned grid are determined as follows [refer to Fig. 8.5(c)]. First, point A is selected at random. The x coordinate at point A is then used with three new random y coordinates to determine points

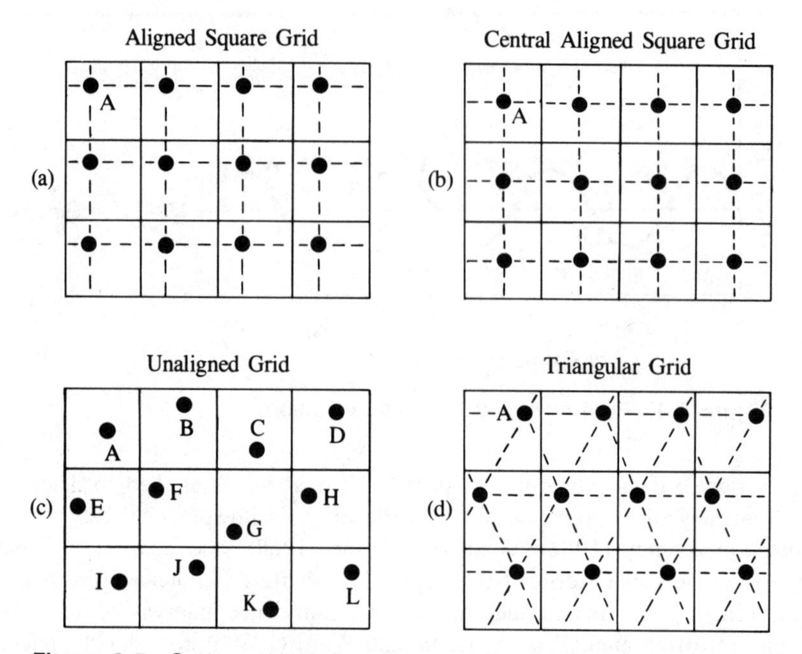

Figure 8.5 Systematic designs for sampling in space.

B, *C*, and *D*. The *y* coordinate of *A* is then used with two new random *x* coordinates to locate points *E* and *I*. The *x* coordinate of *E* and the *y* coordinate of *B* are then used as the coordinates of point *F*. Similarly, the *x* and *y* coordinates of *E* and *C*, respectively, are the *x* and *y* coordinates of point *G*, and so on, for the remaining cells.

The triangular systematic grid shown in Figure 8.5(d) is a variation on the aligned square grid. After the random point *A* is chosen, the other sampling points are fixed by the imposed triangular arrangement. Studies by McBratney, Webster, and Burgess (1981) suggest the triangular grid is slightly superior to the square grid if the spatial correlation structure varies with direction. Olea (1984) evaluated 14 different patterns of spatial sampling for estimating spatial processes. He found the triangular grid best but almost matched by the square pattern. Clustered patterns performed poorly relative to regular patterns. Section 8.7 discusses the estimation of spatial distributions of pollutants, which is usually best accomplished using one of the grid designs in Figure 8.5.

8.3 COMPARING SYSTEMATIC WITH RANDOM SAMPLING

Systematic sampling is easy to use in practice. But, we may ask, will it give an estimated mean, $\bar{x}$, with a variance at least as small as would be obtained if simple random sampling or stratified random sampling had been used? This question has been studied by examining environmental data sets. Results from 13 data sets are summarized by Cochran (1977, p. 221). These data suggest systematic sampling is superior to stratified random sampling plans that use one or two data per strata. Another method for comparing sampling plans is to derive mathematical expressions for $\mathrm{Var}(\bar{x})$ for each plan and to see under what conditions this variance is less for one plan than for another. These comparisons have been made by Cochran (1977, pp. 207–223) and Sukhatme and Sukhatme

(1970, pp. 356–369). A brief summary of their results is given in subsections 8.3.1, 8.3.2, and 8.3.3. Application of these results to environmental contaminant data is given in Section 8.4. Section 8.6 shows how Var($\bar{x}$) can be estimated with data obtained from a single systematic sample.

The performance of systematic sampling relative to simple random and stratified random sampling depends on the particular trends, patterns, hot spots, and correlations present in the population. For some populations systematic sampling will give very precise estimates of μ; for others, simple random sampling will do better. The structure of the population must be known to devise an efficient systematic sampling plan. This fact should be clear from the discussion in Section 8.1.2. Four types of population structures have been studied in the literature: populations in random order, populations with linear trends, populations with periodicities, and populations with correlations between values in close proximity. The use of systematic sampling for these structures is now discussed.

8.3.1 Populations in Random Order

Suppose the following conditions apply:

1. There are no trends in concentrations.
2. There are no natural strata—that is, regions within the population where concentrations are locally elevated.
3. Concentrations for all population units are uncorrelated.

Then a systematic sample of n observations will, *on the average*, estimate the true mean with the same precision as a simple random sample of size n and with the same precision as a stratified random sample with one observation in each of n strata.

Conditions 1–3 are summarized by saying the population is in "random order." In practice, even if conditions 1–3 are fulfilled, Var($\bar{x}$), obtained by systematic sampling, depends on the particular value chosen for k. But, on the average—that is, if the study were repeated many times with different values of k—systematic, stratified random (with one observation per stratum), and simple random sampling methods have equal Var($\bar{x}$). This equality suggests that for populations in random order we might as well use systematic sampling because of its convenience.

An example of a random order population might be radioactive fallout from atmospheric nuclear weapons tests that is uniformly distributed over large areas of land. In the absence of redistribution and concentration of these radionuclides by humans or nature, the "random" order population model could apply. If so, systematic sampling on a square grid would be convenient and should give an estimate of μ as precise as simple random sampling. It would also provide a uniform coverage of the area that would provide information on any unsuspected trends that may be present. This information could be used to plan future sampling designs.

8.3.2 Populations with Linear Trend

In general, contaminants over space and time cannot be considered to be in random order. If a pollutant comes from a point source, concentrations will usually decrease with distance from the source. If the population being sampled consists entirely of a linear trend, Cochran (1977) shows that (1) systematic sampling, will on the average, give a smaller Var($\bar{x}$) than simple random sampling, and (2) stratified random sampling will, on the average, give a smaller Var($\bar{x}$) than either systematic sampling or simple random sampling. However, the performance of systematic sampling for estimating a mean can be improved by using a *weighted* estimate of the mean (see Section 8.5).

If a linear or monotone trend is suspected, our first concern may be to estimate that trend rather than the mean. If unsuspected periodicities are not present, systematic sampling should work well for either objective. Methods for estimating and testing for trends are given in Chapters 16 and 17.

8.3.3 Correlated Populations

Stratified random sampling will, on the average, give a smaller Var($\bar{x}$) than simple random sampling will if the population has the following structure:

1. Measurements at two points a distance u apart are correlated; the correlation is denoted by ρ_u.
2. This correlation is solely a function of u and decreases as u increases.
3. There is no change in mean level or variability over the population.

However, for this correlation structure, no general result holds for systematic sampling. That is, random start systematic sampling might be better or worse than stratified random sampling or simple random sampling, depending on the chosen value of k. Suppose the population has such a correlation structure and the correlogram (graph of the correlations ρ_u versus distance u) is concave upward ($\rho_{u+1} + \rho_{u-1} - 2\rho_u \geq 0$ for all distances u). Then, on the average (over repetitions of the study), Var($\bar{x}$) for a centered systematic sample such as in Figure 8.5(b) is less than or equal to Var($\bar{x}$) for a random start systematic sample, a stratified random sample, or a simple random sample.

8.4 ESTIMATING THE MEAN AND VARIANCE

In Chapters 4–7 the mean μ and total $N\mu$ were estimated using measurements made on randomly selected population units. In this chapter, even though systematic sampling is used in place of random sampling, the same computing formulas are used to estimate μ and $N\mu$. However, different formulas must usually be used when estimating Var($\bar{x}$). Unless the population of units is in random order, it is difficult to obtain a completely valid estimate of this variance from a single systematic sample of size n because the random start position fixes all population units that will be measured. The following three designs that use systematic sampling *do* allow one to obtain an unbiased estimate of Var($\bar{x}$):

1. *Multiple systematic sampling* using a randomly determined starting position for each
2. *Systematic stratified sampling* where two or more systematic samples (each with a different random start position) are taken within each stratum

3. *Two-stage sampling* where the subsamples are collected according to a systematic sampling design
4. *Complementary systematic and random sampling* where a systematic sample is supplemented by a random sample of size m from the remaining population units

If a single systematic sample of size n has already been collected or if the foregoing designs are not feasible, $\text{Var}(\bar{x})$ might be approximated by one of the methods in Section 8.6. The first three options are considered in the next three sections. Option 4 is discussed by Zinger (1980).

8.4.1 Multiple Systematic Sampling

Suppose $J > 1$ systematic samples each of size n are obtained, each with an independently determined random start position. Let $\bar{x}_j$ be the arithmetic mean of the n data from the jth systematic sample. Then

$$\bar{x} = \frac{1}{J} \sum_{j=1}^{J} \bar{x}_j \qquad \textbf{8.1}$$

and $N\bar{x}$ are unbiased estimates of μ and $N\mu$, respectively. Also

$$s^2(\bar{x}) = \frac{(1 - J/k)}{J(J - 1)} \sum_{j=1}^{J} (\bar{x}_j - \bar{x})^2 \qquad \textbf{8.2}$$

and $N^2 s^2(\bar{x})$ are unbiased estimates of $\text{Var}(\bar{x})$ and $\text{Var}(N\bar{x})$, respectively. Note that $1 - J/k$ is a finite population correction factor that approximately equals 1 if J is much smaller than k, the total number of possible systematic samples.

If N cannot be factored exactly into nk parts, then all J systematic samples may not have the same number of observations. Then $\bar{x}$ is estimated as the weighted mean

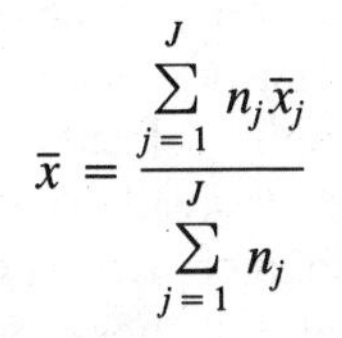

$$\bar{x} = \frac{\sum_{j=1}^{J} n_j \bar{x}_j}{\sum_{j=1}^{J} n_j} \qquad \textbf{8.3}$$

where $\bar{x}_j$ is the arithmetic mean of the n_j observations in the jth systematic sample. Equation 8.2 may still be used to estimate $\text{Var}(\bar{x})$.

Multiple systematic samples can be used for both line and area sampling. Example 8.1 illustrates line sampling. For area sampling, J systematic grids could be used. Referring to Figure 8.5, we see that J random starting points would be chosen in the upper left-hand square. A systematic grid sample would correspond to each such point. Then Eqs. 8.1–8.3 would be used, where $\bar{x}_j$ is the mean of the observations of the jth grid.

The conclusions in Sections 8.3.1 and 8.3.2 remain unchanged if we compare collecting J independent random start systematic samples with stratified random sampling (J units selected at random from each of n strata) with a simple random sample of Jn units. If the population is in random order, then, on the average, an equally precise estimate of μ will be obtained whether we use a single random start systematic sample of size Jn or J independent random start systematic samples each of size n. However, the former plan will be more precise, on the average, for a population with a linear trend or with a positive correlation between the elements where the correlogram is concave upward.

EXAMPLE 8.1

We want to estimate the average maximum daily ozone concentration at an air monitoring station from May 1 to September 30. Prior data indicate 1-week cycles are expected. Five systematic samples are taken, so an unbiased estimate of $\mathrm{Var}(\bar{x})$ can be computed. The total number of population units (days) in the target population is $N = 153$. A sampling interval of $k = 10$ days is chosen. Five random numbers between 1 and 10 (excluding 7) are then selected: for example, 1, 3, 4, 6, and 8. Table 8.2 gives the data (from Chambers et al., 1983). An unbiased estimate of μ is the weighted average of the 5 means, which are computed by Eq. 8.3. Thus $\bar{x} = 56.8$. Equation 8.2 gives $s^2(\bar{x}) = 3.264$ or $s(\bar{x}) = 1.8$.

8.4.2 Systematic-Stratified Sampling

If two or more systematic samples are taken in each stratum, the estimated stratum mean and its variance for each stratum may be obtained by Eqs. 8.1 (or 8.3) and 8.2, respectively. Then the techniques in Chapter 5 can be used to estimate the mean and the standard error for all strata combined. In some sampling programs it may not be necessary to estimate $\mathrm{Var}(\bar{x})$ in every stratum. In that case most strata would need only a single systematic sample of size n. But two or more systematic samples could be taken in a few randomly selected strata to estimate $\mathrm{Var}(\bar{x})$ for those strata.

8.4.3 Two-Stage Sampling

If systematic sampling is used to obtain subsamples in a two-stage sampling plan, a valid estimate of $\mathrm{Var}(\bar{x})$ can be obtained. The only condition is that the number of first-stage (primary) units selected for subsampling should be very small relative to the total number N available for selection. In this section we revert to the notation used in Chapter 6 for two-stage sampling. Hence, n = number of primary units selected for subsampling rather than the number of observations in a single systematic sample.

Suppose a single systematic sample of size m is taken in each of the n randomly selected primary units. Let $\bar{x}_i$ be the arithmetic mean of the m data in the ith primary unit. If all N primary units in the population have the same number of possible subunits, then

$$\bar{x} = \frac{1}{n} \sum_{i=1}^{n} \bar{x}_i \qquad \textbf{8.4}$$

is an unbiased estimate of the mean μ of the population. An unbiased estimator of $\mathrm{Var}(\bar{x})$ is

$$s^2(\bar{x}) = \frac{1}{n(n-1)} \sum_{i=1}^{n} (\bar{x}_i - \bar{x})^2 \qquad \textbf{8.5}$$

Table 8.2 Five Systematic Samples of Maximum Daily Ozone Concentrations (ppb) at Yonkers, N.Y. in 1974

	Systematic Samples									
	1		*2*		*3*		*4*		*5*	
Month	*Day*	*Max. Ozone*	*Day*	*Max. Ozone*	*Day*	*Max. Ozone*	*Day*	*Max. Ozone*	*Day*	*Max. Ozone*
May	1	47	3	27	4	37	6	42[a]	8	52
	11	27	13	40[a]	14	55	16	132	18	106
	21	80	23	21	24	50	26	37	28	33
	31	45								
June	10	106	2	24	3	52	5	111	7	31
	20	107	12	64	13	83	15	79	17	51
	30	119	22	68	23	19	25	20	27	30
July	10	50	2	108	3	85	5	48	7	54
	20	25	12	37	13	47	15	46	17	49
	30	70	22	78	23	40	25	25	27	62
Aug.	9	70	1	66	2	82	4	28	6	55
	19	75	11	67	12	127	14	56	16	100
	29	87	21	70	22	53	24	117	26	27
			31	114						
Sept.	8	67	10	74	1	66	3	25	5	27
	18	34	20	35	11	75	13	42	15	38
	28	15	30	18	21	24	23	17	25	14
$\bar{x}_i$		64.0		56.9		59.7		55.0		48.6
n		16		16		15		15		15

Source: After Chambers et al., 1983, p. 346.
[a] No datum for this day is given by Chambers et al. (1983). The datum shown here is the average of the adjacent preceding and following concentrations.

If the primary units have unequal numbers of subunits, then Eqs. 8.4 and 8.5 may be replaced by Eq. 6.12 and the first term in Eq. 6.13, respectively.

EXAMPLE 8.2

Suppose the average ^{238}Pu concentration in herbaceous vegetation over a large area is to be estimated. The area is divided into, say, $N = 500$ square first-stage units of equal size, each of which is further divided into $M = 100$ smaller second-stage units of equal size. The population sampling unit of interest is a second-stage unit because the entire plant material in selected second-stage units will be ashed and analyzed for ^{238}Pu. Suppose $n = 20$ first-stage units are selected at random. Within each such unit a single systematic sample of 10 secondary units is selected, and the mean $\bar{x}_i$ for that unit is computed. Then Eq. 8.4 is used to compute $\bar{x}$, and Eq. 8.5 is used to estimate $\text{Var}(\bar{x})$.

8.5 POPULATIONS WITH TRENDS

One disadvantage of a random start systematic sample is that any pronounced trend in the population will result in a substantial reduction in the accuracy of

$\bar{x}$ as an estimate of μ as compared to stratified random sampling. However, the loss of accuracy can be avoided by the use of end corrections, which were investigated by Yates (1948). Other types of end corrections have also been proposed, as discussed by Cochran (1977, p. 216) and compared by Bellhouse and Rao (1975). The methods perform similarly and are superior to ordinary systematic sampling (without end corrections) when a linear or parabolic trend is present.

End corrections for area sampling similar to those for line sampling could be developed. However, a simple solution is to use a central aligned square grid [Fig. 8.5(b)] so that each datum gets the same weight. This approach could also be used for line sampling.

8.6 ESTIMATING Var($\bar{x}$) FROM A SINGLE SYSTEMATIC SAMPLE

If a single systematic sample of size n is collected, no completely valid estimate of Var($\bar{x}$) can be obtained from the n data values without some assumptions about the population. This section illustrates several approximate methods. Wolter (1984) provides guidance on which of these and others are preferred when sampling establishments and people. But he cautions that his conclusions could be somewhat different when sampling environmental populations, because the data may be more highly correlated.

8.6.1 Population in Random Order

If the population is in random order, as defined in Section 8.3.1, then an unbiased estimate of Var($\bar{x}$) is obtained by

$$s^2(\bar{x}) = \frac{1 - (n/N)}{n(n-1)} \sum_{i=1}^{n} (x_i - \bar{x})^2 \qquad \mathbf{8.6}$$

the same estimator used with simple random sampling (Eq. 4.7). Similarly, if systematic sampling is used in more complicated designs such as stratified random or multistage sampling, the formulas for Var($\bar{x}$) given in Chapters 5-7 may be used, if one assumes the random order hypothesis is true within strata and primary units. The random order hypothesis is not tenable for environmental contaminant populations that contain trends, periodicities, or correlated data. Then Var($\bar{x}$) may be approximated by the method in the next section.

8.6.2 Balanced Differences Along a Line

Yates (1948) showed how to estimate Var($\bar{x}$) by computing ''balanced differences'' involving segments of nine consecutive and overlapping data values in a systematic sample. The first two differences are

$$d_1 = \tfrac{1}{2} x_1 - x_2 + x_3 - x_4 + x_5 - x_6 + x_7 - x_8 + \tfrac{1}{2} x_9$$

$$d_2 = \tfrac{1}{2} x_9 - x_{10} + x_{11} - x_{12} + x_{13} - x_{14} + x_{15} - x_{16} + \tfrac{1}{2} x_{17}$$

There will be g such differences, where $g \cong n/9$, and n is the number of data in the systematic sample. The sampling error is approximated by computing

$$s^2(\bar{x}) = \frac{1 - (n/N)}{7.5\, gn} \sum_{u=1}^{g} d_u^2 \qquad \textbf{8.7}$$

where 7.5 is the sum of squares of the nine coefficients in each difference d_u. In natural populations studied by Yates (1948), he found Eq. 8.7 overestimated the sampling error.

EXAMPLE 8.4

We return to Example 8.1 and the maximum daily ozone data. The use of Eq. 8.7 requires that n be fairly large so that the number of differences g will not be too small. In Example 8.1, $N = 153$. If $k = 2$, that is, if maximum ozone is measured every other day over the 153 days, then $g = 9$. The data for Yonkers, N.Y., on alternate days from the May 1 through September 30 period of 1974 are given in Table 8.3. The arithmetic mean of the $n = 77$ data is $\bar{x} = 54.9$, our estimate of μ for the 5-month period. The $g = 9$ differences are

$$d_1 = \tfrac{1}{2}(47) - 27 + 38 - 45 + 51 - 27 + 40 - 72 + \tfrac{1}{2}(119) = 41$$

$$d_2 = \tfrac{1}{2}(119) - 42 + 80 - 21 + 31 - 19 + 22 - 45 + \tfrac{1}{2}(24) = 77.5$$

$$\vdots$$

$$d_9 = \tfrac{1}{2}(9) - 67 + 74 - 74 + 40 - 23 + 34 - 35 + \tfrac{1}{2}(27) = -33$$

as given in Table 8.3. Therefore, Eq. 8.7 gives

$$s^2(\bar{x}) = \frac{(1 - 77/153)(37104.5)}{7.5(9)(77)} = 3.546$$

or $s(\bar{x}) = 1.9$. This number agrees well with $s(\bar{x}) = 1.8$ obtained in Example 8.1 by using Eq. 8.2, where 5 independent systematic samples were used, each with a sample spacing of $k = 10$ days.

8.6.3 Differences Over an Area

The preceding methods of estimating the sampling error of $\bar{x}$ when sampling along a line can be extended to systematic samples over space. Two approaches are discussed by Yates (1981, p. 231, 233). However, little is known about their accuracy for environmental contaminant data. If the random order model applies, the usual estimator $(1 - n/N)s^2/n$ appropriate to simple random sampling can be used. If the random order model is inappropriate, one may collect several independent systematic area samples and estimate $s^2(\bar{x})$ as discussed in Section 8.4.1. Koop (1971) shows that estimates of Var($\bar{x}$) can be seriously biased if the systematic sample is split into two halves or into successive pairs of units and each half is treated as an independent replicate.

If the correlation structure of the phenomenon can be accurately estimated, then a weighted moving-average technique called *kriging* may be used to estimate area means. Kriging is discussed further in the next section.

Table 8.3 Maximum Daily Ozone Concentrations (ppb) on Alternate Days at Yonkers, N.Y.

Date	i	*Max. Ozone* x_i	u	d_u	*Date*	i	*Max. Ozone* x_i	u	d_u
May 1	1	47	1	41	July 14	38	71		
3	2	27			16	39	41		
5	3	38			18	40	59		
7	4	45			20	41	25	6	−47.5
9	5	51			22	42	78		
11	6	27			24	43	13		
13	7	40[a]			26	44	46		
15	8	72			28	45	80		
17	9	119[a]	2	77.5	30	46	70		
19	10	42			Aug. 1	47	66		
21	11	80			3	48	47		
23	12	21			5	49	44	7	37
25	13	31			7	50	34		
27	14	19			9	51	70		
29	15	22			11	52	67		
31	16	45			13	53	96		
June 2	17	24	3	144.5	15	54	54		
4	18	88			17	55	44		
6	19	117			19	56	75		
8	20	37			21	57	70	8	−53.5
10	21	106			23	58	36		
12	22	64			25	59	43		
14	23	97			27	60	77		
16	24	36			29	61	87		
18	25	75	4	−5.5	31	62	114		
20	26	107			Sept. 2	63	18		
22	27	68			4	64	14		
24	28	67			6	65	9	9	−33
26	29	35			8	66	67		
28	30	31			10	67	74		
30	31	119			12	68	74		
July 2	32	108			14	69	40[a]		
4	33	96	5	−30.5	16	70	23		
6	34	60			18	71	34		
8	35	71			20	72	35		
10	36	50			22	73	27		
12	37	37			24	74	21		
					26	75	32		
					28	76	15		
					30	77	18		

Source: After Chambers et al., 1983, p. 346.
[a]No datum for this day is given by Chambers et al. (1983). The datum shown here is the average of the adjacent preceding and following concentrations.

8.7 ESTIMATING SPATIAL DISTRIBUTIONS

If a region is known to be contaminated with pollutants, there may be a need to estimate the spatial distribution (pattern) and total amount of pollution present over the region to assess the situation and plan for possible remedial actions. When estimating spatial distributions, it is usually best if the data are collected on a central aligned or triangular grid system (shown in Fig. 8.5) to ensure that all areas of the region are represented.

Data from the grid may be used in one or more ways to estimate the spatial distribution. Two commonly used methods are trend surface analysis and weighted moving averages. Trend surface analysis is described by Koch and Link (1980, Vol. 2), Davis (1973), and Krumbein and Graybill (1965). The technique can work well if the change in mean levels over the region is relatively small, if data are collected on a fine-enough grid over the region, and the goal is to show only the broad features of the spatial pattern. Delfiner and Delhomme (1975), Whitten (1975), and Davis (1973) discuss problems with trend surface analysis.

Weighted moving-average methods use the notion that data values closest to the point or block where the mean is being estimated contain more information than data farther away and, hence, get more weight. A problem is knowing what weight to assign to each datum. *Kriging*, a weighted moving-average estimation technique based on geostatistics, determines the weights by using the spatial correlation structure of the pollutant, which is estimated from the grid data. In this way a mean is estimated at each node of a grid laid over the region of interest. Kriging permits one to estimate the variance of each estimated mean and hence to assess whether additional data are needed in any portion of the region.

A difficulty in using kriging is estimating the correlation structure, particularly when data are not normally distributed. If the data are lognormal, correlation estimation and kriging may be done on the log-transformed data. However, the resulting estimated means and variances in the original scale will be biased if the correlation structure is not modeled correctly (see Gilbert and Simpson, 1985). This problem is difficult if the correlation structure is changing over the region.

Examples of applications of kriging to pollution problems are Bromenshenk et al. (1985), Gilbert and Simpson (1985), Flatman and Yfantis (1984), Flatman (1984), Journel (1984), Eynon and Switzer (1983), Sophocleous (1983), Hughes and Lettenmaier (1981), and Delhomme (1978, 1979). An introduction to geostatistics and kriging is given by Clark (1979). Journel and Huijbregts (1978) give a comprehensive discussion and provide computer codes. Recent developments are discussed by Verly et al. (1984).

8.8 SUMMARY

The periodic features of a population should be known before systematic sampling is adopted. If unsuspected periodicities are present, biased estimates of means and standard errors can occur. Tests of significance, confidence intervals about means, and other statistical analyses based on the data would also be biased. If periodicities have been quantified before the study begins, then sampling intervals that are even multiples of the period can be avoided.

Valid estimates of $\mathrm{Var}(\bar{x})$ can be obtained when systematic sampling is used if one of the following designs is utilized: (1) multiple systematic sampling, each with a randomly determined starting point, (2) stratified sampling where two or more systematic samples are taken in each stratum, and (3) two-stage sampling where subsamples are obtained by systematic sampling, and the fraction

Table 8.4 Five Systematic Samples of Maximum Daily Ozone Concentrations (ppb) at Yonkers, N.Y., in 1974

	Systematic Samples									
	1		*2*		*3*		*4*		*5*	
Month	*Day*	*Ozone*	*Day*	*Ozone*	*Day*	*Ozone*	*Day*	*Ozone*	*Day*	*Ozone*
May	2	37	5	38	7	45	9	51	10	22
	12	25	15	72	17	119[a]	19	42	20	45
	22	107	25	31	27	19	29	22	30	67
June	1	36	4	88	6	117	8	37	9	93
	11	49	14	97	16	36	18	75	19	104
	21	56	24	67	26	35	28	31	29	81
July	1	76	4	96	6	60	8	71	10	60.5[a]
	11	27	14	71	16	41	18	59	20	53
	21	45	24	13	26	46	28	80	30	39
	31	74								
Aug.	10	41	3	47	5	44	7	34	8	60
	20	86	13	96	15	54	17	44	18	44
	30	47	23	36	25	43	27	77	28	75
Sept.	9	74	2	18	4	14	6	9	7	16
	19	58	12	74	14	40[a]	16	23	17	50
	29	21	22	27	24	21	26	32	27	51

Source: After Chambers et al., 1983, p. 346.
[a]No datum for this day is given by Chambers et al. (1983). The datum given here is the average of the adjacent preceding and following concentrations.

of first-stage (primary) units that are subsampled is near zero. For long-term environmental studies it may be permissible to estimate $\mathrm{Var}(\bar{x})$ only occasionally. At those times multiple systematic samples could be used. At other times a single systematic sample may suffice.

Estimating $\mathrm{Var}(\bar{x})$ from a single systematic sample can be accomplished by using segments of nine consecutive data values, as illustrated in Section 8.6.2, but the estimated variance will tend to be too large. Methods for estimating $\mathrm{Var}(\bar{x})$ developed for random sampling plans can be used with confidence only when the population is in random order.

Estimating spatial distributions of pollutants is best achieved if measurements are taken on a grid system and if allowance is made for the correlation structure in the data. References to geostatistical and trend surface methods are provided.

EXERCISES

8.1 Suppose there is a need to estimate the total amount of radioactive fallout per square mile that has occurred from atmospheric testing within a defined geographical region. The total will be estimated by collecting surface soil samples with the assumption that no fallout radioactivity has migrated below the standard depth to which soil will be collected. Which of the following types of areal sampling designs over the region of interest would you choose to estimate the total: simple random, grid sampling, stratified random, or a combination of these? Why? Take into account that radioactive particles in the soil can be redistributed by weathering processes such as runoff from rain storms and human activities such as road building.

8.2 The data in Table 8.4 are $J = 5$ systematic samples of maximum ozone concentrations at a station in Yonkers, N.Y., drawn from the same population of $N = 153$ days as used in Example 8.1 and Table 8.2. The 5 systematic samples in Table 8.2 begin on days May 1, 3, 4, 6, and 8. The systematic samples in Table 8.4 begin on days May 2, 5, 7, 9, and 10. Estimate the average maximum ozone concentration and its standard error at this station for the period from May 1 to September 30, using the data in Table 8.4. Do these values of $\bar{x}$ and $s(\bar{x})$ agree well with those in Example 8.1?

8.3 Using the data in Table 8.4, estimate $\bar{x}$ and $s(\bar{x})$ as if the data were obtained by using simple random sampling. Do $\bar{x}$ and $s(\bar{x})$ computed in this way differ very much from $\bar{x}$ and $s(\bar{x})$ in Exercise 8.2?

ANSWERS

8.1 Discussion.

8.2 $\bar{x}_1 = 53.69$, $\bar{x}_2 = 58.07$, $\bar{x}_3 = 48.93$, $\bar{x}_4 = 45.80$, $\bar{x}_5 = 57.37$, $n_1 = 16$, $n_2 = n_3 = n_4 = n_5 = 15$. From Eq. 8.3, $\bar{x} = 4011.59/76 = 53$. From Eq. 8.2, $s(\bar{x}) = [(1 - 5/10)5.6711]^{1/2} = 1.7$. Yes.

8.3 From Eqs. 4.3 and 4.7, $\bar{x} = 53$ and $s(\bar{x}) = 2.1$.

9 Double Sampling

Frequently, two or more techniques may be available for measuring the amount of pollutant in an environmental sample or unit. For example, a mobile laboratory situated at the study site may use an analytical technique different from what might be used if the sample were shipped to a laboratory for analysis. The mobile laboratory procedure may be less accurate (''fallible'') but also less time consuming and costly. The shortened turnaround time may be a significant advantage in the field. For example, if a cleanup operation is underway, data may be needed quickly to avoid costly delays in taking remedial action.

More generally, if the goal of sampling is to estimate an average concentration or total amount, one approach is to use both techniques on a relatively small number of samples. Then this information is supplemented with a larger number of samples measured only by the fallible method. This aproach, called *double sampling*, will be cost effective if the linear correlation between measurements obtained by both techniques on the same samples is sufficiently near 1 and if the fallible method is substantially less costly than the more accurate method.

This chapter

- Describes how to use double sampling to estimate the population mean and total
- Shows how to determine whether double sampling is more cost effective than simple random sampling
- Gives methods for determining the optimum number of accurate and fallible measurements
- Gives a detailed case study to illustrate these methods.

This chapter is based on discussions by Cochran (1977), Sukhatme and Sukhatme (1970), and Tenenbein (1971, 1974).

9.1 LINEAR REGRESSION DOUBLE SAMPLING

9.1.1 Method

Following are the steps required to estimate μ and the standard error of that estimate by linear regression double sampling. The assumption is made that there is an underlying linear relationship between the accurate and the fallible methods. This assumption must be verified by conducting one or more special studies before double sampling can be considered for use. Once this verification is done, the following steps are taken to estimate μ:

1. Select a random sample of n' units from the target population.
2. Select a random sample of n units from among the n' units and measure each, using *both* the accurate and the fallible methods. Let x_{Ai} and x_{Fi} denote the accurate and fallible measurements, respectively, on the ith unit.
3. Measure the remaining $n' - n$ units, using only the fallible method.
4. The estimate of μ, denoted by $\bar{x}_{\text{lr}}$ (lr stands for *l*inear *r*egression), is obtained by computing

$$\bar{x}_{\text{lr}} = \bar{x}_A + b(\bar{x}_{n'} - \bar{x}_F) \qquad \textbf{9.1}$$

   where $\bar{x}_A$ and $\bar{x}_F$ are the means of the n accurate and fallible measurements, respectively, $\bar{x}_{n'}$ is the mean of the n' fallible values, and b is the slope of the estimated linear regression of accurate on fallible values. The estimate of the total, $N\mu$, is $N\bar{x}_{\text{lr}}$.
5. The estimated variance of $\bar{x}_{\text{lr}}$ is computed as

$$s^2(\bar{x}_{\text{lr}}) = s^2_{A \cdot F}\left[\frac{1}{n} + \frac{(\bar{x}_{n'} - \bar{x}_F)^2}{(n-1)s^2_F}\right] + \frac{s^2_A - s^2_{A \cdot F}}{n'} - \frac{s^2_A}{N} \qquad \textbf{9.2}$$

   where s^2_A and s^2_F are the variances of the n accurate and fallible values, respectively, and $s^2_{A \cdot F}$ is the residual variance about the estimated linear regression line. The standard error of $\bar{x}_{\text{lr}}$ is the square root of Eq. 9.2.

Computing formulas for all terms in Eqs. 9.1 and 9.2 are given in Table 9.1. The terms $\bar{x}_A$, $\bar{x}_F$, s^2_A, s^2_F, and $s^2_{A \cdot F}$ may also be obtained by a linear regression computer program from one of the standard statistical packages, such as the Biomedical Computer Programs, P-Series, (BMDP, 1983, programs P6D, P1R, or P2R). The last term in Eq. 9.2 is negligibly small if the number of population units N is very large relative to s^2_A.

Although the method is reasonably straightforward to apply in practice (see Section 9.3 for an example), it may not be cost effective unless the conditions given in the next section are met. Also, it can be shown mathematically that $\bar{x}_{\text{lr}}$ is a statistically biased estimate of μ unless n and n' are reasonably large (Sukhatme and Sukhatme, 1970, p. 211, give the equation for the bias).

We note that for populations with no unsuspected periodicities, systematic sampling (Chapter 8) may be preferred to random sampling in steps 1 and 2. However, if periodicities are present in the population, then $\bar{x}_{n'}$, $\bar{x}_A$, and $\bar{x}_F$ could be biased, resulting in a biased estimate of μ.

9.1.2 Comparison to Simple Random Sampling

Suppose the cost equation

$$C = c_A n + c_F n' = \text{total dollars available for measuring collected samples} \qquad \textbf{9.3}$$

Table 9.1 Quantities Needed for Use in Equations 9.1 and 9.2 to Estimate the Mean and Its Variance Using Linear Regression Double Sampling

$$\bar{x}_A = \frac{1}{n}\sum_{i=1}^{n} x_{Ai} = \text{mean of } n \text{ accurate values}$$

$$\bar{x}_F = \frac{1}{n}\sum_{i=1}^{n} x_{Fi} = \text{mean of } n \text{ fallible values}$$

$$s_A^2 = \frac{1}{n-1}\sum_{i=1}^{n} (x_{Ai} - \bar{x}_A)^2 = \text{variance of the } n \text{ accurate values}$$

$$s_F^2 = \frac{1}{n-1}\sum_{i=1}^{n} (x_{Fi} - \bar{x}_F)^2 = \text{variance of the } n' \text{ fallible values}$$

$$\bar{x}_{n'} = \frac{1}{n'}\sum_{i=1}^{n'} x_{Fi} = \text{mean of the } n \text{ fallible values}$$

$$b = \frac{\sum_{i=1}^{n} (x_{Ai} - \bar{x}_A)(x_{Fi} - \bar{x}_F)}{\sum_{i=1}^{n} (x_{Fi} - \bar{x}_F)^2}$$

= slope of the estimated linear regression of the accurate on the fallible values

$$s_{A \cdot F}^2 = \frac{n-1}{n-2}(s_A^2 - b^2 s_F^2) = \text{residual variance about the estimated linear regression line}$$

$$\hat{\rho} = \frac{\sum_{i=1}^{n} (x_{Ai} - \bar{x}_A)(x_{Fi} - \bar{x}_F)}{\left[\sum_{i=1}^{n} (x_{Ai} - \bar{x}_A)^2 \sum_{i=1}^{n} (x_{Fi} - \bar{x}_F)^2\right]^{1/2}}$$

= estimated correlation coefficient between the accurate and fallible values

applies, where c_A and c_F are the costs per unit of making an accurate and a fallible measurement, respectively. (Note that C does not include costs of collecting samples.) Also, let $R = c_A/c_F$. Then under the following three conditions, linear regression doubling sampling will, on the average, yield a more precise estimate of μ (smaller variance of the estimated mean) than would be achieved by the accurate method on C/c_A units selected by simple random sampling:

Condition 1. The underlying relationship between the accurate and fallible values is linear.

Condition 2. The optimum values of n and n' are used to estimate μ [those that minimize $\text{Var}(\bar{x}_{\text{lr}})$].

Condition 3.

$$R > \frac{(1 + \sqrt{1 - \rho^2})^2}{\rho^2} \tag{9.4}$$

or, equivalently,

$$\rho^2 > \frac{4R}{(1 + R)^2} \tag{9.5}$$

Data from prior or pilot studies may be used to evaluate the linear hypothesis in condition 1. The optimum n and n' in condition 2 depend on the true correlation coefficient ρ (see next section) as do Eqs. 9.4 and 9.5. In practice, since ρ is seldom known a priori, data from prior studies or from special pilot studies are usually needed to estimate ρ and hence to evaluate conditions 1, 2, and 3. The next section shows how to determine n and n' if ρ is known a priori and suggests what can be done when ρ must be estimated.

9.1.3 Number of Samples

Assume that the cost function given by Eq. 9.3 applies. Suppose we want to find n and n' that will minimize $\text{Var}(\bar{x}_{\text{lr}})$ for a given fixed budget of C dollars (the *fixed-cost* case). These optimum n and n' are given by the following equations (from Eqs. 2.4 and 2.5 of Tenenbein, 1971):

$$n = \frac{Cf_0}{c_A f_0 + c_F} \qquad \textbf{9.6}$$

and

$$n' = \frac{C - c_A n}{c_F} \qquad \textbf{9.7}$$

where

$$f_0 = \left(\frac{1 - \rho^2}{\rho^2 R}\right)^{1/2} \qquad \textbf{9.8}$$

and f_0 is set equal to 1 if Eq. 9.8 gives an f_0 greater than 1. Hence, the procedure is to first determine f_0 by Eq. 9.8, then compute n and n' from Eqs. 9.6 and 9.7, respectively.

Alternatively, we may consider a *fixed-variance* case, where the goal is to choose n and n' such that cost is minimized subject to the constraint that $\text{Var}(\bar{x}_{\text{lr}})$ is no greater than the variance of a mean computed by a simple random sample of n_v accurate measurements. In that case the optimum n and n' are obtained as follows (from Eqs. 2.6 and 2.7 in Tenenbein, 1971):

$$n = n_v(1 - \rho^2 + \rho^2 f_0) \qquad \textbf{9.9}$$

and

$$n' = \frac{n n_v \rho^2}{n - n_v(1 - \rho^2)} \qquad \textbf{9.10}$$

where f_0 is computed as in Eq. 9.8.

Now, as previously noted, the true ρ is never known in practice. This means that Eqs. 9.6–9.10 will give only estimates of the optimum n and n'. Hence, there is no complete assurance in practice that linear regression double sampling will do better than simple random sampling, even if a linear model is appropriate and Eqs. 9.4 and 9.5 are satisfied by $\hat{\rho}$. The problem is that $\text{Var}(\bar{x}_{\text{lr}})$ may be much larger than its minimum value if the estimates of n and n' are greatly different from their optimum values.

Tenenbein (1974) gives guidance on how much $\text{Var}(\bar{x}_{\text{lr}})$ can be increased when nonoptimum n and n' are used. For the fixed-cost case he shows that if the true $\rho < 0.95$, then $[\text{Var}(\bar{x}_{\text{lr}})]^{1/2}$ will not exceed about 15% of its minimum value if Eqs. 9.6 and 9.7 are replaced by

$$n = \frac{C\sqrt{R}}{(\sqrt{R} + 1)c_A} \qquad 9.11$$

$$n' = \frac{C}{c_F(\sqrt{R} + 1)} \qquad 9.12$$

In other words, as long as $\rho < 0.95$, the use of Eqs. 9.11 and 9.12 instead of Eqs. 9.6 and 9.7 will not greatly increase $\text{Var}(\bar{x}_{lr})$ over its minimum possible value. (Equations 9.11 and 9.12 are obtained by using $\rho^2 = 0.5$ in Eqs. 9.6–9.8.) However, if the true ρ is believed to be greater than 0.95, then Eqs. 9.11 and 9.12 should be avoided and Eqs. 9.6–9.8 should be used instead. For example, if $\rho = 0.98$, then the percentage increase in $[\text{Var}(\bar{x}_{lr})]^{1/2}$, when using Eqs. 9.11 and 9.12 to obtain n and n', will be 57% and 104% when $R = 5$ and 10, respectively (computed from Eq. 4.1 in Tenenbein, 1971).

9.1.4 Number of Samples from a Pilot Study

Note that n and n' can be determined by Eqs. 9.11 and 9.12 without having to first estimate ρ or evaluate from data the validity of the linear regression assumption. This practice, of course, is not recommended. Although Eqs. 9.11 and 9.12 can give a rough guide to the n and n' that might be required to use double sampling, it is always necessary to conduct pilot studies or to use suitable data from prior studies to estimate ρ and to evaluate the linear regression hypothesis.

If a pilot study is needed, a three-stage sampling procedure recommended by Tenenbein (1974) may be used to estimate ρ and then n and n'. Consider first the *fixed-cost* case. The first stage consists of randomly selecting m units from the population and making both accurate and fallible measurements on each unit to estimate ρ. Then assuming condition 3 (Eq. 9.4 or 9.5) is satisfied, one obtains estimates $\hat{n}$ and $\hat{n}'$ by Eqs. 9.6–9.8. If $m < \hat{n}$ then $\hat{n} - m$ additional units are randomly selected for measurement by both methods. This selection is the second stage of the sampling plan. At this point ρ can be estimated by using all $\hat{n}$ data, and condition 3 can be rechecked before proceeding. The third stage consists of using $\hat{n}$ for n in Eq. 9.7 to reestimate n'. Once the data are in hand, $\bar{x}_{lr}$ and $s^2(\bar{x}_{lr})$ may be computed by Eqs. 9.1 and 9.2.

Tenenbein (1974) recommends the following rule for deciding the number of units, m, to select at the first stage for this fixed-cost case: Use

$$\begin{aligned} m &= 21 \qquad \text{if Eq. 9.11} > 21 \\ &= \text{Eq. 9.11} \qquad \text{otherwise.} \end{aligned} \qquad 9.13$$

He found that estimating ρ on the basis of m accurate-fallible data pairs as determined by this rule gave values for n and n' sufficiently close to the optimum values that there was a less than a 1 in 20 chance that the increase in $\text{Var}(\bar{x}_{lr})$ would exceed 10%. This rule establishes an upper bound of $m = 21$, which limits the losses that could occur by making too many accurate measurements at the first stage—that is, if it should happen that the second stage gives an $\hat{n}$ less than m.

If the *fixed-variance* approach is used, then the three-stage sampling plan remains unchanged except that (1) n and n' are estimated by Eqs. 9.9 and 9.10

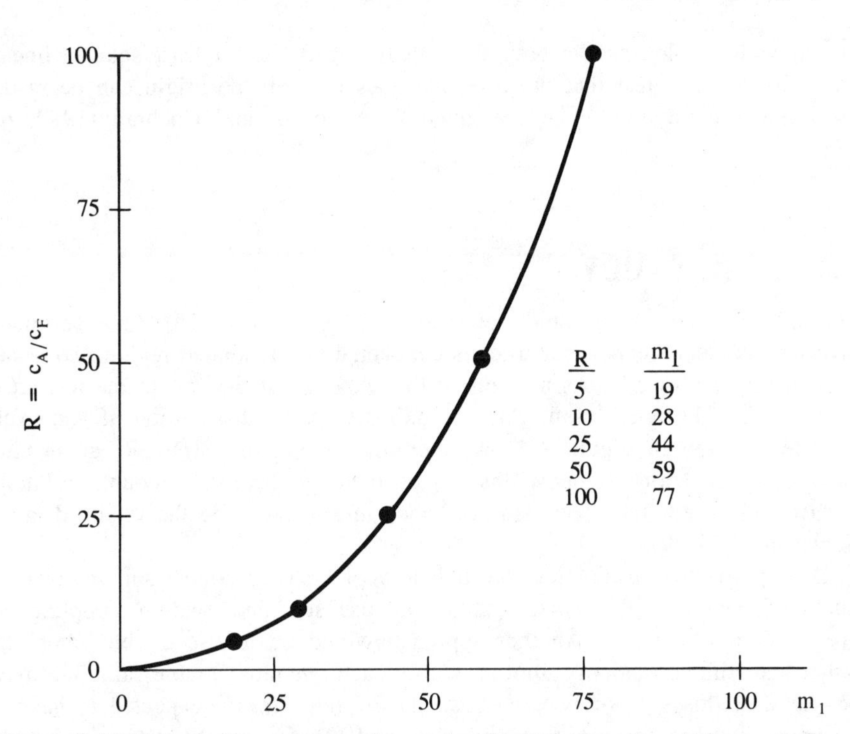

Figure 9.1 Values of m_1 versus the cost ratio R for the Fixed Variance Case. (The plotted points are from Table 1 of Tenenbein, 1974.)

instead of Eqs. 9.6 and 9.7, respectively, and (2) the following rule is recommended by Tenenbein (1974) for determining the number m of first stage units:

$$m = n_v\left[0.3 + \left(\frac{0.21}{R}\right)^{1/2}\right] \quad \text{if} \quad m_1 > n_v\left[0.3 + \left(\frac{0.21}{R}\right)^{1/2}\right]$$

$$= m_1 \quad \text{otherwise,} \qquad \mathbf{9.14}$$

where m_1 depends on R and is plotted versus R in Figure 9.1. Recall that n_v was defined in the paragraph preceding Eq. 9.9. This rule also establishes an upper bound for m.

The rules given by Eqs. 9.13 and 9.14 should be used to *guide* not dictate the number of units to collect at the first stage. Also, a minimum of $m = 10$ or 15 seems generally advisable in cases where very little information on ρ is available prior to sampling.

9.2 RATIO DOUBLE SAMPLING

Thus far we have assumed that the regression of the accurate method on the fallible method is linear. If the additional assumption can be made that there is a ratio relationship between the two methods (equivalent to a linear regression through the origin), then a ratio double sampling estimator may be considered for estimating μ and $N\mu$. This method is discussed by Cochran (1963, 1977) and Sukhatme and Sukhatme (1970).

In practice, it is unwise to assume the ratio model (regression through the origin) without plotting the data and calculating the usual least squares linear regression line. A test that the true line goes through the origin can be made by a linear regression procedure given by Snedecor and Cochran (1980, p. 173).

9.3 CASE STUDY

This example is based on a study reported by Gilbert et al. (1975). An experiment involving nuclear weapon materials is conducted in an isolated region (Tonopah Test Range) in Nevada. As a result of this experiment the soil in the test area contains $^{239,240}Pu$ and ^{241}Am. The approximate spatial distribution of Am over the site is given in Figure 9.2 (as determined using the FIDLER, an in situ gamma ray detector). Suppose that an estimate of the total amount of Pu in surface soil is required. The area is divided into strata using the in situ data as shown in Figure 9.2.

One approach to estimating the inventory of Pu is to collect soil samples at random locations within each stratum and use stratified random sampling as discussed in Chapter 5. Another approach would be to use double sampling rather than simple random sampling within each stratum. Double sampling may be useful at this site for two reasons: (1) Pu and Am are expected to have a linear relationship and a high correlation, and (2) Am can be determined with less expense (using gamma ray spectrometry) than can Pu since the latter requires expensive radiochemical techniques. Sections 9.3.1 and 9.3.2 show how to set up a double sampling program in stratum 1. The extension to stratified double sampling is discussed in Section 9.3.3.

9.3.1 Fixed Cost

An estimate of the total Pu in surface soil within stratum 1 is needed. It will be obtained by first estimating the average Pu concentration on a per m^2 basis. Then this mean is multiplied by the land area (m^2) in stratum 1 to estimate the total Pu.

Suppose data from similar field experiments suggest that a straight-line regression model is likely to be applicable and that correlation between the accurate and fallible measurements is very high, about 0.99. Also, suppose \$5000 is available to measure the soil samples collected in stratum 1, and that the cost of Pu and Am measurements are $c_A = \$200$ and $c_F = \$100$, respectively. Hence, $R = c_A/c_F = 2$. Therefore $\hat{\rho}^2 = 0.98$ and $4R/(1 + R)^2 = 8/9 = 0.89$. Hence, by Eq. 9.5 (since $0.98 > 0.89$), it appears that double sampling may be more cost effective than simple random sampling.

If double sampling is used and we want the estimate of μ to have the minimum possible variance for a fixed cost of \$5000, then Eqs. 9.6–9.8 are used to determine the optimum n and n' as follows:

$$f_0 = \left[\frac{1 - 0.98}{0.98(2)}\right]^{1/2} = 0.10$$

$$n = \frac{5000(0.1)}{200(0.1) + 100} = 4.17 \quad \text{which is rounded up to 5}$$

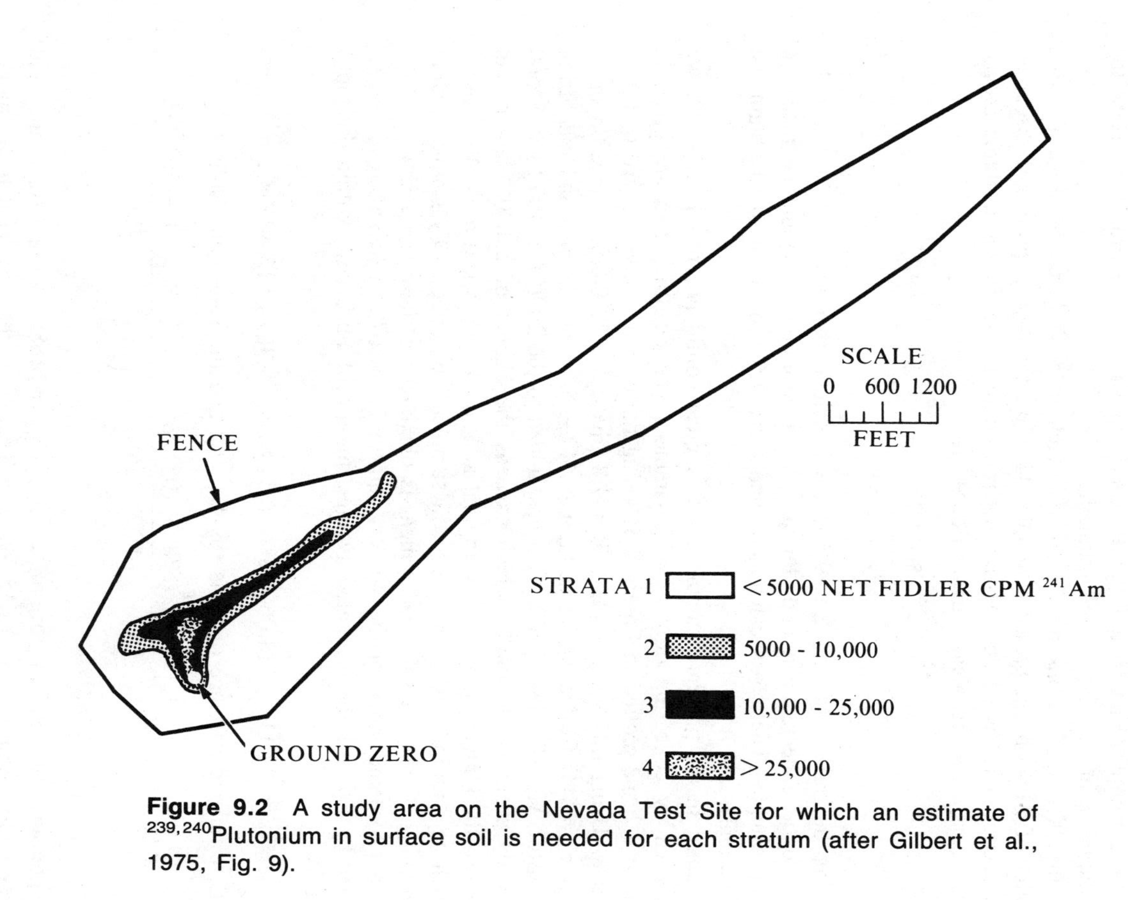

Figure 9.2 A study area on the Nevada Test Site for which an estimate of ^{239,240}Plutonium in surface soil is needed for each stratum (after Gilbert et al., 1975, Fig. 9).

and

$$n' = \frac{5000 - 200(5)}{100} = 40$$

Hence, if indeed $\rho = 0.99$, then 40 samples should be collected within stratum 1 by simple random sampling. All are measured for Am, and a random subset of five samples is measured for Pu. Since at this point no samples have been collected within stratum 1, the recommended procedure is to first collect the $n = 5$ samples and to use the Pu and Am values to estimate ρ. Then the optimum n and n' can be recomputed.

In other applications there may be much more uncertainty about the true value of ρ than in the present example. Then the three-stage procedure discussed in Section 9.1.4 is needed. Using that procedure in the present example for the fixed-cost case, we obtain, using Eq. 9.11,

$$n = \frac{5000\sqrt{2}}{(\sqrt{2} + 1)200} = 14.6 \text{ or } 15$$

Using the rule given by Eq. 9.13, we conclude that $m = 15$ soil samples should be collected and both Pu and Am measurements should be made on each sample. Then $\hat{\rho}$ is calculated, and Eqs. 9.6–9.8 are used to obtain the optimum n and n'.

Since in the present example there is great confidence that $\rho \geq 0.99$, we proceed by collecting only $n = 5$ samples. The resulting data are given in Table 9.2 and plotted in Figure 9.3. Using the data in Table 9.2, we find $\hat{\rho} = 0.998$. Using this value in Eqs. 9.6–9.8 gives $f_0 = 0.0448$, $n = 2.06$ or 3, and $n' = 44$. Collecting an additional $44 - 5 = 39$ samples and measuring each for Am will cost \$3900. This cost added to the $200(5) + 100(5) = \$1500$ already spent gives \$5400, which exceeds the budget of \$5000. Rather than exceed the budget, suppose the original allocation $n = 5$ and $n' = 40$ is used. Then an additional $n' - n = 35$ soil samples are collected by simple random sampling. Only the Am measurement is obtained for these samples.

Suppose the mean of the 40 Am data is $\bar{x}_{n'} = 678$ (a hypothetical value). Then by Eq. 9.1 the estimated average amount of Pu per m^2 within the top 5 cm is

$$\bar{x}_{lr} = 22{,}112 + 18.06(678 - 1051.8) = 15{,}361 \text{ or } 15{,}400 \text{ nCi/m}^2$$

The estimated variance of $\bar{x}_{lr}$ is (by Eq. 9.2 and the results in Table 9.2)

$$s^2(\bar{x}_{lr}) = (562{,}176)\left[\frac{1}{5} + \frac{(678 - 1051.8)^2}{4(327{,}336)}\right] + \frac{107{,}186{,}720 - 562{,}176}{40}$$

$$= 2{,}838{,}041$$

Therefore, the standard error of $\bar{x}_{lr}$ is $s^2(\bar{x}_{lr}) = 1680 \text{ nCi/m}^2$. Now, the land area within stratum 1 is 1,615,000 m^2. Therefore, the estimated total amount of Pu present in surface soil is

$$(1{,}615{,}000 \text{ m}^2)(15{,}361 \text{ nCi/m}^2) = 2.48 \times 10^{10} \text{ nCi} = 24.8 \text{ or } 25 \text{ Ci}$$

The estimated standard error of this estimate is

$$(1{,}615{,}000 \text{ m}^2)(1685 \text{ nCi/n}^2) = 2.72 \times 10^9 \text{ nCi} = 2.7 \text{ Ci}$$

Table 9.2 Concentrations of $^{239,240}Pu$ and ^{241}Am in Surface Soil (nCi/m^2)[a]

Accurate Pu	15,860	30,200	8,500	33,900	22,100
Fallible Am	719	1,540	310	1,690	1,000

$\bar{x}_A = 22{,}112,\ s_A^2 = 107{,}186{,}720,\ \bar{x}_F = 1051.8,\ s_F^2 = 327{,}336$
$b = 18.06,\ s_{A \cdot F}^2 = 562{,}176,\ \hat{\rho} = 0.998,\ \hat{\rho}^2 = 0.996$

[a]These data are part of a larger data set given in Figure B12(b) of Gilbert et al. (1975). Data used by permission of the Nevada Applied Ecology Group, U.S. Department of Energy, Las Vegas.

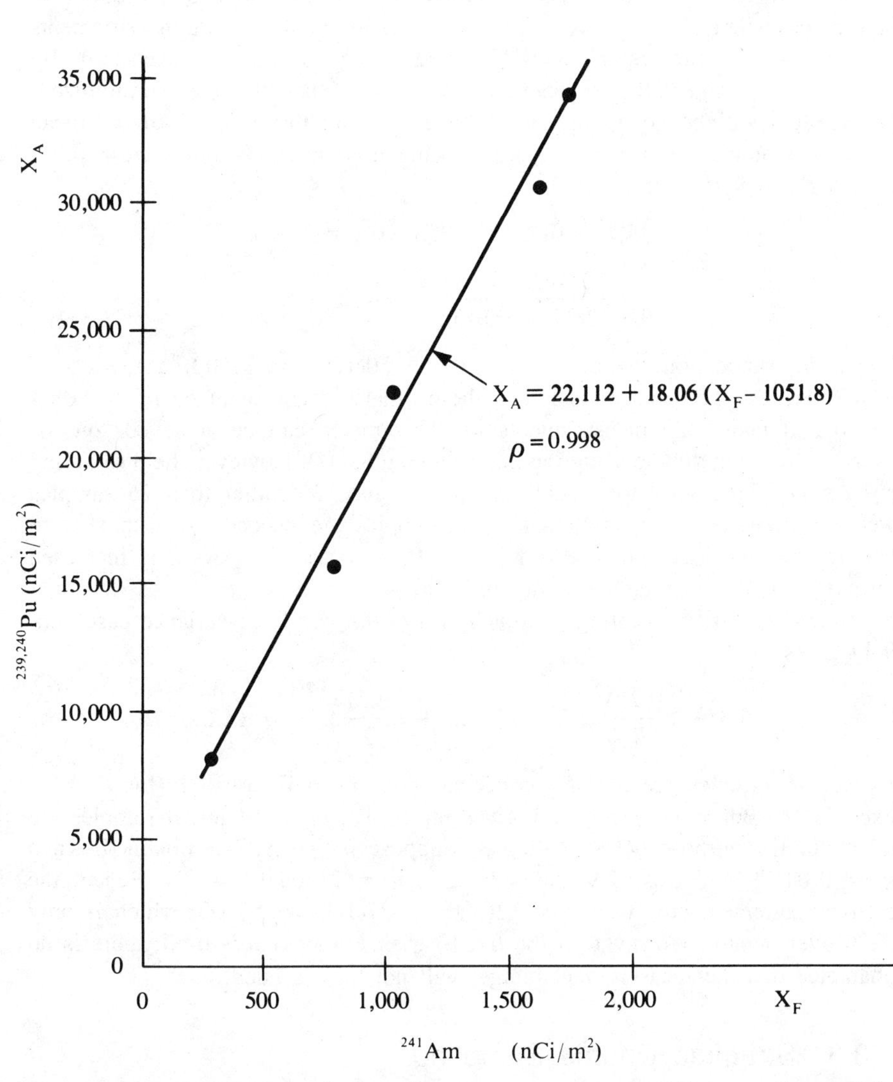

Figure 9.3 Regression of the accurate measurement (Pu) on the fallible measurement (Am) for five randomly selected soil samples (data from Gilbert et al., 1975, Fig. B12(b)).

It may be impractical to collect field samples in two or three stages. Therefore data from prior studies at the same or similar areas should be used to the maximum extent so that the initial estimates of ρ, n, and n' are as accurate as possible. As a last resort one can use Eqs. 9.11 and 9.12 to approximate n and n' since these equations do not require an advanced estimate of ρ. For the present case study these equations yield $n = 15$ and $n' = 21$. These compare with $n = 3$ and $n' = 44$, obtained by using $\hat{\rho} = 0.998$, the best available estimate of ρ.

9.3.2 Fixed Variance

In the previous section the optimum n and n' were obtained to estimate μ with the smallest possible variance, given that the total measurement budget is \$5000. However, there may be a need to minimize costs to allow part of the \$5000 to be used elsewhere. If the entire \$5000 were used to obtain only Pu measurements on each sample, then $n_v = 5000/200 = 25$ samples could be measured. By using Eqs. 9.9 and 9.10, we obtain n and n' such that the cost is minimized, yet $\text{Var}(\bar{x}_{\text{lr}})$ will be no greater on the average than the variance of the mean of the 25 samples. For the case study being used here, assuming $\rho = 0.99$, Eqs. 9.9 and 9.10 give

$$n = 25[1 - 0.98 + 0.98(0.10)] = 2.95 \text{ or } 3$$

$$n' = \frac{3(25)(0.98)}{3 - 25(1 - 0.98)} = 29.4 \text{ or } 30$$

For this allocation the cost is $3(200) + 30(100) = \$3600$, a savings of \$1400. Hence, the same precision in the estimated mean obtained by spending \$5000 and taking Pu measurements on 25 samples can be achieved, on the average, by using double sampling and spending \$3600. However, field sampling costs would be higher for double sampling, since 30 rather than 25 samples were required. Also, more than $n = 3$ samples are needed to estimate ρ to confirm that the assumed value of $\rho = 0.99$ is reasonable. As n is increased over $n = 3$, the cost savings will diminish.

If the three-stage sampling approach is used for the fixed-variance case, Eq. 9.14 gives

$$n_v\left[0.3 + \left(\frac{0.21}{R}\right)^{1/2}\right] = 25\left[0.3 + \left(\frac{0.21}{2}\right)^{1/2}\right] = 15.6 \text{ or } 16$$

Hence, 16 samples are initially collected if m_1 from Figure 9.1 (for $R = 2$) exceeds 16. But since Figure 9.1 gives $m_1 \cong 12$, $m = 12$ initial samples are taken for the purpose of estimating ρ. Suppose $\hat{\rho} = 0.998$ is obtained. Then $f_0 = 0.04479$ and Eqs. 9.9 and 9.10 give $n = 2$ and $n' = 27$. Hence, the total measurement cost would be $12(200) + 27(100) = \$5100$, which is only \$100 over budget. However, if the fixed-variance approach is used, there is no guarantee that the measurement budget will not be exceeded.

9.3.3 Stratified Double Sampling

Double sampling could be applied separately and independently to each of the strata in Figure 9.2. Let $\bar{x}_{\text{lr},h}$ and $s^2(\bar{x}_{\text{lr},h})$ denote the estimated mean and its variance in the hth stratum. Then, in accordance with Eq. 5.3 in Chapter 5, the estimated mean concentration over the L strata is

$$\bar{x}_{\text{lr}} = \sum_{h=1}^{L} W_h \bar{x}_{\text{lr},h}$$

Since the sampling is independent in different strata, $\text{Var}(\bar{x}_{\text{lr}})$ is estimated by computing

$$s^2(\bar{x}_{\text{lr}}) = \sum_{h=1}^{L} W_h^2 s^2(\bar{x}_{\text{lr},h})$$

The estimated total for the hth stratum is $N_h \bar{x}_{\text{lr},h}$, where in this example, N_h is the total area in m^2 for the hth stratum. Hence, the total estimated inventory over all four strata in Figure 9.2 is $\Sigma_{h=1}^{4} N_h \bar{x}_{\text{lr},h}$. The estimated variance of this estimate is $\Sigma_{h=1}^{4} N_h^2 s^2(\bar{x}_{\text{lr},h})$.

9.4 SUMMARY

If a linear relationship exits between an accurate but expensive measurement technique and a less accurate and inexpensive technique, then double sampling may be cost effective compared with simple random sampling. In practice, it is necessary to know with some assurance the correlation ρ between the two measurement techniques and their relative costs before cost effectiveness can be evaluated. Methods are given here for deciding on the optimum number of each type of measurement to make with methods developed by Tenenbein (1971, 1974). Once the data are collected, the estimated population mean and its estimated standard error are obtained by using linear regression or ratio methods. Double sampling may also be used in conjunction with other sampling plans such as stratified random sampling.

EXERCISES

9.1 Suppose there is a linear relationship between accurate and fallible types of measurement for a pollutant. Suppose the correlation between them is $\rho = 0.80$ and the costs of the accurate and fallible methods are $c_A = \$500$ and $c_F = \$350$, respectively. Is it likely that linear regression double sampling will yield an estimate of the mean concentration that is more precise than if simple random sampling is used? Assume equal dollars are available for both approaches and that appropriate n and n' are used in double sampling. What is your conclusion if $\rho = 0.98$?

9.2 Continuing on with Exercise 9.1, suppose $R = c_A/c_F = 500/350 = 1.43$ and $\rho = 0.98$. (a) Find the optimum n and n' that will minimize $\text{Var}(\bar{x}_{\text{lr}})$ for a fixed budget of $C = \$50{,}000$. (b) Find the optimum n and n' that ensure that $\text{Var}(\bar{x}_{\text{lr}})$ will be no larger than the variance of the mean obtained by simple random sampling where each sample is analyzed with the accurate method. (c) How much would this latter optimum allocation cost? (d) Suppose $\rho = 0.98$ is only an estimate of the true correlation based on 4 observations. Would that affect your decision whether to use double sampling? What should you do in this situation?

9.3 Suppose a pilot study is conducted to estimate ρ. Suppose $m = 10$ units are measured by both accurate and fallible methods and that the estimate of ρ is 0.80. Estimate the optimum n and n', assuming $c_A = \$500$, $c_F = \$350$, and the total cost is fixed at \$50,000. Do additional units need to be measured? If new units are collected and analyzed, should they be combined and used to reestimate ρ?

ANSWERS

9.1 Using Eq. 9.4, R should exceed 2.5 for double sampling to be more efficient. Since $R = 1.43$, simple random sampling may be best. If $\rho = 0.98$, R must exceed 1.25. Since R exceeds 1.25, double sampling is preferred.

9.2 (a) By Eq. 9.8, $f_0 = 0.1698$. Equations 9.6 and 9.7 give $n = 20$ and $n' = 114$. (b) From Eqs. 9.8, 9.9, and 9.10, $f_0 = 0.1698$, $n = 21$, and $n' = 119$. (c) From Eq. 9.3, total cost equals \$52,150. (d) Yes. The true ρ could be much smaller than 0.98, so double sampling is actually less efficient than simple random sampling. Take more measurements to obtain a better estimate of ρ.

9.3 $R = 1.43$, $\hat{\rho} = 0.80$. From Eqs. 9.6, 9.7, and 9.8, $f_0 = 0.63$, $n = 48$, and $n' = 75$. Since $m = 10$, which is smaller than 48, this implies 38 additional units should be measured by the accurate and fallible methods. Yes.

10 Locating Hot Spots

Chapters 4 through 9 have discussed sampling designs for estimating average concentrations or total amounts of pollutants in environmental media. Suppose, however, that the objective of sampling is not to estimate an average but to determine whether "hot spots," or highly contaminated local areas are present. For example, it may be known or suspected that hazardous chemical wastes have been buried in a land fill but its exact location is unknown. This chapter provides methods for answering the following questions when a square, rectangular, or triangular systematic sampling grid is used in an attempt to find hot spots:

What grid spacing is needed to hit a hot spot with specified confidence?
For a given grid spacing, what is the probability of hitting a hot spot of specified size?
What is the probability that a hot spot exists when no hot spots were found by sampling on a grid?

This discussion is based on an approach developed by Singer (1972, 1975) for locating geologic deposits by sampling on a square, rectangular, or triangular grid. He developed a computer program (ELIPGRID) that was used by Zirschky and Gilbert (1984) to develop nomographs for answering the preceding three questions. These nomographs are given in Figures 10.3, 10.4, and 10.5. We concentrate here on single hot spots. Some approaches for finding multiple hot spots are discussed by Gilbert (1982) and Holoway et al. (1981).

The methods in this chapter require the following assumptions:

1. The target (hot spot) is circular or elliptical. For subsurface targets this applies to the projection of the target to the surface (Fig. 10.1).
2. Samples or measurements are taken on a square, rectangular, or triangular grid (Fig. 10.2).
3. The distance between grid points is much larger than the area sampled, measured, or cored at grid points—that is, a very small proportion of the area being studied can actually be measured.
4. The definition of "hot spot" is clear and unambiguous. This definition implies that the types of measurement and the levels of contamination that constitute a hot spot are clearly defined.

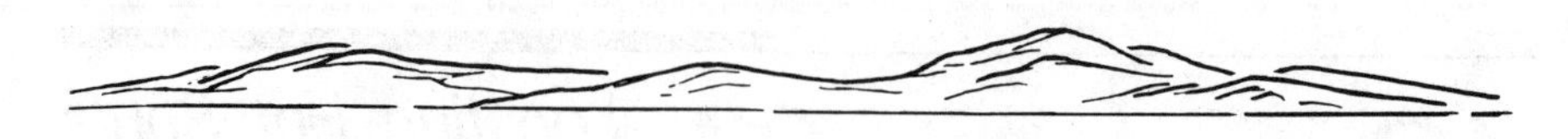

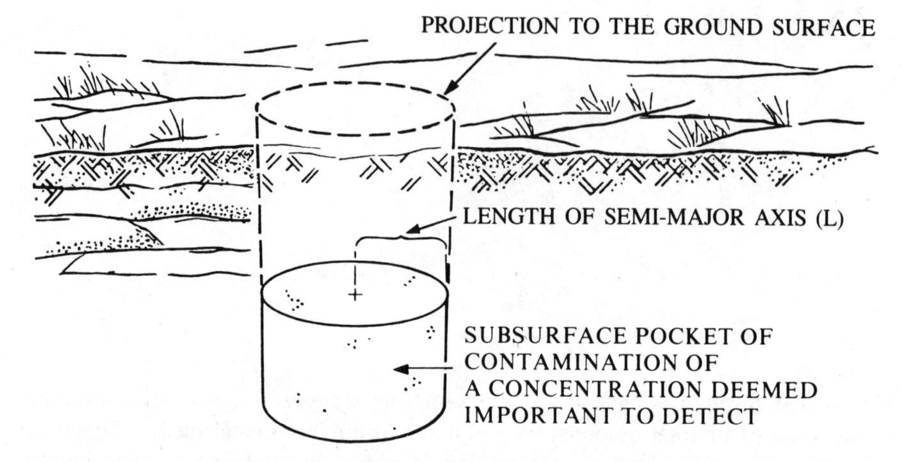

Figure 10.1 Hypothetical subsurface pocket of contamination (after Gilbert, 1982, Fig. 1).

5. There are no measurement misclassification errors—that is, no errors are made in deciding when a hot spot has been hit.

Parkhurst (1984) compared triangular and square grids when the objective is to obtain an unbiased estimate of the density of waste clusters in a hazardous waste site. He showed that the triangular grid was more likely to provide more

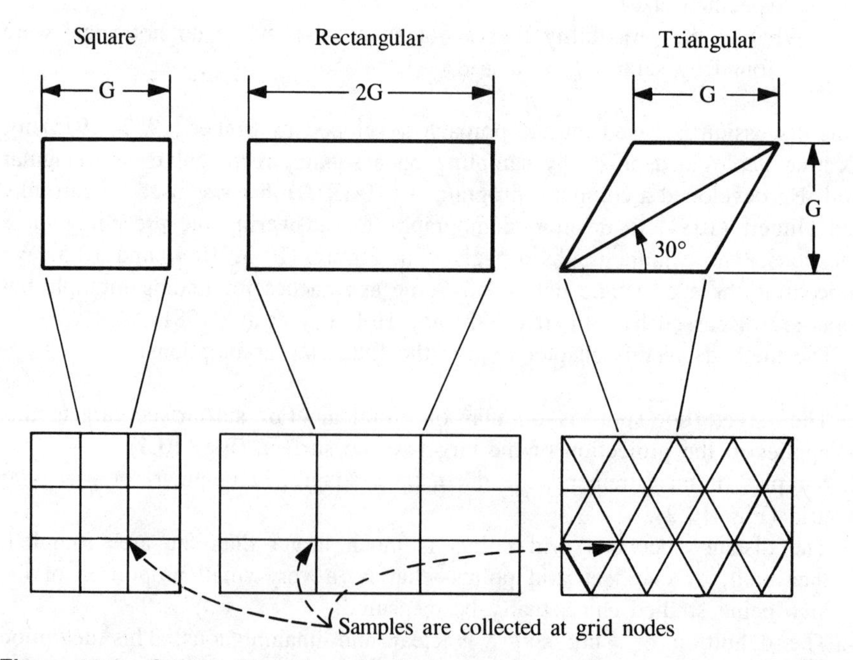

Figure 10.2 Grid configurations for finding hot spots (after Zirschky and Gilbert, 1984, Fig. 1).

information than the square grid. He also concluded that if the waste clusters are expected to follow an unknown but regular pattern, the wells should be drilled at randomly selected locations. But for randomly located clusters, a triangular or square grid is preferred.

10.1 DETERMINING GRID SPACING

The grid spacing required to find a hot spot of prespecified size and shape with specified confidence may be determined from the following procedure:

1. Specify L, the length of the semimajor axis of the smallest hot spot important to detect (see Fig. 10.1). L is one half the length of the long axis of the ellipse.
2. Specify the expected shape (S) of the elliptical target, where

$$S = \frac{\text{length of short axis of the ellipse}}{\text{length of long axis of the ellipse}}$$

   Note that $0 < S \leq 1$ and that $S = 1$ for a circle. If S is not known in advance, a conservative approach is to assume a rather skinny elliptical shape, perhaps $S = 0.5$, to give a smaller spacing between grid points than if a circular or "fatter" ellipse is assumed. That is, we sample on a finer grid to compensate for lack of knowledge about the target shape.
3. Specify an acceptable probability (β) of not finding the hot spot. The value β is known as the "consumer's risk." To illustrate, we may be willing to accept a $100\beta\% = 20\%$ chance of not finding a small hot spot, say one for which $L = 5$ cm. But if L is much larger, say $L = 5$ m, a probability of only $\beta = 0.01$ (1 chance in 100) may be required.
4. Turn to Figures 10.3, 10.4, or 10.5 for a square, rectangular, or triangular grid, respectively. These nomographs give the relationship between β and the ratio L/G, where G is the spacing between grid lines (Fig. 10.2). Using the curve corresponding to the shape (S) of interest, find L/G on the horizontal axis that corresponds to the prespecified β. Then solve L/G for G, the required grid spacing. The total number of grid points (sampling locations) can then be found because the dimensions of the land area to be sampled are known.

For elliptical targets ($S < 1$) the curves in Figures 10.3, 10.4, and 10.5 are average curves over all possible orientations of the target relative to the grid. Singer (1975, Fig. 1) illustrates how the orientation affects the probability of not hitting the target. If the orientation is known, Singer's (1972) program will give the curves for that specific orientation.

EXAMPLE 10.1

Suppose a square grid is used and we want to take no more than a $100\beta\% = 10\%$ chance of not hitting a circular target of radius $L = 100$ cm or larger. Using the curve in Figure 10.3 for $S = 1$, we find $L/G = 0.56$ corresponds to $\beta = 0.10$. Solving for G yields $G = L/0.56 = 100\text{ cm}/0.56 \cong 180$ cm. Hence, if cores are taken

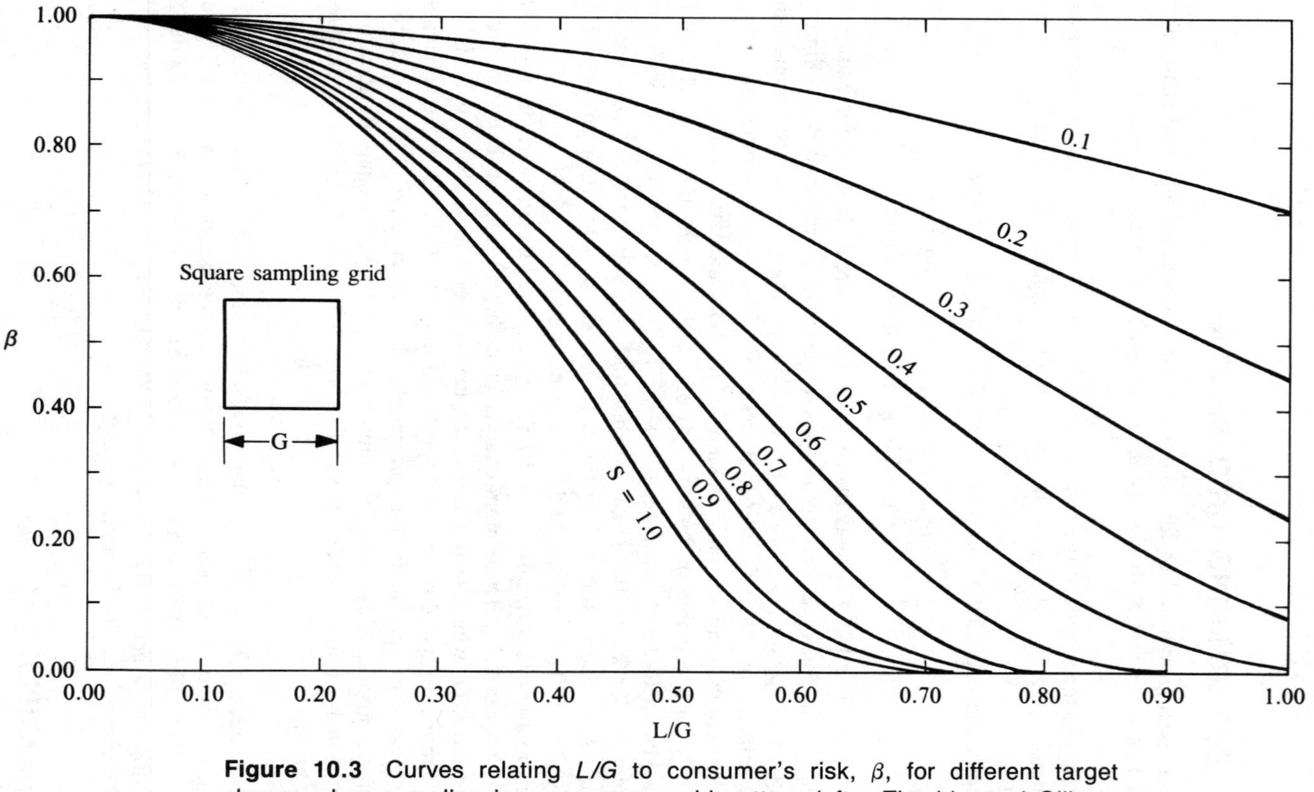

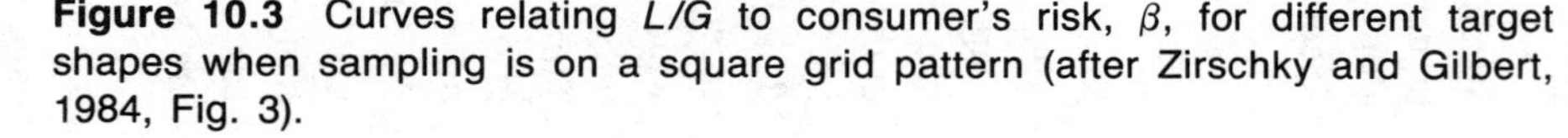

Figure 10.3 Curves relating L/G to consumer's risk, β, for different target shapes when sampling is on a square grid pattern (after Zirschky and Gilbert, 1984, Fig. 3).

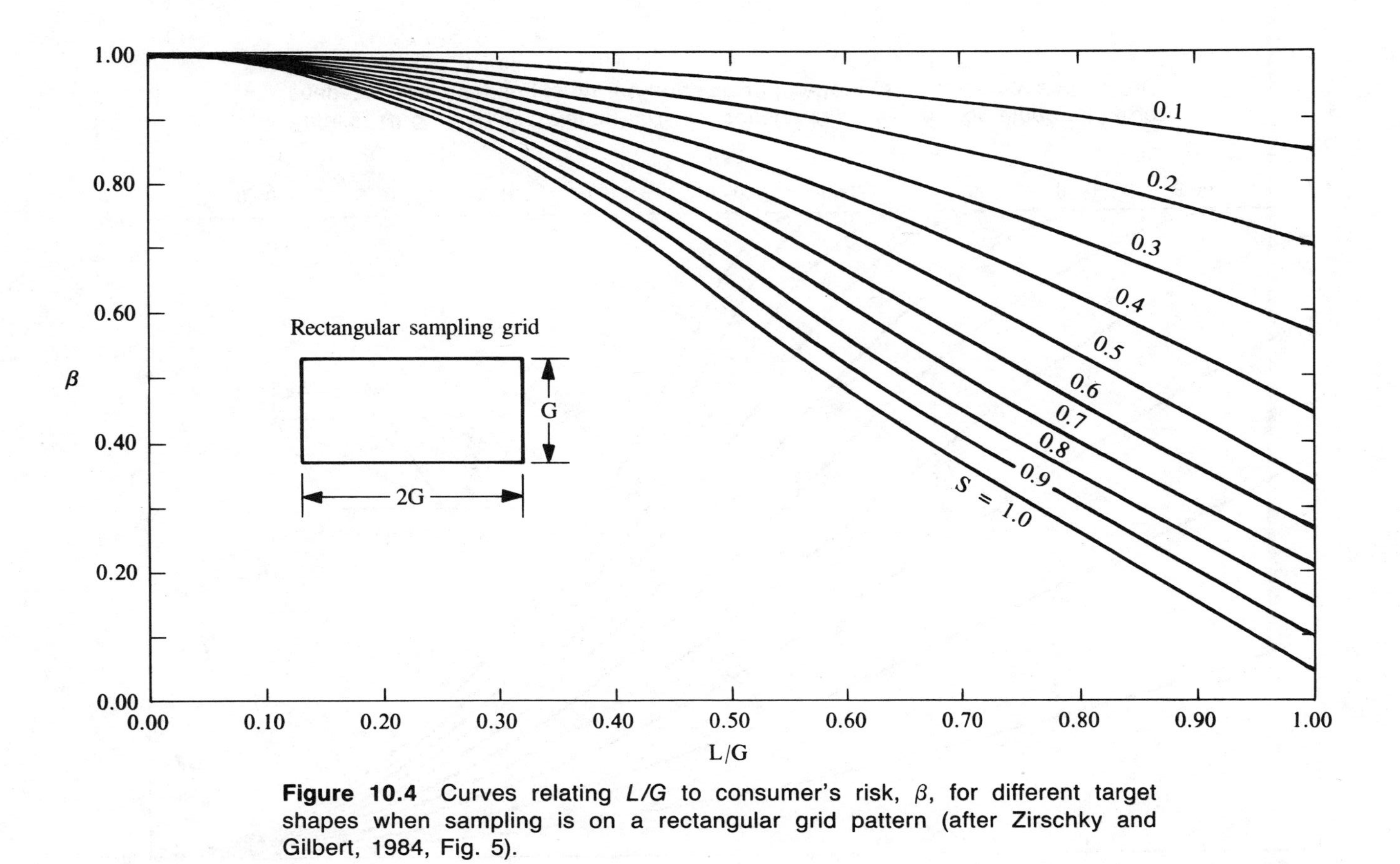

Figure 10.4 Curves relating *L/G* to consumer's risk, β, for different target shapes when sampling is on a rectangular grid pattern (after Zirschky and Gilbert, 1984, Fig. 5).

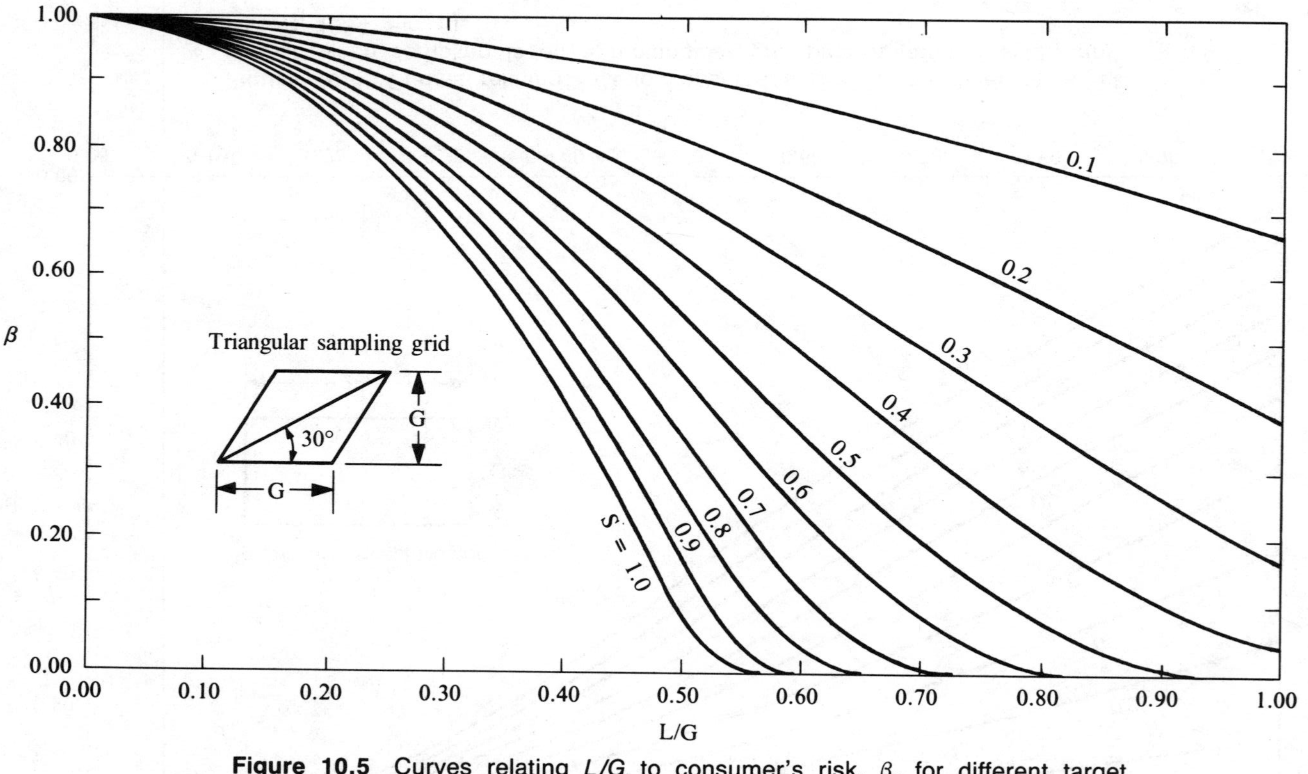

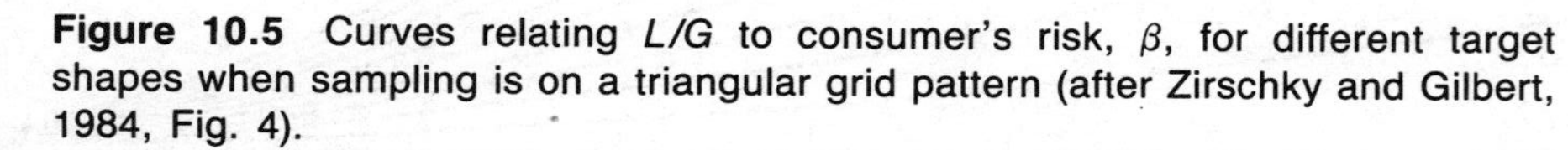

Figure 10.5 Curves relating L/G to consumer's risk, β, for different target shapes when sampling is on a triangular grid pattern (after Zirschky and Gilbert, 1984, Fig. 4).

on a square grid with spacing of 180 cm, we are assured the probability is only 0.10 (1 chance in 10) of not hitting a circular target that is 100 or more cm in radius.

10.2 SIZE OF HOT SPOT LIKELY TO BE HIT

Figures 10.2, 10.3, and 10.4 can also be used to find the maximum size hot spot that can be located for a given cost and consumer's risk. Suppose, for example, we can afford to take measurements at no more than 25 locations on a square grid system. What size elliptical target (characterized by L) can we expect to find with confidence $1 - \beta$ (the probability of hitting a target at least once)? The general procedure is to specify β, G, and S, then use the curves to solve for L.

EXAMPLE 10.2

Suppose our budget allows taking measurements at $n = 25$ locations on a square grid pattern. Suppose also that a grid spacing of $G =$ 200 cm covers the area of interest. What size circular target can we be at least 90% confident of detecting—that is, for which the probability of not hitting the target is $\beta = 0.10$ or less? Using $S =$ 1 in Figure 10.3, we find $L/G = 0.56$ for $\beta = 0.10$. Hence, $L =$ (200 cm) (0.56) = 112 cm. Therefore, we estimate that a circle with a radius of 112 cm or larger has no more than a 10% chance of not being hit when using a square grid spacing of 200 cm. If the circular target has a radius L less than 112 cm, the probability of not locating it will exceed 0.10. Conversely, if $L >$ 112 cm, the probability of not locating it will be less than 0.10. If we require only a 50% chance of hitting the target (i.e., $\beta = 0.50$), the curve for $S = 1$ gives $L/G = 0.4$ or $L =$ (200 cm) (0.4) = 80 cm.

By computing L as in Example 10.2 for different values of β and G, we can generate curves that give the probability of hitting a circular or elliptical target of any size. These curves for grid spacings of 100, 200, and 300 distance units for two target shapes, $S = 1$ and 0.5, are given in Figure 10.6.

For example, suppose the target is circular ($S = 1$) and the grid spacing is $G = 100$ units. Then the probability β that we do not hit a circular target of radius $L = 50$ units (same units as G) is about 0.2. If the target is smaller, say $L = 20$ units, then β is larger, about 0.87.

10.3 PROBABILITY OF NOT HITTING A HOT SPOT

Figures 10.3–10.5 can also be used to estimate the consumer's risk β of not hitting a hot spot of given size and shape when using a specified grid size.

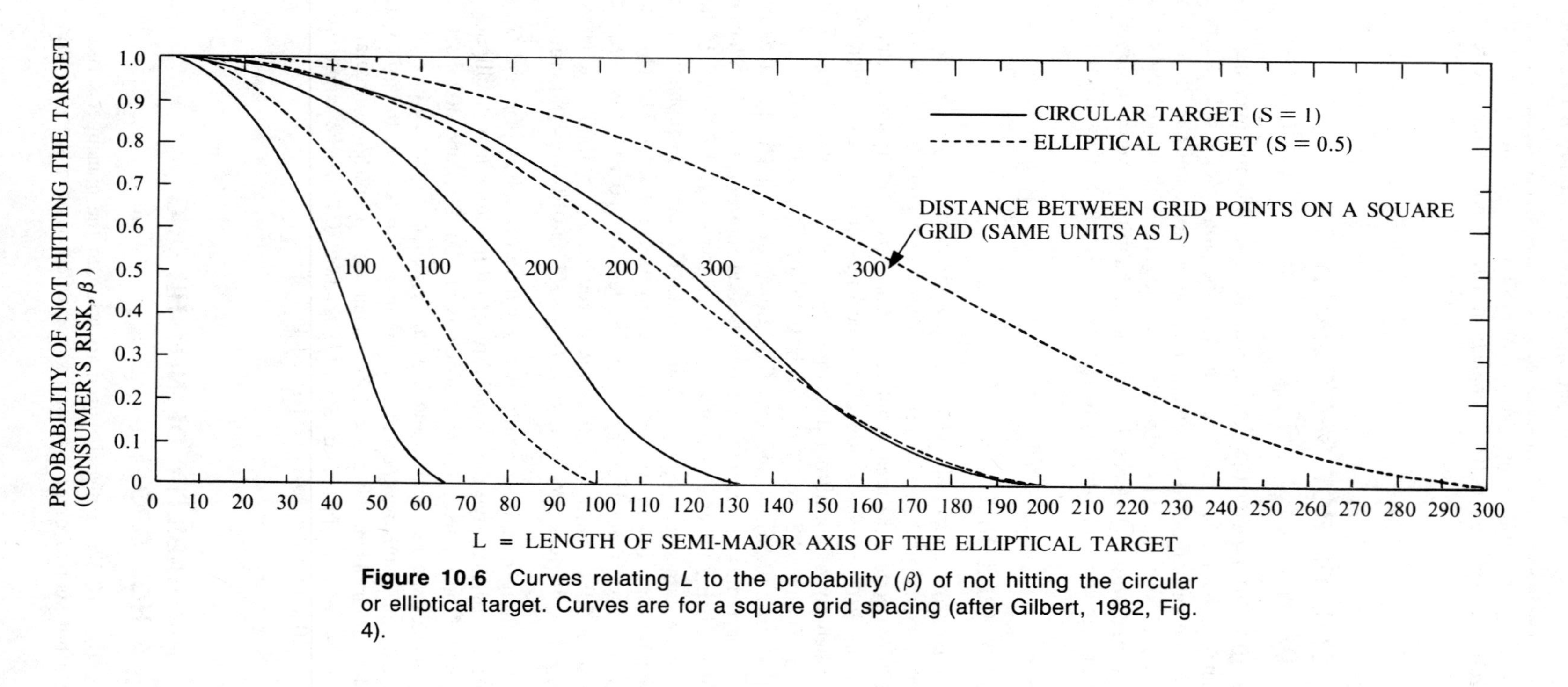

Figure 10.6 Curves relating L to the probability (β) of not hitting the circular or elliptical target. Curves are for a square grid spacing (after Gilbert, 1982, Fig. 4).

EXAMPLE 10.3

What is the average probability of not finding an elliptical hot spot that is twice as long as it is wide and for which the semimajor axis (L) is 40% as long as the spacing G between grid points? Suppose a rectangular sampling grid is used. Using $S = 0.5$ and $L/G = 0.40$ in Figure 10.4, we find β to be 0.87. Hence, there is about an 87% chance that this size and shape target would not be found by sampling at the grid points. The actual β could be somewhat smaller or larger than 0.87, depending on the orientation of the target relative to the grid.

10.4 TAKING PRIOR INFORMATION INTO ACCOUNT

Thus far we have assumed that a hot spot does actually exist. In practice, no such assurance may be warranted. Now we consider how prior information about the probability that a hot spot exists can be used to obtain a more realistic estimate of β. Let

A = event that a hot spot of size L or larger exists

B = event that a hot spot of size L or larger is hit by taking measurements on a grid.

Then the law of conditional probabilities (see, e.g., Fisz, 1963, p. 20) says that

$$P(B|A) = \frac{P(A, B)}{P(A)}$$

= probability that a hot spot of size L or larger is hit, *given* such a hot spot exists **10.1**

where

$P(A, B)$ = probability that a hot spot of size L or larger exists *and* is discovered by sampling on a grid

and

$P(A)$ = probability that a hot spot of size L or larger exists

Whenever there is doubt whether a hot spot of size L or larger exists, then $P(A, B)$ is of interest. From Eq. 10.1 we have

$$P(A, B) = P(B|A)P(A) \qquad \textbf{10.2}$$

Now $P(B|A)$ is just $1 - \beta$. Hence, $P(A, B)$ can be estimated by using Figures 10.3–10.5 and by specifying a value for $P(A)$. In many situations a hot spot of size L or larger will be known to exist so that $P(A) = 1$ and $P(A, B) = P(B|A)$. Then if a square grid is used, Figure 10.3 gives the final result. In other situations an informed guess at $P(A)$ may be made based on prior surveys and other knowledge. Then Eq. 10.2 may be used to approximate $P(A, B)$.

EXAMPLE 10.4

Suppose $L = 100$ cm and $\beta = 0.10$ for a circular hot spot as in Example 10.1. Use of Figure 10.3 gives a grid spacing of $G = 180$ cm. Suppose also that prior information about the site suggests $P(A)$ is very low, say $P(A) = 0.01$ (1 chance in 100). Then, since $P(B|A) = 1 - \beta = 0.90$, Eq. 10.2 gives $P(A, B) = (0.9)(0.01) = 0.009$. Hence, if 180-cm grid spacing is used, the probability that a hot spot of size $L = 100$ cm or larger exists *and* is found is 0.009, assuming $P(A) = 0.01$.

10.5 PROBABILITY THAT A HOT SPOT EXISTS WHEN NONE HAS BEEN FOUND

Suppose samples are taken on a grid spacing determined by S, L, and β, but no hot spot of size L or larger is found. Then it is natural to ask the question, What is the probability that a hot spot of size L or larger exists even though it was not found? A procedure for answering that question is now given. Let

A = event a hot spot of size L or larger exists

$\bar{A}$ = event a hot spot of size L or larger does not exist

B = event a hot spot of size L or larger is hit

$\bar{B}$ = event a hot spot of size L or larger is not hit

Using Eq. 10.1 gives

$P(A|\bar{B})$ = probability that a hot spot of size L or larger exists given that our sampling effort on a grid did not find it

$$= \frac{P(A, \bar{B})}{P(\bar{B})} \qquad 10.3$$

But using Eq. 10.1 again, we find that the numerator of Eq. 10.3 is $P(\bar{B}|A)P(A)$. Also, since either A or $\bar{A}$ must occur, the denominator of Eq. 10.3 can be written as

$$P(\bar{B}) = P(\bar{B}|A)\, P(A) + P(\bar{B}|\bar{A})\, P(\bar{A})$$

Hence

$$P(A|\bar{B}) = \frac{P(\bar{B}|A)\, P(A)}{P(\bar{B}|A)\, P(A) + P(\bar{B}|\bar{A})\, P(\bar{A})} \qquad 10.4$$

Equation 10.4 is known as *Bayes' formula* (Fisz, 1963, p. 23). Since $P(\bar{B}|\bar{A}) = 1$ and $P(\bar{A}) = 1 - P(A)$, Eq. 10.4 becomes

$$P(A|\overline{B}) = \frac{P(\overline{B}|A)\, P(A)}{P(\overline{B}|A)\, P(A) + 1 - P(A)}$$

$$= \frac{\beta P(A)}{\beta P(A) + 1 - P(A)} \quad \textbf{10.5}$$

Hence, $P(A|\overline{B})$ for a given grid spacing can be estimated by Eq. 10.5 if β and $P(A)$ are specified.

EXAMPLE 10.5

Suppose we can tolerate a consumer's risk of no more than 10% of not hitting a circular target of radius $L = 100$ cm or greater. This leads to a grid spacing of 180 cm (Example 10.1). Also, suppose $P(A) = 0.01$ is our best guess for the probability that a circular hot spot of size L or greater exists at the site. If no hot spot of size L or greater is found by taking measurements on the 180-cm grid, the probability that such a hot spot exists at the site is estimated to be (by Eq. 10.5)

$$P(A|\overline{B}) = \frac{(0.10)(0.01)}{(0.10)(0.01) + 1 - 0.01} = 0.001$$

The probability $P(A|\overline{B})$, computed by Eq. 10.5, is plotted in Figure 10.7 for a range of values of β and $P(A)$. Figure 10.7 shows that $P(A)$ has a major impact on the value of $P(A|\overline{B})$. Figure 10.7 also shows the importance of choosing a small value for β if we want high confidence that a hot spot has not been missed.

EXAMPLE 10.6

Suppose we set $P(A) = 0.50$ and $\beta = 0.10$ for $L = 100$ cm or larger for a circular hot spot. Then Eq. 10.5 gives $P(A|\overline{B}) = 0.091$. Hence, for these values of $P(A)$ and β the chances are about 1 in 10 that a circular hot spot of size $L = 100$ cm or greater exists even though it was not found. For $\beta = 0.50$, $P(A|\overline{B})$ increases to 0.33. $P(A|\overline{B})$ increases as β increases because larger β's result in wider grid spacing. Hence, there is less chance of finding the hot spot.

If grid spacing is determined for a circular hot spot, but the target is actually an ellipse, then β is actually larger than expected since a smaller grid spacing should have been used. Looking at Figure 10.7, we see then that $P(A|\overline{B})$ is actually larger than expected. When in doubt about the shape of the target, the conservative approach is to assume a skinnier ellipse (smaller value of S) than expected, which will result in the use of a smaller grid spacing and a conservative (larger) estimate of $P(A|\overline{B})$.

10.6 CHOOSING THE CONSUMER'S RISK

Figure 10.7 can be used to help decide on a value for β. Suppose $P(A|\overline{B})$ must be no larger than some prespecified value, say 0.01. That is, we want to be

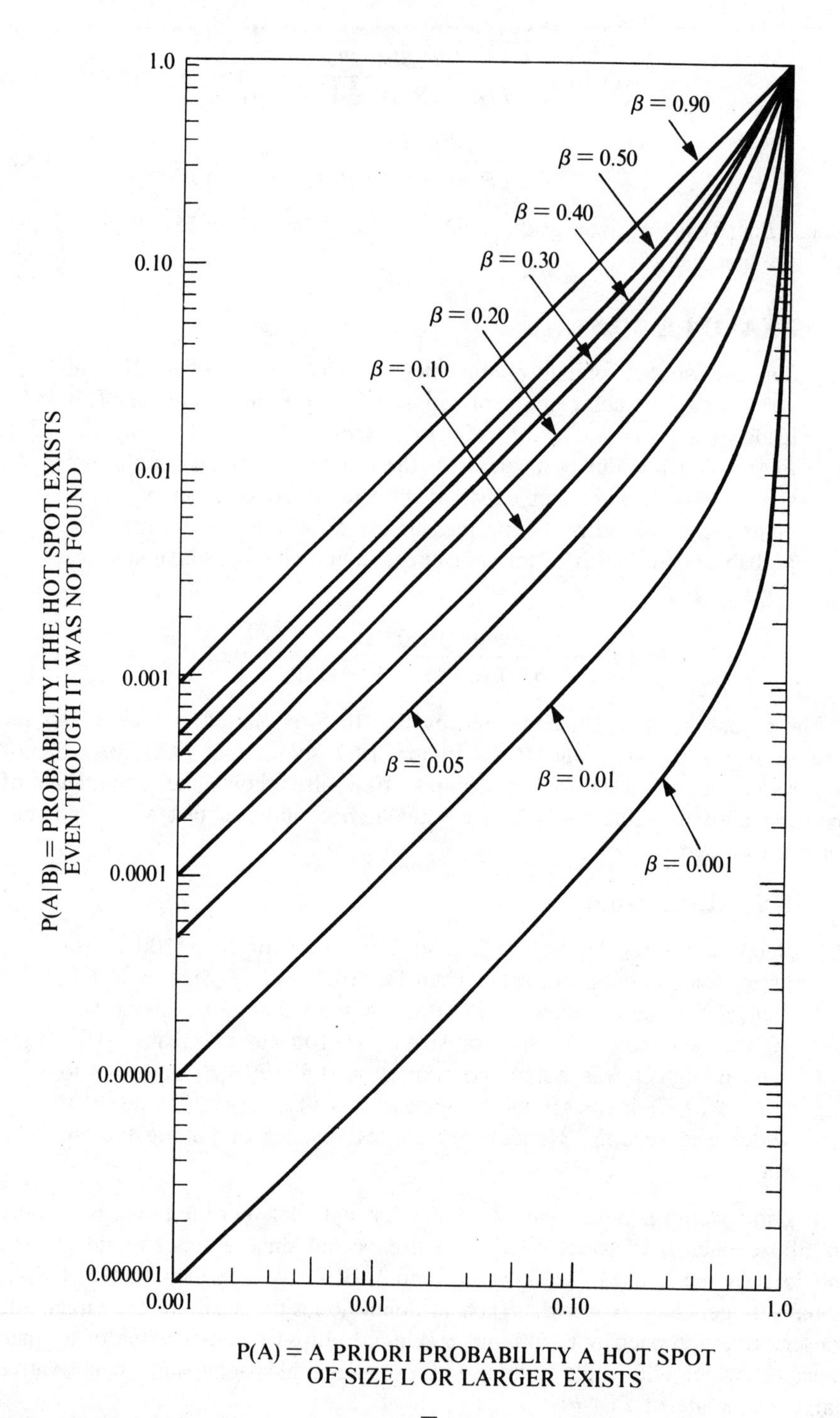

Figure 10.7 Relationship between $P(A|\overline{B})$, $P(A)$ and the consumer's risk β (after Gilbert, 1982, Fig. 5).

99% confident that a hot spot does not exist, given that no hot spots have been found. If at the planning stages of the survey effort some reasonable value for $P(A)$ can be determined, then Figure 10.7 can be used to determine β. For example, for $P(A|\bar{B}) = 0.01$ and $P(A) = 0.50$, we find $\beta = 0.01$. This value of β may then be used to determine grid spacing.

10.7 SUMMARY

This chapter gives methods for determining grid spacing when the primary objective is to search for circular or elliptical hot spots. The grid spacings are obtained so that the consumer's risk is held to an acceptable level. The nomographs presented for this purpose can also be used to determine the consumer's risk for a given grid spacing that has been used.

Since grid spacing must be small to have a high probability of finding small hot spots, the cost of sampling and analyses can be high. For that reason judgment is necessary to decide in advance where hot spots are most likely to lie and to concentrate sampling in those areas. Larger grid spacing can be used in areas where hot spots are less likely to be present.

EXERCISES

10.1 Find the required square grid spacing to achieve a consumer's risk no greater than $\beta = 0.10$ of not hitting the target if the target is expected to be twice as long as it is wide and if $L \geq 100$ cm.

10.2 What size circular hot spot can we be 80% sure of detecting if a triangular grid of spacing $G = 10$ m is used?

10.3 Determine the probability that a circular target of radius $L = 30$ units will not be hit when a square grid spacing of 200 units is used.

10.4 In Example 10.4 suppose that the probability that a circular hot spot of radius $L = 100$ cm exists is 0.90 instead of 0.01. Using a consumer's risk of $\beta = 0.25$, determine $P(A, B)$. State your conclusions.

10.5 In Example 10.5 suppose that $\beta = 0.20$ and $P(A) = 0.60$. Find $P(A|\bar{B})$. State your conclusions.

ANSWERS

10.1 Using Figure 10.3 when $S = 0.5$, we obtain $L/G = 0.84$ or $G = 100/0.84 = 119$ cm.

10.2 Using Figure 10.5, we obtain $L/G = 0.47$, so $L = 0.47$ (10 m) = 4.7-m radius circle.

10.3 Using Figure 10.7, we obtain $\beta = 0.93$.

10.4 $P(A, B) = P(B|A)\,P(A) = (1 - \beta)\,P(A) = 0.75(0.90) = 0.675$.

10.5 $$P(A|\bar{B}) = \frac{0.20(0.60)}{0.2(0.60) + 1 - 0.60} = 0.23.$$

11 Quantiles, Proportions, and Means

This chapter discusses the normal (Gaussian) distribution and shows how to estimate confidence limits on quantiles, proportions, and means. It then

Discusses estimators of the true mean and variance that are appropriate if the data are nonnormal or if outliers or trace data are present
Gives nonparametric (distribution-free) methods for estimating quantiles and confidence limits on quantiles and proportions
Shows how to put confidence limits on the mean when data are correlated
Discusses advantages and disadvantages of nonlinear transformations to achieve normality

The normal distribution is important because many statistical procedures such as tests of significance, confidence limits, and estimation procedures are strictly valid only for normally distributed data. Even though most pollutants are not normally distributed, the data can often be transformed to be approximately normal. Also, inferences about population means of nonnormal populations are still possible if n is sufficiently large, since in that case the sample mean, $\bar{x}$, is approximately normally distributed.

11.1 BASIC CONCEPTS

The normal distribution is a bell-shaped, symmetric distribution. It is described mathematically by its *probability density function*

$$f(x) = \frac{1}{\sigma\sqrt{2\pi}} \exp\left[-\frac{1}{2\sigma^2}(x - u)^2\right]$$

$$-\infty < x < \infty, \quad -\infty < \mu < \infty, \quad \sigma > 0$$

where $f(x)$ is the height (ordinate) of the curve at the value x. The density function is completely specified by two parameters, μ and σ^2, which are also the mean and variance, respectively, of the distribution. We use the notation $N(\mu, \sigma^2)$ to denote a normal probability density function (in short, normal distribution) with mean μ and variance σ^2.

Figure 11.1 shows two normal distributions. The solid curve is $N(0, 1)$; the dashed curve is $N(1, 2.25)$. There is a different normal distribution for each

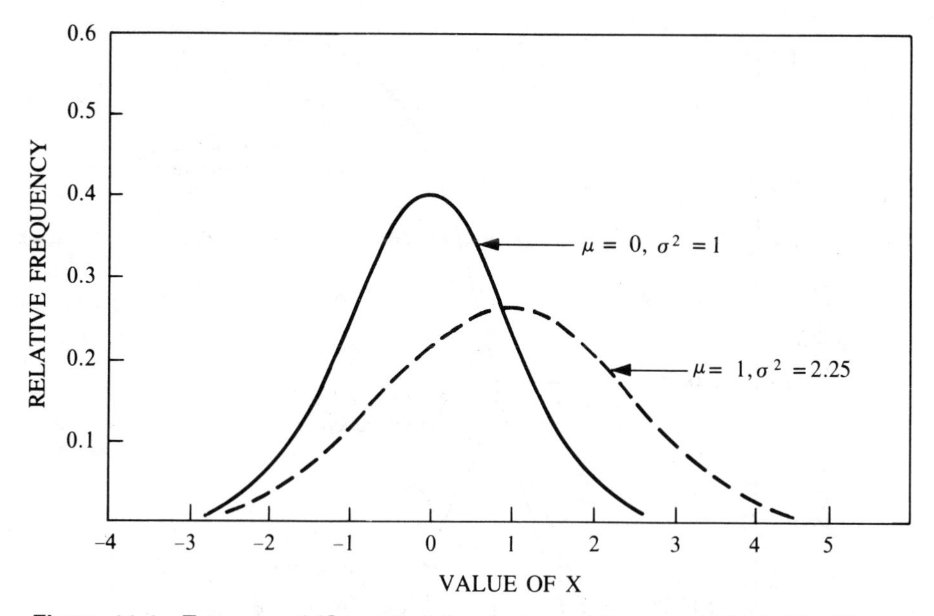

Figure 11.1 Two normal (Gaussian) distributions: $N(0, 1)$ and $N(1, 2.25)$. $N(0, 1)$ is the "standard normal" distribution.

combination of μ and σ^2. However, they can all be transformed to the $N(0, 1)$ distribution by the transformation

$$Z = \frac{X - \mu}{\sigma}$$

That is, the random variable Z has the $N(0, 1)$ distribution shown in Figure 11.1 if the random variable X is $N(\mu, \sigma^2)$. Z is commonly called a *standard normal deviate*.

Figure 11.2 shows the density function $f(x)$ and the *cumulative distribution function* (CDF) for a $N(\mu, \sigma^2)$ distribution. The CDF is denoted by $F(x)$ and is defined as follows:

$$F(x) = \text{Prob}\ [X \leq x]$$

$$= \text{probability that the random variable } X \text{ will take on a value less than or equal to a specified value } x$$

In other words, $F(x)$ gives the cumulative percentage of the normal density function that lies between $-\infty$ and the point x on the abscissa.

From Figure 11.2 we see, for example, that 2.15% of the density function lies between $\mu + 2\sigma$ and $\mu + 3\sigma$, 2.28% lies below $\mu - 2\sigma$, and 2.28% lies above $\mu + 2\sigma$. Stated another way, $\mu - 2\sigma$ is the 0.0228 quantile of the $N(\mu, \sigma^2)$ distribution, or $x_{0.0228}$ is the quantile of order 0.0228. Similarly, $\mu + 2\sigma$ is the 0.9772 quantile of the $N(\mu, \sigma^2)$ distribution. More formally, the pth quantile, x_p (where $0 < p < 1$), is the value such that the probability is p that a unit in the population will have an observed value less than or equal to x_p, and the probability is $1 - p$ that a units value will be larger than x_p. The median is the 0.5 quantile, and $x_{0.25}$ and $x_{0.75}$ are the lower and upper *quartiles*, respectively. Quantiles are also called percentiles.

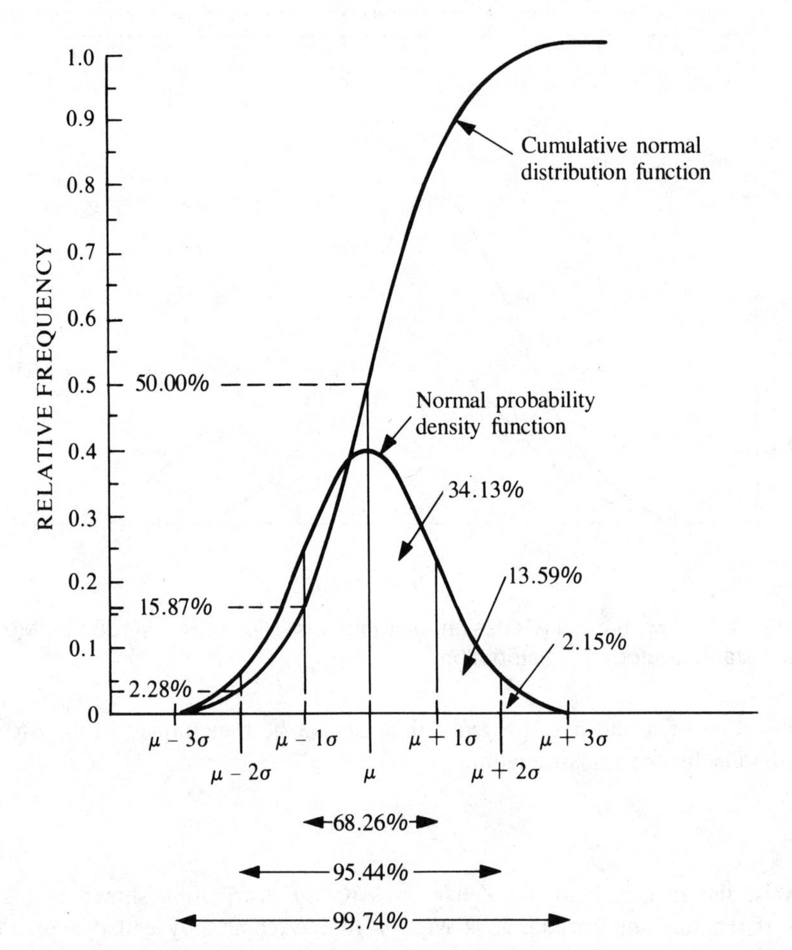

Figure 11.2 Areas under the normal probability density function and the cumulative normal distribution function (after Sokal and Rohlf, 1981, p. 103.)

11.2 ESTIMATING QUANTILES (PERCENTILES)

Quantiles of distributions are frequently estimated to determine whether environmental pollution levels exceed specified limits. For example, a regulation may require that the true 0.98 quantile of the population, $x_{0.98}$, must not exceed 1 ppm. In practice, x_p must be estimated from data. This section gives two methods for estimating x_p when the underlying distribution is normal. The following section shows how to put an upper confidence limit on x_p when the distribution is normal. Methods for estimating quantiles of a lognormal distribution are given in Section 13.6.

Quantiles of a normal distribution can be estimated by using the sample mean, $\bar{x}$, and standard deviation, s, as computed by Eq. 4.3 and the square root of Eq. 4.4, respectively. Suppose the n data are a simple random sample from a normal distribution. Then x_p is estimated by computing

$$\hat{x}_p = \bar{x} + Z_p s \qquad \textbf{11.1}$$

where Z_p is the pth quantile of the standard normal distribution. Table A1 gives values of p that correspond to Z_p. For example, $\hat{x}_{0.9772} = \bar{x} + 2s$ is an estimate of the 0.9772 quantile of the distribution because Table A1 gives $Z_{0.9772} = 2$. Similarly, $\hat{x}_{0.0228} = \bar{x} - 2s$ is an estimate of the 0.0228 quantile because $Z_{0.0228} = -2$.

Saltzman (1972) gives a nomograph (his Fig. 2) for finding $Z_p s$ without looking up Z_p in Table A1. The user supplies s and p, and the resulting value of $Z_p s$ from the nomograph is added (by the user) to $\bar{x}$ to obtain $\hat{x}_p$.

Another method of estimating normal quantiles is to use probability plotting. The procedure is to first order the untransformed data from smallest to largest. Let $x_{[1]} \leq x_{[2]} \leq \cdots \leq x_{[n]}$ denote the ordered data. The $x_{[i]}$ are called the *order statistics* of the data set. Then plot $x_{[i]}$ versus $(i - 0.5)100/n$ on normal probability paper. If the data are from a normal distribution, the plotted points should lie approximately on a straight line. If so, a best-fitting straight line is drawn subjectively by eye. Then quantiles can be easily approximated from the plot. An objective method for fitting a unique straight line to the points was developed by Mage (1982a, 1982b).

EXAMPLE 11.1

Figure 11.3 shows a normal probability plot for the concentration of ^{241}Am (pCi/g) for 20 soil samples collected near a nuclear facility (Price, Gilbert, and Gano, 1981). A straight line (fit by eye) fits the plotted points reasonably well, suggesting the underlying distribution may be normal. Using the line, we estimate the 0.9 quantile to be 0.065 pCi/g. Similarly, we estimate the 0.5 quantile to be 0.038

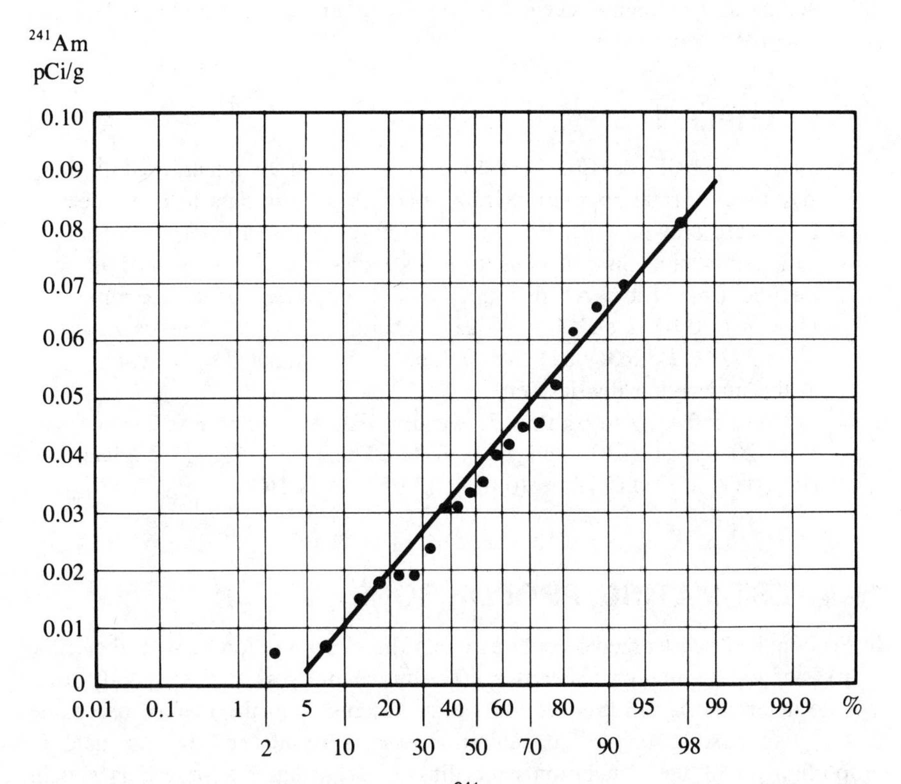

Figure 11.3 Normal probability plot of ^{241}Am pCi/g soil data (after Price et al., 1981).

pCi/g. Since the normal distribution is symmetrical, the true mean and median of the distribution are identical. Hence, the 0.5 quantile estimates both the mean and median of the normal distribution. The standard deviation, σ, may be estimated from the probability plot by computing $(\hat{x}_{0.84} - \hat{x}_{0.16})/2$, where $\hat{x}_{0.84}$ and $\hat{x}_{0.16}$ are the 0.84 and 0.16 quantiles as read from the plot. From Figure 11.3 we find $\hat{x}_{0.84} = 0.059$ and $\hat{x}_{0.16} = 0.0165$. Hence, the estimated standard deviation is $(0.059 - 0.0165)/2 = 0.021$. For this data set, $\bar{x} = 0.0372$ and $s = 0.0211$, which agree well with the estimates obtained from the probability plot.

11.3 CONFIDENCE LIMITS FOR QUANTILES

An upper $100(1 - \alpha)\%$ confidence limit for the true pth quantile, x_p, can be easily obtained if the underlying distribution is normal. This upper limit, denoted by $UL_{1-\alpha}(x_p)$, is

$$UL_{1-\alpha}(x_p) = \bar{x} + sK_{1-\alpha,p} \qquad \textbf{11.2}$$

where $K_{1-\alpha,p}$ is obtained from Table A3 for specified α and p. Note that Eq. 11.2 is identical to Eq. 11.1 except that $K_{1-\alpha,p}$ replaces Z_p and $K_{1-\alpha,p} > Z_p$.

This upper limit could be used to test whether the true x_p for the sampled population actually exceeds a specified x_p. One rule would be to conclude that the specified x_p has been exceeded unless the estimated upper limit $UL_{1-\alpha}(x_p)$ is less than the specified x_p.

EXAMPLE 11.2

First, we shall use Eq. 11.1 to estimate the 0.99 quantile of the (assumed) normal population from which the ^{241}Am data in Example 11.1 were drawn. Then Eq. 11.2 will be used to estimate an upper 90% confidence limit for the true 0.99 quantile. Since $p = 0.99$, we find from Table A1 that $Z_{0.99} = 2.3263$. Also, from Example 11.1, $\bar{x} = 0.0372$ and $s = 0.0211$. Therefore, Eq. 11.1 gives $\hat{x}_{0.99} = 0.0372 + 2.3263(0.0211) = 0.0863$ as the estimated 0.99 quantile of the underlying distribution.

Now, referring to Table A3, we find that $K_{1-\alpha,p} = 3.052$ when $n = 20$, $\alpha = 0.10$, and $p = 0.99$. Therefore, Eq. 11.2 gives $UL_{0.90}(x_{0.99}) = 0.0372 + (0.0211)(3.052) = 0.102$.

11.4 ESTIMATING PROPORTIONS

In Section 11.2 we learned how to estimate the concentration x_p such that $100(1 - p)\%$ of the population exceeds x_p. The procedure was to first specify p and then to determine $\hat{x}_p$. In this section we are interested in the reverse procedure, that is, we first specify a concentration, say x_c, and then we estimate the proportion p_{x_c} of the population exceeding x_c. This latter approach is suitable if regulations specify that the proportion of the population exceeding a specified concentration x_c (upper limit) must be less than some specified value.

If the random variable X is known to be normally distributed with parameters μ and σ^2, then the proportion of the population that exceeds x_c is

$$p_{x_c} = \text{Prob}[X > x_c] = 1 - \phi\left(\frac{x_c - \mu}{\sigma}\right) \qquad \textbf{11.3}$$

where ϕ denotes the cumulative distribution function (CDF) of the $N(0, 1)$ distribution. For example, if $\mu = 1$ and $\sigma^2 = 2.25$ (the dashed-line normal distribution in Fig. 11.1) and we want to determine the proportion of the population exceeding $x_c = 3$, then

$$\phi\left(\frac{x_c - \mu}{\sigma}\right) = \phi\left(\frac{3 - 1}{(2.25)^{1/2}}\right) = \phi(1.33)$$

Using Table A1, we find $\phi(1.33) = 0.9082$. Hence, $p_{x_c} = \text{Prob}[X > 3] = 1 - 0.9082 = 0.0918$. Hence, 9.18% of this normal population exceeds $x_c = 3$. Of course, in practice, μ and σ^2 are almost never known a priori. Then the estimates $\bar{x}$ and s are used in place of μ and σ in Eq. 11.3. Hence, the estimate of p_{x_c} is

$$\hat{p}_{x_c} = 1 - \phi\left(\frac{x_c - \bar{x}}{s}\right)$$

An alternative approach for normally distributed data is to construct a normal probability plot and read the cumulative probability for the specified concentration x_c directly off the plot. For example, using the normal probability plot in Figure 11.3, we see that an estimated 15% of the population exceeds $x_c = 0.06$ pCi ^{241}Am/g.

11.5 TWO-SIDED CONFIDENCE LIMITS FOR THE MEAN

This section shows how to compute two-sided confidence limits for the population mean μ when either the data values x_i or the estimated mean, $\bar{x}$, are normally distributed. Methods appropriate for lognormal data are given in Chapter 13. Two-sided limits give an interval in which the true mean is expected to lie with specified confidence. This interval can be compared with intervals computed for different times and/or areas. One-sided limits can also be computed and used to test for compliance with environmental limits. These limits are discussed in Section 11.6.

Throughout this section we assume the data are independent and therefore uncorrelated. Methods for computing confidence limits about μ when data are correlated are given in Section 11.12, an important topic because pollution data are frequently correlated if collected at short time and/or space intervals.

11.5.1 Known Variance

If n data have been drawn by simple random sampling from a normal distribution, then $\bar{x}$ is also normally distributed, no matter how small or large n may be. Hence, if σ^2 is known a priori, a two-sided $100(1 - \alpha)\%$ confidence interval about μ is

$$\bar{x} - Z_{1-\alpha/2} \frac{\sigma}{\sqrt{n}} \leq \mu \leq \bar{x} + Z_{1-\alpha/2} \frac{\sigma}{\sqrt{n}} \qquad \textbf{11.4}$$

where $\sigma/\sqrt{n}$ is the standard error of $\bar{x}$ and $Z_{1-\alpha/2}$ is the value of the standard normal variable that cuts off $(100\alpha/2)\%$ of the upper tail of the $N(0, 1)$ distribution. From our discussion of quantiles in Section 11.2, we know that $Z_{1-\alpha/2}$ is the $1 - \alpha/2$ quantile of the $N(0, 1)$ distribution. Values of $Z_{1-\alpha/2}$ are obtained from Table A1. For example, if $\alpha = 0.05$, then $Z_{0.975}$ equals 1.96 and Eq. 11.4 becomes

$$\bar{x} - 1.96 \frac{\sigma}{\sqrt{n}} \leq \mu \leq \bar{x} + 1.96 \frac{\sigma}{\sqrt{n}}$$

This interval is easily computed, since σ is known by assumption and $\bar{x}$ can be computed from the n data.

11.5.2 Unknown Variance

For the more realistic case where σ^2 is unknown, the two-sided $100(1 - \alpha)\%$ confidence interval about μ is

$$\bar{x} - t_{1-\alpha/2,n-1} \frac{s}{\sqrt{n}} \leq \mu \leq \bar{x} + t_{1-\alpha/2,n-1} \frac{s}{\sqrt{n}} \qquad \textbf{11.5}$$

where s is an estimate of σ computed from n data drawn at random from a normal distribution, and $t_{1-\alpha/2,n-1}$ is the value that cuts off $(100\alpha/2)\%$ of the upper tail of the t distribution that has $n - 1$ degrees of freedom (df). The cumulative t distribution for various degrees of freedom is tabulated in Table A2. The validity of Eq. 11.5 does not require n to be large, but the underlying distribution must be normal.

Returning for a moment to Eq. 11.4 when σ is known, we see that the width of the confidence interval (upper limit minus lower limit) is constant and given by $2(1.96)\sigma/\sqrt{n}$. But when σ is replaced by an estimate s as computed from a particular data set, the width will vary from data set to data set. Hence, even though two data sets are drawn at random from the same population, the width of the estimated confidence limits will be different.

It is instructive to discuss the meaning of a $100(1 - \alpha)\%$ confidence interval. Suppose we repeat many times the process of withdrawing n samples at random from the population, each time computing a $100(1 - \alpha)\%$ confidence interval. Then $100(1 - \alpha)\%$ of the computed intervals will, on the average, contain the true value μ. Hence, when a 95% confidence interval is computed by using n randomly drawn data, that interval may be expected to include the true mean μ unless this interval is one of those that will occur by chance 5% of the time.

11.6 ONE-SIDED CONFIDENCE LIMITS FOR THE MEAN

The upper one-sided $100(1 - \alpha)\%$ confidence limit for μ when σ is known is

$$UL_{1-\alpha} = \bar{x} + Z_{1-\alpha} \frac{\sigma}{\sqrt{n}}$$

Similarly, the upper limit when σ is estimated by s is

$$\mathrm{UL}_{1-\alpha} = \bar{x} + t_{1-\alpha,n-1} \frac{s}{\sqrt{n}} \tag{11.6}$$

The corresponding *lower* one-sided limits are

$$\mathrm{LL}_{\alpha} = \bar{x} - Z_{1-\alpha} \frac{\sigma}{\sqrt{n}}$$

and

$$\mathrm{LL}_{\alpha} = \bar{x} - t_{1-\alpha,n-1} \frac{s}{\sqrt{n}} \tag{11.7}$$

respectively. Note that the upper (or lower) $100(1 - \alpha)\%$ one-sided limit uses $Z_{1-\alpha}$ or $t_{1-\alpha}$, whereas $100(1 - \alpha)\%$ two-sided limits use $Z_{1-\alpha/2}$ or $t_{1-\alpha/2}$. For example, if $\alpha = 0.05$, $Z_{0.95} = 1.645$ is used for computing a one-sided limit on μ, whereas $Z_{0.975} = 1.96$ is used for the two-sided limits.

Upper limits on the true mean μ may be computed to test for compliance with regulations that specify some mean value, say μ_L, as an upper limit. If $\mathrm{UL}_{1-\alpha} > \mu_L$, this might be taken as evidence that μ may exceed μ_L.

EXAMPLE 11.3

First we use Eq. 11.5 to compute a 90% confidence interval ($\alpha = 0.10$), using the ^{241}Am data in Example 11.1. Since $\bar{x} = 0.0372$, $s = 0.0211$, $n - 1 = 19$, and $t_{0.95,19} = 1.729$ (from Table A2), Eq. 11.5 gives

$$0.0372 \pm \frac{1.729(0.0211)}{\sqrt{20}} \quad \text{or} \quad 0.0290 \le \mu \le 0.0454$$

Second, a one-sided upper 90% confidence limit for μ is computed by Eq. 11.6. Since $t_{0.90,19} = 1.328$, we obtain

$$\mathrm{UL}_{0.90} = 0.0372 + \frac{1.328(0.0211)}{\sqrt{20}} = 0.0435$$

11.7 APPROXIMATE CONFIDENCE LIMITS FOR THE MEAN

In the previous two sections we learned how to compute confidence limits for the true mean of an underlying normal distribution. But suppose the distribution is not normal, or suppose we are unwilling to make that assumption. Then if n is sufficiently large, a two-sided $100(1 - \alpha)\%$ confidence inteval for the mean μ is approximated by

$$\bar{x} - Z_{1-\alpha/2} \frac{s}{\sqrt{n}} \le \mu \le \bar{x} + Z_{1-\alpha/2} \frac{s}{\sqrt{n}} \tag{11.8}$$

Similarly, for large n, approximate one-sided upper and lower $100(1 - \alpha)\%$ confidence limits are

$$\bar{x} + Z_{1-\alpha}\frac{s}{\sqrt{n}} \qquad \textbf{11.9}$$

and

$$\bar{x} - Z_{1-\alpha}\frac{s}{\sqrt{n}} \qquad \textbf{11.10}$$

respectively.

In practice, there appears to be no simple rule for determining how large n should be for Eqs. 11.8, 11.9, and 11.10 to be used. It depends on the amount of bias in the confidence limits that can be tolerated and also on the shape of the distribution from which the data have been drawn. If the distribution is highly skewed, an n of 50 or more may be required.

11.8 ALTERNATIVE ESTIMATORS FOR THE MEAN AND STANDARD DEVIATION

Thus far in this chapter we have used $\bar{x}$ and s^2 to estimate the true mean μ and variance σ^2 of a normal distribution. It can be shown that for data drawn at random from a normal distribution, $\bar{x}$ and s^2 are the minimum variance unbiased (MVU) estimators of μ and σ^2. That is, of all unbiased estimators of μ, the mean $\bar{x}$ has the smallest error that arises because only a portion of the population units are measured. (Unbiased estimators were defined in Section 2.5.3.) Nevertheless, $\bar{x}$ and s^2 are not always well suited for environmental data. For example, data sets are often highly skewed to the right, so a few data are much larger than most. In this case, even though $\bar{x}$ and s^2 are unbiased estimators of the mean and variance of the distribution, they may not be accurate. If the skewed data set is believed to be drawn from a lognormal distribution, the methods for estimating μ and σ^2 illustrated in Chapter 13 may be used. That chapter also gives methods for estimating the population median when the lognormal distribution is applicable.

A related problem is the frequent occurrence of outliers, where Hunt et al. (1981) define an outlier to be "an observation that does not conform to the pattern established by other observations." If the underlying distribution is believed to be symmetric (but not necessarily normal) the median, trimmed mean, or Winsorized mean can be used to estimate μ, as discussed in Chapter 14.

Another problem is that environmental data sets are often censored—that is, the actual measured values for some population units are not available. Censoring may occur when the pollutant concentration is very near or below the measurement limit of detection (LOD) and the datum is reported to the data analyst as "trace," the letters ND ("not detected"), or the LOD itself. In this case the median, trimmed mean, and Winsorized mean may be useful because these estimators do not use data in the tails of the data set. Alternatively, if the censored data are believed to be from normal or lognormal distributions, the efficient estimators illustrated in Chapter 14 may be used.

Finally, environmental data are often correlated in time and/or space. The sample mean $\bar{x}$ is still an unbiased estimator of the mean μ, but correlation should be taken into account when estimating $\mathrm{Var}(\bar{x})$. Methods for estimating $\mathrm{Var}(\bar{x})$ are discussed in Sections 4.5 and 4.6.

11.9 NONPARAMETRIC ESTIMATORS OF QUANTILES

Until now we have used methods appropriate when data have been drawn at random from normal distributions. However, quantiles, proportions, and confidence limits on means can also be estimated when the underlying distribution is either unknown or nonnormal. These methods are called *nonparametric* or *distribution-free* techniques, since their validity does not depend on the data being drawn from any particular distribution.

This section gives a nonparametric procedure for estimating quantiles that may be used in place of the method in Section 11.2. Suppose we are unwilling or unable to assume the distribution is normal, and we wish to estimate the pth quantile, x_p, where p is some specified proportion ($0 < p < 1$). The procedure is as follows: Draw n data at random from the underlying population and order them to obtain the sample order statistics $x_{[1]} \leq x_{[2]} \leq \cdots \leq x_{[n]}$. To estimate x_p, we first compute $k = p(n + 1)$. If k is an integer, the estimated pth percentile, $\hat{x}_p$, is simply the kth order statistic $x_{[k]}$, that is, the kth largest datum in the data set. If k is not an integer, $\hat{x}_p$ is obtained by linear interpolation between the two closest order statistics. This procedure is used by Snedecor and Cochran (1967, p. 125) and by Gibbons, Olkin, and Sobel (1977, p. 195).

For example, if we want to estimate $x_{0.97}$, the 0.97 quantile, and $n = 100$ data are obtained by random sampling, then $k = (0.97)101 = 97.97$. Since k is not an integer, $\hat{x}_{0.97}$ is found by linear interpolation between the 97th and 98th largest of the $n = 100$ data.

11.10 NONPARAMETRIC CONFIDENCE LIMITS FOR QUANTILES

Suppose we want to estimate the lower and upper $100(1 - \alpha)\%$ confidence limits for the true pth quantile, x_p, of an unknown distribution. If $n \leq 20$, it can be done with the procedure described by Conover (1980, p. 112) in conjunction with his Table A3 (pp. 433–444), which gives the cumulative distribution function of the binomial distribution.

If $n > 20$, the following method may be used. First compute

$$l = p(n + 1) - Z_{1-\alpha/2}[np(1 - p)]^{1/2} \qquad \textbf{11.11}$$

and

$$u = p(n + 1) + Z_{1-\alpha/2}[np(1 - p)]^{1/2} \qquad \textbf{11.12}$$

Since l and u are usually not integers, the limits are obtained by linear interpolation between the closest order statistics. For example, if 95% limits are desired about the $p = 0.90$ quantile, and if $n = 1000$, then

$$l = 0.9(1001) - 1.96[1000(0.9)(0.1)]^{1/2} = 882.306$$

and

$$u = 0.9(1001) + 1.96[1000(0.9)(0.1)]^{1/2} = 919.494$$

Then the lower limit is obtained by linear interpolation between the 882nd and 883rd order statistic. Similarly, interpolating between the 919th and 920th order statistics gives the upper limit.

One-sided confidence limits for the true pth quantile are also easily obtained. Suppose an upper $100(1 - \alpha)\%$ limit is required. If $n > 20$, this limit is obtained by computing

$$u = p(n + 1) + Z_{1-\alpha}[np(1 - p)]^{1/2} \qquad \textbf{11.13}$$

and interpolating between the two closest order statistics. For example, if the upper 95% limit for the 0.90 quantile is desired and if $n = 1000$, then

$$u = 0.9(1001) + 1.645[1000(0.9)(0.1)]^{1/2} = 916.506$$

Therefore, the estimated upper limit is the value that is 50.6% of the way between the 916th and 917th largest values. A one-sided lower limit is obtained in a similar manner by using a negative sign in front of $Z_{1-\alpha}$ in Eq. 11.13.

11.11 NONPARAMETRIC CONFIDENCE LIMITS FOR PROPORTIONS

The two approaches for estimating proportions given in Section 11.4 are appropriate for normal distributions. The following approach is valid for any distribution, as long as the data are uncorrelated and were drawn by random sampling. To estimate p_{x_c}, the proportion of the population exceeding x_c, we compute

$$\hat{p}_{x_c} = \frac{u}{n} \qquad \textbf{11.14}$$

where n is the number of observations and u is the number of those that exceed x_c.

A confidence interval for p_{x_c} can easily be obtained. If $n \leq 30$, then 95% and 99% confidence intervals can be read directly from Table A4 (from Blyth and Still, 1983). For example, suppose $n = 30$ observations are drawn at random from the population and that $u = 1$ of these exceeds a prespecified concentration x_c. Then Eq. 11.14 gives $\hat{p}_{x_c} = \frac{1}{30} = 0.033$. Also, from Table A4 we find that 0 and 0.16 are the lower and upper 95% confidence limits for the true proportion of the population exceeding x_c.

If $n > 30$, Blyth and Still (1983) recommend that the lower and upper limits of a two-sided $100(1 - \alpha)\%$ confidence interval be computed as follows.

$$\text{Lower Limit} = \frac{1}{n + Z^2_{1-\alpha/2}} \cdot \left\{(u - 0.5) + \frac{Z^2_{1-\alpha/2}}{2} - Z_{1-\alpha/2}\left[(u - 0.5) - \frac{(u - 0.5)^2}{n} + \frac{Z^2_{1-\alpha/2}}{4}\right]^{1/2}\right\} \qquad \textbf{11.15}$$

except that the lower limit equals 0 if $u = 0$.

$$\text{Upper Limit} = \frac{1}{n + Z^2_{1-\alpha/2}} \cdot \left\{ (u + 0.5) + \frac{Z^2_{1-\alpha/2}}{2} + Z_{1-\alpha/2}\left[(u + 0.5) - \frac{(u + 0.5)^2}{n} + \frac{Z^2_{1-\alpha/2}}{4}\right]^{1/2} \right\} \qquad \mathbf{11.16}$$

except that the upper limit equals 1 if $u = n$.

Confidence limits about a proportion may also be obtained from data charts given, for example, by Conover (1980, Table A4). Finally, if np_{x_c} and $np_{x_c}(1 - p_{x_c})$ are both greater than 5, the following upper and lower limits give an approximate two-sided $100(1 - \alpha)\%$ confidence interval for p_{x_c}:

$$\hat{p}_{x_c} \pm Z_{1-\alpha/2}\left[\frac{\hat{p}_{x_c}(1 - \hat{p}_{x_c})}{n}\right]^{1/2} \qquad \mathbf{11.17}$$

EXAMPLE 11.4

Table 11.1 gives CO vehicle emissions data [in units of grams/mile (g/mi)] obtained on $n = 46$ randomly selected vehicles (reported by Lorenzen, 1980). We shall use these data to estimate the proportion of the population (from which the 46 vehicles were drawn) that exceeds 15 g/mi. We also compute a 90% confidence interval about the true proportion exceeding 15 g/mi using the nonparametric procedure.

From Table 11.1 we find $u = 4$ data that exceed $x_c = 15$ g/mi. Therefore, Eq. 11.14 gives $\hat{p}_{15} = u/n = \frac{4}{46} = 0.087$ as the estimated proportion of the population exceeding 15 g/mi. The confidence interval is obtained from Eqs. 11.15 and 11.16. Table A1 gives $Z_{1-\alpha/2} = Z_{0.95} = 1.645$. Therefore, the lower limit obtained from Eq. 11.15 is

$$\frac{\{(4 - 0.5) + 2.706/2 - 1.645[(4 - 0.5) - (4 - 0.5)^2/46 + 2.706/4]^{1/2}\}}{46 + 2.706} = 0.033$$

Similarly, using Eq. 11.16, we find that the upper limit is 0.194. Therefore, the 90% confidence interval about the true proportion of the population greater than 15 g/mi is from 0.033 to 0.19. A point estimate of that proportion is 0.087.

Since $n\hat{p}_{x_c} = 4$ and $n\hat{p}_{x_c}(1 - \hat{p}_{x_c}) = [46(4)/46](1 - \frac{4}{46}) = 3.65$, it is possible that np_{x_c} and $np_{x_c}(1 - p_{x_c})$ are not greater than 5. Hence, we should not use Eq. 11.17 to obtain confidence limits. However, for illustration's sake, Eq. 11.17 gives

$$0.087 \pm 1.645\left[\frac{0.087(0.913)}{46}\right]^{1/2}$$

or 0.019 and 0.16 for the lower and upper limits.

Table 11.1 Vehicle Emission Carbon Monoxide (CO) Data (grams/mile) for $n = 46$ Randomly Chosen Vehicles

Vehicle	*CO*	*Vehicle*	*CO*	*Vehicle*	*CO*
1	5.01	17	15.13	33	5.36
2	14.67	18	5.04	34	14.83
3	8.60	19	3.95	35	5.69
4	4.42	20	3.38	36	6.35
5	4.95	21	4.12	37	6.02
6	7.24	22	23.53	38	5.79
7	7.51	23	19.00	39	2.03
8	12.30	24	22.92	40	4.62
9	14.59	25	11.20	41	6.78
10	7.98	26	3.81	42	8.43
11	11.53	27	3.45	43	6.02
12	4.10	28	1.85	44	3.99
13	5.21	29	4.10	45	5.22
14	12.10	30	2.26	46	7.47
15	9.62	31	4.74		
16	14.97	32	4.29		

Source: After Lorenzen, 1980, Table 2.

11.12 CONFIDENCE LIMITS WHEN DATA ARE CORRELATED

The methods given in Sections 11.5.2, 11.6, and 11.7 for placing confidence limits about the true mean μ are appropriate when the data are not correlated. The same formulas may be used when data are correlated except that the estimates of the standard error of the mean, $s/\sqrt{n}$, must be modified as discussed in Sections 4.5 and 4.6. The confidence intervals given in subsections 11.12.1 and 11.12.2 require that estimates of serial and/or spatial correlation coefficients be obtained from the data. Since these estimates will not be accurate if they are based on only a few data, the formulas in this section should not be used unless n is large, preferably $n \geq 50$.

11.12.1 Single Station

Consider first the case of time (serial) correlation but no spatial correlation. Suppose data are collected at equal intervals sequentially in time at a monitoring station, and we wish to compute a confidence interval for the true mean over that period of time at that station. The two-sided and one-sided confidence intervals for μ given by Eqs. 11.5–11.10 can be used for this purpose if $s/\sqrt{n}$ in these equations is replaced by

$$s\left[\frac{1}{n}\left(1 + 2\sum_{l=1}^{n-1}\hat{\rho}_l\right)\right]^{1/2} \tag{11.18}$$

where s is the square root of Eq. 4.4 and the $\hat{\rho}_l$ are the estimated serial correlation coefficients of lag $l = 1, 2, \ldots, n-1$ computed by Eq. 4.22. For example, by Eq. 11.5 the approximate $100(1-\alpha)\%$ confidence interval for the true mean for the station is

$$\bar{x} \pm t_{1-\alpha/2,n-1}s\left[\frac{1}{n}\left(1 + 2\sum_{l=1}^{n-1}\hat{\rho}_l\right)\right]^{1/2} \tag{11.19}$$

If the serial correlation coefficients with lags greater than, say, $l = m$ are all zero, then Eq. 11.18 is the same as given by Albers (1978b, Eq. 2.4), who developed a t test for dependent data.

A somewhat more accurate confidence interval may be obtained by using

$$s\left\{\frac{1}{n}\left[1 + \frac{2}{n}\sum_{l=1}^{n-1}(n - l)\hat{\rho}_l\right]\right\}^{1/2} \qquad \textbf{11.20}$$

instead of Eq. 11.18. We have encountered Eqs. 11.18 and 11.20 before in connection with determining the number of measurements needed for estimating a mean when data are correlated; see Eqs. 4.18 and 4.21.

EXAMPLE 11.5

In Exercise 4.3 there was serial correlation between the $n = 100$ measurements taken along a line in space. The serial correlations for lags $l = 1, 2, \ldots, 14$ were nonzero and summed to $\Sigma_{l=1}^{14} \hat{\rho}_l = 4.612$. Also, the sample mean and standard deviation were $\bar{x} = 18{,}680$ and $s = 4030$. Therefore, by Eqs. 11.5 and 11.18 a 90% confidence interval about the true mean along the transect is given by

$$\bar{x} \pm t_{0.95,99}s\left[\frac{1 + 2(4.612)}{100}\right]^{1/2}$$

or $18{,}680 \pm 1.661(4030)(0.3197)$ or 16,500 and 20,800.

11.12.2 Regional Means

Suppose n observations over time are collected at each of n_s monitoring stations. The arithmetic mean of the observations at the ith station is

$$\bar{x}_i = \frac{1}{n}\sum_{j=1}^{n} x_{ij}$$

Then an estimate of the true regional mean is

$$\bar{x} = \frac{1}{nn_s}\sum_{i=1}^{n_s}\sum_{j=1}^{n} x_{ij} = \frac{1}{n_s}\sum_{i=1}^{n_s}\bar{x}_i$$

If there is no correlation between stations, a confidence interval about the true regional mean is

$$\bar{x} \pm t_{1-\alpha/2,n_s-1}\left[\frac{1}{n_s(n_s - 1)}\sum_{i=1}^{n_s}(\bar{x}_i - \bar{x})^2\right]^{1/2} \qquad \textbf{11.21}$$

Equation 11.21 is valid even if there is serial correlation between the measurements at each station. The quantity under the square-root sign in Eq. 11.21 is the standard error given previously in Section 4.6. Note that this estimator does not use individual data points at the stations but only the station means.

If there is a spatial correlation ρ_c between n_s stations but no time correlation between the n measurements at each station, the confidence interval about the true regional mean is (by Eq. 4.14)

$$\bar{x} \pm t_{1-\alpha/2,nn_s-1}s\left[\frac{1 + \hat{\rho}_c(n_s - 1)}{nn_s}\right]^{1/2} \qquad \textbf{11.22}$$

where s is approximated by the square root of $nn_s s^2(\bar{x})$, where $s^2(\bar{x})$ is Eq. 4.27. Example 4.6 shows how to obtain $\hat{\rho}_c$.

Finally, if there are both spatial and temporal correlation, the confidence interval is (from Eq. 4.23)

$$\bar{x} \pm t_{1-\alpha/2, nn_s - 1} s \left[\frac{1}{nn_s} \left(1 + \frac{2}{n} \sum_{l=1}^{n-1} (n - l)\hat{\rho}_l \right) (1 + \hat{\rho}_c(n_s - 1)) \right]^{1/2} \qquad \mathbf{11.23}$$

where s is obtained as above for Eq. 11.22. Equations 11.18, 11.19, and 11.20, are approximate, since s in these equations is computed by Eq. 4.4, which is appropriate only when the data are independent. An alternative estimator for s is suggested in Exercise 4.4.

11.13 RANK VON NEUMANN TEST FOR SERIAL CORRELATION

Sections 4.5.2 and 11.12.1 gave methods for estimating $\text{Var}(\bar{x})$ and for deciding how many measurements to take to compute $\bar{x}$ when the data are collected sequentially over time and are serially correlated. This section discusses the rank von Neumann test (Bartels, 1982), which tests the null hypothesis H_0 that $\rho_1 = 0$ versus the alternative hypothesis that $\rho_1 > 0$, where ρ_1 is the lag 1 serial correlation coefficient. If this test is nonsignificant (null hypothesis not rejected) and at least 25 measurements are used in the test, we may tentatively conclude that the formulas for $\text{Var}(\bar{x})$ given in Sections 4.5.2 and 11.12.1 may not be needed; the simpler formula s^2/n is sufficient. Additional tests can be carried out to help decide whether serial correlations of lags greater than 1 are equal to zero by the method given by Box and Jenkins (1976, p. 35).

The von Neumann test will also detect trends and/or cycles in a sequence of data. Hence, if the test statistic gives a significant result, it could be due to trends, cycles, and/or autocorrelation. Also, for the test to be meaningful, the data should be collected at equal or approximately equal intervals.

Assume there are no trends or cycles present. Then the null hypothesis being tested is that $\rho_1 = 0$. The alternative hypothesis of most interest is that $\rho_1 > 0$, since positive correlation is the usual case with pollution data. Let $x_1, x_2, \ldots, x_n$ be a sequence (time series) of n observations obtained over time at a monitoring station. Then do the following:

1. Assign the rank of 1 to the smallest observation, the rank of 2 to the next smallest, . . . , and the rank of n to the largest observation. Let R_1 be the rank of x_1, R_2 be the rank of x_2, . . . , and R_n be the rank of x_n.
2. Compute the rank von Neumann statistic, R_v, as follows:

$$R_v = \frac{12}{n(n^2 - 1)} \sum_{i=1}^{n-1} (R_i - R_{i+1})^2$$

where: R_i = rank of the ith observation in the sequence,
R_{i+1} = rank of the $(i + 1)$st observation in the sequence (the following observation).

3. If $10 \leq n \leq 100$, reject the null hypothesis that $\rho_1 = 0$ at the α significance level and conclude that $\rho_1 > 0$ if R_v is less than $R_{v,\alpha}$, the α quantile for R_v given in Table A5 for the appropriate n. This table gives quantiles for

$10 \leq n \leq 100$. Tests for serial correlation are not recommended if $n \leq 10$, but such tests are possible with Table 1 in Bartels (1982).

4. If $n > 100$, compute

$$Z_R = \frac{\sqrt{n}(R_v - 2)}{2}$$

and reject H_0 and accept that $\rho_1 > 0$ if Z_R is negative and $|Z_R| > Z_{1-\alpha}$, where $Z_{1-\alpha}$ is obtained from Table A1.

The rank von Neumann test is not exact if some measurements are tied (have the same magnitude). If the number of tied values is small, one may assign to each observation in a tied set the midrank, that is, the average of the ranks that would be assigned to the set. When this procedure is used, the critical values in the table ''may provide a reasonable approximation when the number of ties is small'' (Bartels, 1982, p. 41).

EXAMPLE 11.6

Hakonson and White (1979) conducted a field study in which soil samples were collected on a grid at 1-m intervals both before and after the site was rototilled. Cesium concentrations (^{137}Cs) were obtained for each sample. The authors have kindly provided the data for our use. We shall test at the $\alpha = 0.05$ significance level that there is no serial correlation between Cs concentrations along grid lines. The 13 ^{137}Cs concentrations and their ranks obtained in order along one line of the rectangular grid before rototilling are

datum,	rank
2.20	10
2.74	13
0.42	4
0.63	6
0.82	7
0.86	8
0.31	2
2.33	12
0.50	5
2.22	11
1.10	9
0.32	3
0.01	1

Since $n(n^2 - 1)/12 = 13(168)/12 = 182$, the rank von Neumann statistic is

$$[(10 - 13)^2 + (13 - 4)^2 + (4 - 6)^2 + \ldots + (3 - 1)^2]/182 = \frac{361}{182} = 1.98$$

From Table A5 the critical value at the $\alpha = 0.05$ level is 1.14. Since $1.98 > 1.14$, we cannot reject the null hypothesis that $\rho_1 = 0$.

When testing for serial correlations over space, it is necessary to test along lines in several directions, for example, north-south, east-west, and intermediate angles, since correlations can exist in some directions but not in others.

11.14 DATA TRANSFORMATIONS

Pollution data are frequently transformed before statistical or graphical analyses. This section briefly discusses when it is desirable to transform and some of the pitfalls that can arise. Our interest is in nonlinear transformations—that is, those that change the scale of measurement. We pay particular attention to the logarithmic transformation, since it is useful for pollution data and is used in later chapters of this book, primarily Chapters 13 and 14.

11.14.1 Reasons for Using Transformations

Here are some reasons for using nonlinear transformations.

1. To obtain a more accurate and meaningful graphical display of the data
2. To obtain a straight-line relationship between two variables
3. To better fulfill the assumptions underlying statistical tests and procedures, such as normality, additive statistical models, and equal spreads (variance) in different data sets
4. To efficiently estimate quantities such as the mean and variance of a lognormal distribution, as illustrated in Chapters 13 and 14

As an example of items 1 and 2, suppose we wish to study the relationship between two pollutants. We measure both on the same set of n samples and plot the untransformed data on a scatter plot. It would not be unusual to obtain several data pairs with very high concentrations relative to the bulk of the data. In this case the low concentrations pairs will be clumped close to zero, and the high concentrations will appear as isolated values far from zero. If a straight line is fit to these data, the fit will be determined mainly by the position of the lower clump and the few high values. That is, the full set of n data pairs will not be efficiently used to examine the relationship between the two variables. If the logarithms of the data are plotted, the points will be more evenly spread out over the coordinate scales and the entire n pairs effectively used and displayed.

Since pollution data tend to be skewed, the scatter plot of log-transformed data will tend to be linear. This behavior is fortunate because statistical methods for linear relationships are relatively simple. For example, as pointed out by Hoaglin, Mosteller, and Tukey (1983), departures from a linear fit are more easily detected, and interpolation and extrapolation are easier than if the relationship is not linear.

Concerning item 3, most statistical methods books, such as Snedecor and Cochran (1980, pp. 282–297), discuss data transformations that are useful before performing analysis of variance (AOV) computations. AOV tests of hypotheses assume that the effects of different factors are additive and that the residual errors have the same variance and are normally distributed. Cochran (1947), Scheffé (1959), and Glass, Peckham, and Sanders (1972) discuss the effects on AOV procedures when these assumptions are not fulfilled. Since pollutant data are often approximately lognormal, it is common practice to use a logarithmic transformation before conducting an AOV. The desired characteristics of additivity, constant variance, and normality are frequently achieved at least approximately when this is done. For the same reason, the logarithmic transformation is frequently used before doing t tests to look for significant differences between two means.

11.14.2 Potential Problems with Transformations

Following are three problems that may arise when using a nonlinear transformation.

1. Estimating quantities such as means, variances, confidence limits, and regression coefficients in the transformed scale typically leads to biased estimates when they are transformed back into the original scale.
2. It may be difficult to understand or apply results of statistical analyses expressed in the transformed scale.
3. More calculations are required.

We may illustrate the bias referred to in item 1 by considering the lognormal distribution. Let x represent an untransformed lognormal datum, and let $y = \ln x$. An unbiased estimator of the mean of the log-transformed distribution is $\bar{y}$, the arithmetic mean of the y's. But if $\bar{y}$ is transformed back to the original scale by computing $\exp(\bar{y})$, the geometric mean, we do not obtain an unbiased estimate of the mean of the untransformed (lognormal) distribution. A similar problem arises when estimating confidence limits for the mean of a lognormal distribution. Chapter 13 gives unbiased methods for estimating lognormal means and confidence limits. Heien (1968) and Agterberg (1974, p. 299) discuss similar bias problems when conducting linear regression and trend surface analysis on transformed data.

Koch and Link (1980, Vol. 1, p. 233) suggest that transformations may be useful "when the conclusions based on the transformed scale can be understood, when biased estimates are acceptable, or when the amount of bias can be estimated and removed because the details of the distribution are known." Hoaglin, Mosteller, and Tukey (1983) point out that we lose some of our intuitive understanding of data in a transformed scale, and that a judgment must be made as to when the benefits justify the "costs." They indicate that a transformation is likely to be useful only when the ratio of the largest datum to the smallest datum in a data set is greater than about 20.

11.15 SUMMARY

The normal distribution plays an important role in the analyses of pollution data even though many environmental data sets are usually not normally distributed. This importance occurs because of the close relationship with the lognormal distribution and because the sample mean, $\bar{x}$, is normally distributed if n is sufficiently large.

This chapter provides a set of tools for characterizing normal distributions. Methods for estimating the mean, variance, proportions, and quantiles and for putting confidence intervals on proportions, quantiles, and the mean are given. Nonparametric (distribution-free) methods that may be used when the distribution is nonnormal are provided. Formulas are given for computing approximate confidence limits about μ when data are correlated over time and/or space. The last section discusses the benefits and potential pitfalls of nonlinear data transformations.

EXERCISES

11.1 What percent of the standard normal distribution falls (a) above $\mu + 1\sigma$, (b) between $\mu - 2\sigma$ and $\mu + 1\sigma$?

11.2 Listed here are $n = 10$ carbon monoxide data from Table 11.1:

6.02	5.79	2.03	4.62	6.78
8.43	6.02	3.99	5.22	7.47

Assume these data were drawn at random from a normal distribution. Estimate (a) the mean, μ, (b) the variance, σ^2, and (c) the 90th percentile of the normal distribution from which these data are assumed drawn. Also, estimate the 90th percentile by the nonparametric method. Can the 95th percentile be estimated by the nonparametric method when only 10 samples are collected?

11.3 Estimate the upper 90% confidence limit for the 90th percentile of the (assumed) normal distribution from which the data in Exercise 11.2 were drawn.

11.4 Using the data in Exercise 11.2, estimate the proportion of the assumed normal distribution that exceeds the value 4.0.

11.5 In Example 11.3 estimate (a) a two-sided 99% confidence interval for μ, (b) a one-sided upper 80% confidence limit for μ, and (c) a one-sided lower 95% confidence limit for μ.

11.6 Use the carbon monoxide data in Table 11.1 to estimate upper and lower 80% confidence limits on the 60th percentile (0.60 quantile) of the population from which these data were drawn.

11.7 Use the data in Table 11.1 to estimate the proportion of the population that exceeds 20 g/mi. Estimate the lower and upper 95% confidence limits for the true proportion greater than 20 g/mi.

11.8 In Example 11.5 we found that when serial correlation was present, the 90% confidence interval for the true mean was from 16,540 to 20,820. Recompute the 90% confidence interval using the same data but without taking the serial correlation into account.

ANSWERS

11.1 (a) 15.87%, (b) 81.85%.

11.2 (a) $\bar{x} = 5.64$, (b) $s^2 = 3.30$, (c) the 90th percentile computed by Eq. 11.1 is $\hat{x}_{0.90} = 5.64 + 1.282(1.817) = 7.97$, and by the nonparametric method, is 90% of the way between the 9th and 10th largest of the 10 data, or 8.33. No.

11.3 From Eq. 11.2, $UL_{.90}(x_{.90}) = 5.64 + 2.066(1.817) = 9.39$.

11.4 $\hat{p}_{4.0} = 1 - \phi\left(\frac{4 - 5.64}{1.817}\right) = 1 - \phi(-0.9026) = 0.817$

11.5 (a) From Eq. 11.5, $0.0372 \pm 2.861(0.0211)/\sqrt{20}$, or 0.0237 and 0.0507.
(b) From Eq. 11.6, $0.0372 + 0.861(0.0211)/\sqrt{20} = 0.0413$.
(c) From Eq. 11.7, $0.0372 - 1.729(0.0211)/\sqrt{20} = 0.0290$.

11.6 Using Eqs. 11.11 and 11.12 gives

$$(l, u) = 0.6(47) \pm 1.282[46(0.6)(0.4)]^{1/2}$$

Therefore, $l = 23.94$, $u = 32.46$, so lower limit $= 6.01$, upper limit $= 8.51$.

11.7 $\hat{p}_{20} = \frac{2}{46} = 0.0435$. By Eq. 11.15,

$$\text{Lower 95\% Limit: } \frac{1}{49.8}\left\{1.5 + \frac{3.84}{2} - 1.96\left[1.5 - \frac{(1.5)^2}{46} + \frac{3.84}{4}\right]^{1/2}\right\}$$

$$= 0.0076$$

By Eq. 11.16,

$$\text{Upper 95\% Limit: } \frac{1}{49.8}\left\{2.5 + \frac{3.84}{2} + 1.96\left[2.5 - \frac{(2.5)^2}{46} + \frac{3.84}{4}\right]^{1/2}\right\}$$

$$= 0.16$$

11.8 $18{,}680 \pm 1.661(4030)/10$, or 18,011 to 19,349. This interval is about one-third the length of the interval computed in Example 11.5.

12 Skewed Distributions and Goodness-of-Fit Tests

In many cases pollution data sets are skewed (asymmetrical) so that the symmetric normal distribution discussed in Chapter 11 is not a suitable model for estimating quantiles, proportions, or means. In that case the nonparametric procedures given in Chapter 11 may be used. Another approach is to find a distribution model that adequately fits the skewed data set. Then statistical methods for that distribution can be used. This chapter describes the lognormal distribution and several methods for testing whether a data set is likely to have arisen from a normal distribution or a lognormal distribution.

12.1 LOGNORMAL DISTRIBUTION

The lognormal distribution is used to model many kinds of environmental contaminant data: for example, air quality data (see the reviews by Mage, 1981; Georgopoulos and Seinfeld, 1982), radionuclide data sets (Pinder and Smith, 1975; McLendon, 1975; and Horton et al., 1980), trace metals in fish (Giesy and Weiner, 1977), and strontium-90 and other fission-product concentrations in human tissues (Schubert, Brodsky, and Tyler, 1967).

Two-, three-, and four-parameter lognormal distributions can be defined. The *two-parameter lognormal* density function is given by

$$f(x) = \frac{1}{x\sigma_y\sqrt{2\pi}} \exp\left[-\frac{1}{2\sigma_y^2}(\ln x - \mu_y)^2\right] \qquad x > 0, \; -\infty < \mu_y < \infty, \; \sigma_y > 0 \tag{12.1}$$

where μ_y and σ_y^2, the two parameters of the distribution, are the true mean and variance, respectively, of the *transformed* random variable $Y = \ln X$. Some authors refer to the true geometric mean [exp (μ_y)] and the true geometric standard deviation [exp (σ_y)] as the parameters of the distribution. We shall use $\Lambda(\mu_y, \sigma_y^2)$ to denote a two-parameter lognormal distribution with parameters μ_y and σ_y^2.

Some two-parameter lognormal distributions are shown in Figure 12.1. The distribution is described in detail by Aitchison and Brown (1969) and Johnson and Kotz (1970*a*), who give several methods for estimating the parameters μ_y and σ_y^2. Mage and Ott (1984) evaluate several methods and demonstrate that

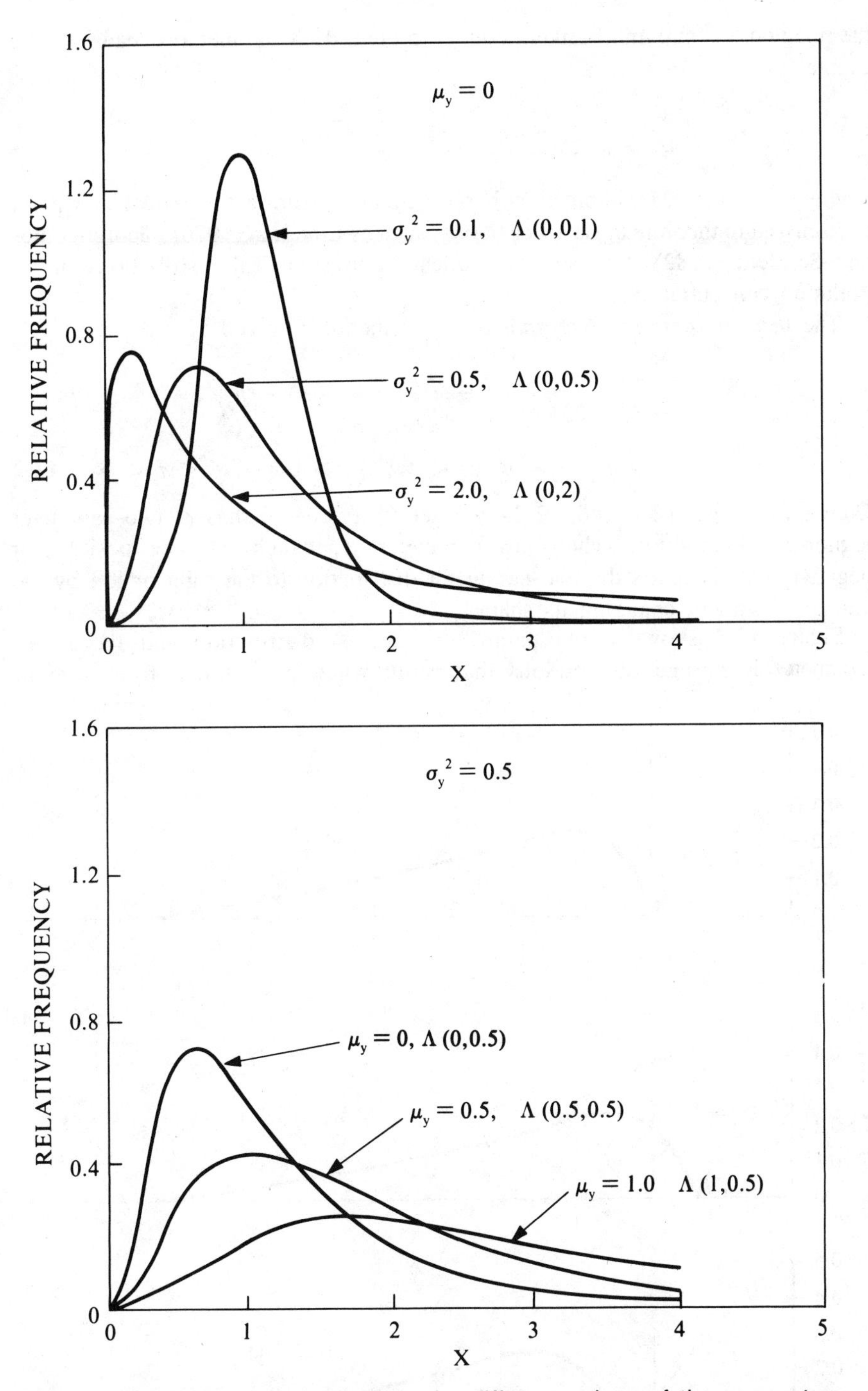

Figure 12.1 Lognormal distributions for different values of the parameters μ_y and σ_y^2, the mean and variance, respectively, of the log-transformed variate (after Aitchison and Brown, 1969, Figs. 2.2 and 2.3).

the method of maximum likelihood is preferred. This method leads to the estimators

$$\bar{y} = \frac{1}{n} \sum_{i=1}^{n} y_i \quad \text{and} \quad s_y^2 = \frac{1}{n} \sum_{i=1}^{n} (y_i - \bar{y})^2$$

where $y_i = \ln x_i$. Maximum likelihood estimation from a theoretical viewpoint is discussed in theoretical statistics books, such as Lindgren (1976). Georgopoulos and Seinfeld (1982) illustrate its application to statistical distributions of air pollution concentrations.

The *three-parameter lognormal* density function is given by

$$f(x) = \frac{1}{(x - \tau)\, \sigma_y \sqrt{2\pi}} \exp\left\{ -\frac{1}{2\sigma_y^2} [\ln (x - \tau) - \mu_y]^2 \right\}$$

$$x > \tau, \quad -\infty < \mu_y < \infty, \quad \sigma_y > 0, \quad -\infty < \tau < \infty \qquad \textbf{12.2}$$

Comparing Eqs. 12.1 and 12.2, we see that $x - \tau$ has a two-parameter lognormal distribution. The third parameter, τ, which may be positive or negative, simply shifts the two-parameter distribution to the right or left by the amount τ without changing its shape.

Figure 12.2 shows a two-parameter $\Lambda(1, 1)$ distribution and the three-parameter lognormal distributions that result when τ is shifted from zero to

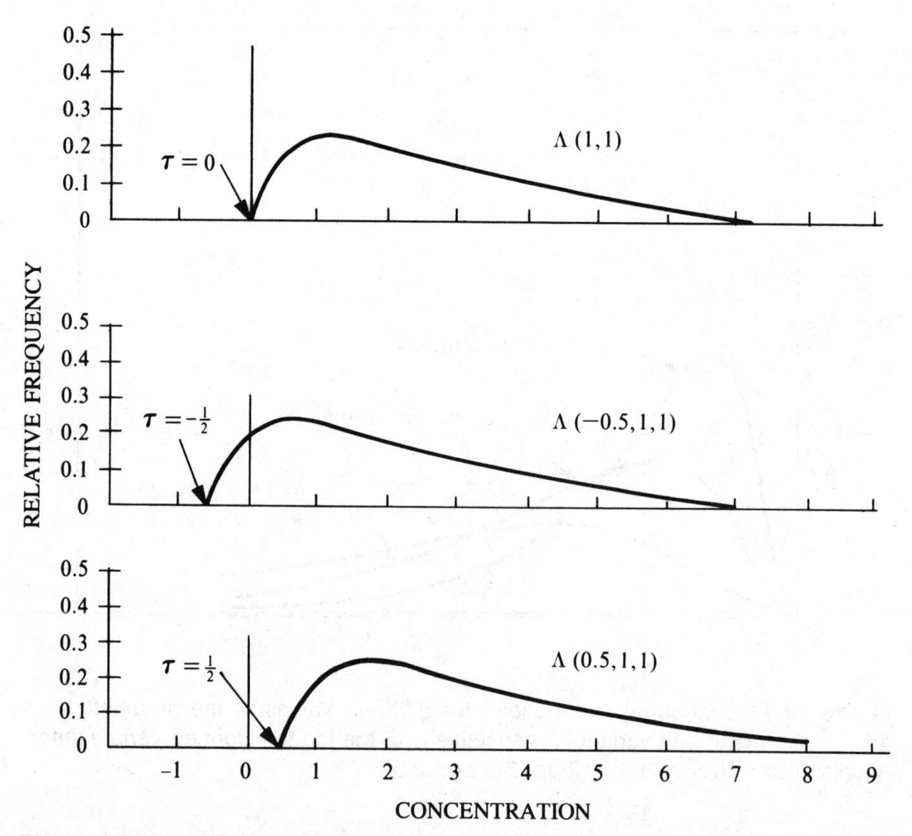

Figure 12.2 The two-parameter lognormal distribution $\Lambda(1, 1)$ and the three-parameter lognormal distributions that result when $\tau = -0.5$ and 0.5 (after Gilbert and Kinnison, 1981, Fig. 1).

-0.5 or to 0.5. The population mean μ and variance σ^2 of the three-parameter lognormal distribution are defined in Table 12.1 along with other characteristics of the distribution. Setting $\tau = 0$ in Table 12.1 gives the appropriate expressions for the two-parameter lognormal distribution.

Maximum likelihood estimates of the parameters μ_y, σ_y^2, and τ may be obtained by the Nelder–Mead simplex procedure (Olsson and Nelson, 1975), as discussed by Holland and Fitz-Simons (1982). One of the conditional maximum likelihood estimates discussed by Cohen and Whitten (1981) could also be used. If the value for τ is known a priori, then μ_y and σ_y^2 are estimated by computing $\bar{y}$ and s_y^2 from the data $y_i = \ln(x_i - \tau)$.

Since pollution concentrations cannot be negative, a three-parameter lognormal with negative τ may seem on first thought to be irrelevant. However, negative measurements can occur due to measurement errors when true concentrations are very near zero. For example, correcting environmental radionuclide measurements by subtracting naturally occurring (background) radiation can give negative data when concentrations are only slightly above background and large measurement errors are present.

The four-parameter lognormal is bounded by a lower bound and an upper bound on the possible values of the variable. Also, both right and left-skewed distribution shapes are possible. This distribution is discussed by Aitchison and Brown (1969) and Mage (1980) and has been applied to air quality data by Mage (1975).

12.2 WEIBULL, GAMMA, AND BETA DISTRIBUTIONS

The Weibull, gamma, and beta distributions are sometimes used to model environmental pollution data. Their density functions are given in Table 12.2 along with the two- and three-parameter lognormal density functions. Plots showing the many shapes these distributions can take are given by Hahn and Shapiro (1967). Georgopoulos and Seinfeld (1982) discuss the application of these distributions to air pollution concentrations.

The parameters γ, α, and β of the *three-parameter Weibull distribution* determine the location, shape, and scale, respectively, of the distribution. The distribution can take on a wide variety of shapes and can be used to model both right- and left-skewed data sets.

The three parameters can be estimated by several methods, including Weibull probability paper (illustrated by King, 1971 and Hahn and Shapiro, 1967) or the maximum likelihood method, discussed, for example, by Johnson and Kotz (1970*a*, p. 255) and Holland and Fitz-Simons (1982).

Pinder and Smith (1975) found that the Weibull distribution fit some radionuclide data sets better than the two-parameter lognormal. Apt (1976) recommends the Weibull distribution as being well suited for describing spatial and temporal distributions of atmospheric radioactivity. He suggests that the estimate of γ would be a reasonably good environmental "background" or "nonimpacted" value, since γ is a threshold or minimum-value parameter. Johnson (1979) reported that ambient ozone data appeared to be better fit by the two-parameter Weibull distribution (i.e., when $\gamma = 0$) than by the two-parameter lognormal.

Table 12.1 Some Characteristics of Normal and Lognormal Populations

	x Is from a Normal Distribution with Parameters μ and σ^2	*x Is from a 3-Parameter Lognormal Distribution[a] with Parameters μ_y, σ_y^2, and τ*
Definitions of Distribution Parameters[b]	$\mu = N^{-1} \sum_{i=1}^{N} x_i$ $\sigma^2 = N^{-1} \sum_{i=1}^{N} (x_i - \mu)^2$	$\mu_y = N^{-1} \sum_{i=1}^{N} \ln(x_i - \tau)$, $\sigma_y^2 = N^{-1} \sum_{i=1}^{N} [\ln(x_i - \tau) - \mu_y]^2$
Mean	μ	$\mu = \exp(\mu_y + \sigma_y^2/2) + \tau$
Geometric mean (GM)	—	$\mu_g = \exp(\mu_y) + \tau$
Median	μ	$\mu_g = \exp(\mu_y) + \tau$
Mode	μ	$\exp(\mu_y - \sigma_y^2) + \tau$
Standard deviation	σ	$\sigma = \sqrt{\exp(2\mu_y + \sigma_y^2)[\exp(\sigma_y^2) - 1]}$
Geometric standard deviation	—	$\sigma_g = \exp(\sigma_y)$
Coefficient of variation	σ/μ	$\eta = \sqrt{\exp(\sigma_y^2) - 1}\left[1 + \frac{\tau}{\exp(\mu_y + \sigma_y^2/2)}\right]^{-1}$
Coefficient of skewness[c]	0	$\eta_1^3 + 3\eta_1$
Coefficient of kurtosis[c]	0	$\eta_1^8 + 6\eta_1^6 + 15\eta_1^4 + 16\eta_1^2$
Central 68% of the distribution	$\mu - \sigma$ to $\mu + \sigma$	μ_g/σ_g to $\mu_g \times \sigma_g$
Central 95% of the distribution	$\mu - 1.96\sigma$ to $\mu + 1.96\sigma$	$\mu_g/\sigma_g^{1.96}$ to $\mu_g \times \sigma_g^{1.96}$

Source: After Miesch, 1976, Table 1.

[a] N = number of population units in the target population.

x_i = datum for the ith population unit.

If $\tau = 0$, the three-parameter lognormal reduces to the two-parameter. Note that the expressions for μ, μ_g, η and the mode simplify when $\tau = 0$, that is, when the variable has the two-parameter lognormal distribution.

[b] Various notations have been used to denote the two parameters and the mean and variance of the lognormal distribution. Care must be taken to avoid confusion. For example, Aitchison and Brown (1969) use α and β^2 instead of our μ and σ^2 for the mean and variance of the untransformed (lognormal) variate. They use μ and σ^2 instead of our μ_y and σ_y^2 for the mean and variance of the log-transformed (normal) variate.

[c] $\eta_1 = \sqrt{\exp(\sigma_y^2) - 1}$

Table 12.2 Probability Density Functions Sometimes Used to Model Environmental Pollutant Concentrations

Distribution	*Probability Density Function (pdf)*
Two-parameter lognormal[a]	$\frac{1}{x\sigma_y\sqrt{2\pi}}\exp\left[-\frac{(\ln x-\mu_y)^2}{2\sigma_y^2}\right]$ $x>0,\quad -\infty<\mu_y<\infty,\quad \sigma_y>0$
Three-parameter lognormal[a]	$\frac{1}{(x-\tau)\,\sigma_y\sqrt{2\pi}}\exp\left\{-\frac{[\ln(x-\tau)-\mu_y]^2}{2\sigma_y^2}\right\}$ $x>\tau,\quad -\infty<\tau<\infty,\quad -\infty<\mu_y<\infty,\quad \sigma_y>0$
Three-parameter Weibull[b]	$\frac{\alpha}{\beta}\left(\frac{x-\gamma}{\beta}\right)^{\alpha-1}\exp\left[-\left(\frac{x-\gamma}{\beta}\right)^{\alpha}\right]$ $-\infty<\gamma<\infty,\ x>\gamma,\ \beta>0,\ \alpha>0$
Three-parameter gamma[b]	$\frac{1}{\beta\Gamma(\alpha)}\left(\frac{x-\gamma}{\beta}\right)^{\alpha-1}\exp\left[-\left(\frac{x-\gamma}{\beta}\right)\right]$ $-\infty<\gamma<\infty,\ x>\gamma,\ \alpha>0,\ \beta>0$
Four-parameter beta[c]	$\frac{\Gamma(\alpha+\beta)}{\Gamma(\alpha)\,\Gamma(\beta)}(\theta-\gamma)^{1-\alpha-\beta}(x-\gamma)^{\alpha-1}(\theta-x)^{\beta-1}$ $\gamma<x<\theta,\quad \alpha>0,\quad \beta>0$

[a] μ_y and σ_y^2 are defined in Table 12.1.
[b] Reduces to two-parameter distributions when $\gamma = 0$.
[c] Reduces to three-parameter beta distribution when $\gamma = 0$.

The parameters γ, α, and β of the *three-parameter gamma distribution* are also location, shape, and scale parameters. The density function (given in Table 12.2) contains the gamma function $\Gamma(\alpha)$, which is defined to be

$$\Gamma(\alpha) = \int_0^\infty x^{\alpha-1}\exp(-x)\,dx$$

where $\Gamma(\alpha) = (\alpha - 1)!$ when α is a positive integer. The maximum likelihood estimates of the three parameters may be obtained as described by Johnson and Kotz (1970*a*, p. 185). Preliminary and easily computed estimates of the parameters can be obtained by using Eqs. 39.1, 39.2, and 39.3 in Johnson and Kotz (1970*a*, p. 186). However, these estimators are not as accurate as the maximum likelihood estimators.

The density function of the *beta distribution* given in Table 12.2 has four parameters: α, β, θ, and γ. The variable X is bounded below by γ and above by θ, a useful feature because such bounds may occur for some types of environmental data. Methods for estimating the parameters are given by Johnson and Kotz (1970*b*, pp. 41–46).

12.3 GOODNESS-OF-FIT TESTS

The previous section presented several density functions that might be used to model environmental contaminant data. The data analyst is faced with deciding on the basis of data which of these probability distributions to use. Common ways to approach this problem are to construct a histogram, stem-and-leaf display, or normal and lognormal probability plots of the data. (See Section 13.1.3 for more on lognormal probability plots.) Histograms and stem-and-leaf displays (the latter are discussed and illustrated by Hoaglin, Mosteller, and Tukey, 1983) will give a visual impression of the shape of the data set, but they are not adequate tools for discrimination. If the normal probability plot is a straight line, it is evidence of an underlying normal distribution. A straight line on the lognormal probability plot suggests the lognormal distribution is a better model. Coefficients of skewness and kurtosis may also be computed from the n data values and used to test for normality (Bowman and Shenton, 1975). Most statistical packages of computer programs contain a code that will plot histograms and compute the coefficients of skewness or kurtosis, and some (e.g., Minitab; see Ryan, Joiner, and Ryan, 1982) will construct probability plots.

This section presents several statistical tests that test the null hypothesis that the distribution is in some specified form. We begin with the W test developed by Shapiro and Wilk (1965), one of the most powerful tests available for detecting departures from a hypothesized normal or lognormal density function. Shapiro and Wilk provided tables that allow the W test to be made if $n \leq 50$. This limitation on n was overcome somewhat by D'Agostino (1971), who developed a related test for when n is between 50 and 1000. D'Agostino's test and the W test are discussed and illustrated in this section. Royston (1982*a*) developed a computational procedure for the W test for n as large as 2000. His procedure is well suited for computation on a computer, and computer codes are available (Royston, 1982*a*, 1982*b*, 1982*c*, 1983; Königer, 1983).

Tests closely related to the W test with similar performance capabilities are those by Shapiro and Francia (1972) and Filliben (1975). Looney and Gulledge (1985) use the correlation coefficient applied to a probability plot to test for normality or lognormality. A table of critical values needed for the test is provided for n between 3 and 100. Their test is simple, and its performance is roughly the same as that of the W test.

The nonparametric Kolmogorov–Smirnov (KS) test and the related Lilliefors test may also be used to evaluate the fit of a hypothesized distribution. These tests, described by Conover (1980) are considered to be more powerful than the chi-square goodness-of-fit tests. The KS test is not valid if the parameters of the hypothesized distribution are estimated from the data set. The Lilliefors test (Lilliefors, 1967, 1969) was developed to surmount this problem when the hypothesized distribution is the normal or lognormal. Iman (1982) developed graphs that simplify the Lilliefors test procedure. Kurtz and Fields (1983*a*, 1983*b*) developed a computer code for computing the KS test. One can also use an IMSL (1982) subroutine as well as SAS (1982, 1985) software, but the Kurtz and Fields' code is valid for smaller n ($n > 3$).

12.3.1 The *W* Test

The W test developed by Shapiro and Wilk (1965) is an effective method for testing whether a data set has been drawn from an underlying normal distribution. Furthermore, by conducting the test on the logarithms of the data, it is an equally effective way of evaluating the hypothesis of a lognormal distribution.

We suppose that $n \leq 50$ data, $x_1, x_2, \ldots, x_n$, have been drawn at random from some population. The null hypothesis to be tested is

H_0: The population has a normal distribution

versus

H_A: The population does not have a normal distribution

If H_0 is rejected, then H_A is accepted. If H_0 is not rejected, the data set is consistent with the H_0 distribution, although a retest using additional data could result in rejecting H_0.

The W test of this H_0 is conducted as follows:

1. Compute the denominator d of the W test statistic, using the n data.

$$d = \sum_{i=1}^{n} (x_i - \bar{x})^2 = \sum_{i=1}^{n} x_i^2 - \frac{1}{n}\left(\sum_{i=1}^{n} x_i\right)^2 \qquad \textbf{12.3}$$

2. Order the n data from smallest to largest to obtain the sample order statistics $x_{[1]} \leq x_{[2]} \leq \cdots \leq x_{[n]}$.
3. Compute k, where

$$k = \frac{n}{2} \qquad \text{if } n \text{ is even}$$

$$= \frac{n-1}{2} \qquad \text{if } n \text{ is odd}$$

4. Turn to Table A6 and for the observed n find the coefficients $a_1, a_2, \ldots, a_k$.
5. Then compute

$$W = \frac{1}{d}\left[\sum_{i=1}^{k} a_i(x_{[n-i+1]} - x_{[i]})\right]^2 \qquad \textbf{12.4}$$

6. Reject H_0 at the α significance level if W is less than the quantile given in Table A7.

To test the null hypothesis

H_0: The population has a lognormal distribution

versus

H_A: The population does not have a lognormal distribution

the preceding procedure is used on the logarithms of the data. That is, we compute d (Eq. 12.3), using $y_1, y_2, \ldots, y_n$, where $y_i = \ln x_i$, and we use the sample order statistics of the logarithms $y_{[1]} \leq y_{[2]} \leq \cdots \leq y_{[n]}$ in place of the $x_{[i]}$ in Eq. 12.4.

EXAMPLE 12.1

Lee and Krutchkoff (1980) list mercury concentrations (ppm) in 115 samples of swordfish. We have selected 10 of these data at random

Table 12.3 Mercury Concentrations (ppm) in Ten Samples of Swordfish

x_i	0.13	0.45	0.60	0.76	1.05
$y_i = \ln x_i$	−2.0402	−0.7985	−0.5108	−0.2744	0.04879
x_i	1.12	1.20	1.37	1.69	2.06
$y_i = \ln x_i$	0.1133	0.1823	0.3148	0.5247	0.7227

Source: After Lee and Krutchkoff, 1980, Table 1.

to illustrate the W test. We test the null hypothesis

H_0: The distribution is lognormal

versus

H_A: The distribution is not lognormal

and we test at the $\alpha = 0.05$ level. The natural logarithms of the 10 randomly selected values are listed from smallest to largest in Table 12.3.

The denominator of the W statistic computed with the 10 y_i data is $d = 5.7865$ (by Eq. 12.3). Since $n = 10$, we have $k = 5$. Using the 5 coefficients $a_1, a_2, \ldots, a_5$ from Table A6 for $n = 10$, we use Eq. 12.4 to obtain

$$\begin{aligned} W = \frac{1}{5.7865} \{ & 0.5739\,[0.7227 - (-2.0402)] \\ & + 0.3291\,[0.5247 - (-0.7985)] \\ & + 0.2141\,[0.3148 - (-0.5108)] \\ & + 0.1224\,[0.1823 - (-0.2744)] \\ & + 0.0399\,[0.1133 - 0.04879]\}^2 \\ = 0.8798 & \end{aligned}$$

From Table A7 we find this calculated W is greater than the 0.05 quantile 0.842. Hence, we cannot reject H_0, and we conclude that, based on the $n = 10$ data, the lognormal distribution may be a reasonable approximation to the true unknown distribution. Of course, if n were much greater than 10, the W test might lead to the opposite conclusion, since the additional data would provide more information about the shape of the target population distribution.

An alternative method of using W to test H_0 is to convert W to a standard normal variable and to use Table A1 to decide whether to reject H_0. This approach is illustrated by Hahn and Shapiro (1967) and by Conover (1980). One attractive feature of this approach is that it can be used to combine several independent W tests into one overall test of normality (or lognormality). This testing procedure is illustrated by Conover (1980, p. 365).

12.3.2 D'Agostino's Test

D'Agostino (1971) developed the D statistic to test the null hypothesis of normality or lognormality when $n \geq 50$. He shows that his test compares favorably with other tests in its ability to reject H_0 when H_0 is actually false. This test complements the W test, since tables needed for the latter test are limited to $n \leq 50$.

Suppose we wish to test the null hypothesis that the underlying distribution is normal. Then the D test is conducted as follows:

1. Draw a random sample $x_1, x_2, \ldots, x_n$ of size $n \geq 50$ from the population of interest.
2. Order the n data from smallest to largest to obtain the sample order statistics $x_{[1]} \leq x_{[2]} \leq \cdots \leq x_{[n]}$.
3. Compute the statistic

$$D = \frac{\sum_{i=1}^{n} [i - \frac{1}{2}(n + 1)]x_{[i]}}{n^2 s}$$

where

$$s = \left[\frac{1}{n} \sum_{i=1}^{n} (x_i - \bar{x})^2\right]^{1/2}$$

4. Transform D to the statistic Y by computing

$$Y = \frac{D - 0.28209479}{0.02998598/\sqrt{n}}$$

[One should aim for five-place numerical accuracy in computing D (step 3), since the denominator of Y is so small.] If n is large and the data are drawn from a normal distribution, then the expected value of Y is zero. For nonnormal distributions Y will tend to be either less than or greater than zero, depending on the particular distribution. This fact necessitates a two-tailed test (step 5).
5. Reject at the α significance level the null hypothesis that the n data were drawn from a normal distribution if Y is less than the $\alpha/2$ quantile or greater than the $1 - \alpha/2$ quantile of the distribution of Y. These quantiles are given in Table A8 for selected values of n between 50 and 1000 (from D'Agostino, 1971).

The Y statistic can also be used to test the null hypothesis of a lognormal population by using $y_i = \ln x_i$ in place of x_i in the calculations.

EXAMPLE 12.2

We test at the $\alpha = 0.05$ significance level that the $n = 115$ mercury swordfish concentrations in Table 1 of Lee and Krutchkoff (1980) have been drawn from a normal distribution. That is, we test

H_0: The distribution is normal

versus

H_A: The distribution is not normal

and we assume that the data were drawn at random from the target population.

The value of s in the denominator of D is computed to be

$$s = \left[\frac{1}{115} \sum_{i=1}^{115} (x_i - \bar{x})^2\right]^{1/2} = 0.4978213$$

Hence the denominator is

$$(115)^2(0.4978213) = 6583.687$$

Since $(n + 1)/2 = 116/2 = 58$, the numerator of D is

$$(1 - 58)x_{[1]} + (2 - 58)x_{[2]} + \cdots + (114 - 58)x_{[114]} + (115 - 58)x_{[115]} = 1833.3$$

Therefore

$$D = \frac{1833.3}{6583.687} = 0.27846099$$

Hence

$$Y = \frac{0.27846099 - 0.28209479}{0.02998598/\sqrt{115}} = -1.30$$

Table A8 contains no quantiles of the Y statistic for $n = 115$. Hence, we must interpolate. If $n = 100$, the $\alpha/2 = 0.05/2 = 0.025$ quantile is -2.552, and the $1 - 0.025 = 0.975$ quantile is 1.303. If $n = 150$, Table A8 gives -2.452 and 1.423 for these quantiles. Linear interpolation between the 0.025 quantiles for $n = 100$ and 150 gives -2.522 as the approximate 0.025 quantile for $n = 115$. The 0.975 quantile when $n = 115$ is similarly approximated to be 1.339. Since $Y = -1.30$ is not less than -2.522 nor greater than 1.339, the null hypothesis of a normal distribution cannot be rejected. Hence, we tentatively accept the hypothesis that the population from which the data were obtained can be approximated by a normal distribution.

12.4 SUMMARY

This chapter introduced the most important frequency distributions used to model environmental data sets. The lognormal distribution is frequently used and will be discussed in more detail in Chapter 13.

Two statistical procedures for testing that a data set has been drawn (at random) from a hypothesized normal or lognormal distribution have also been described and illustrated. One of these, the W test, is recommended as a powerful general-purpose test for normality or lognormality when $n \leq 50$. The other test, by D'Agostino (1971), is appropriate for $n \geq 50$. Two easily used graphical tests are the Kolmogorov–Smirnov (KS) and Lilliefors tests discussed by Conover (1980). If the hypothesized distribution is normal or lognormal, the Lilliefors test is preferred to the KS test because the parameters of the distribution need not be known a priori. The simple correlation coefficient procedure discussed by Looney and Gulledge (1985) is recommended as a test for normal or lognormal distributions if $n \leq 100$.

EXERCISES

12.1 Use the data in Table 12.3 and the W test to test at the $\alpha = 0.05$ level the null hypothesis that mercury concentrations in swordfish are normally distributed. Compare your conclusion to that in Example 12.1, where we tested the null hypothesis that these data are from a lognormal distribution. Does the normal or lognormal seem to be the better choice?

12.2 The following mercury concentrations were drawn at random from the list of 115 values given by Lee and Krutchkoff (1980):

1.00	1.08	1.39	1.89	0.83
0.89	0.13	0.07	1.26	0.92

Combine these data with those in Table 12.3 and use the W test to test at the $\alpha = 0.05$ level that the sampled population is normal. Does the test result differ from that in Exercise 12.1?

ANSWERS

12.1 By Eq. 12.3, $d = 3.0520$. Using the 5 coefficients from Table A6 and Eq. 12.4, we obtain

$$W = \frac{1}{3.0520}[0.5739\,(2.06 - 0.13) + 0.3291\,(1.69 - 0.45)$$
$$+ 0.2141\,(1.37 - 0.60) + 0.1224\,(1.20 - 0.76)$$
$$+ 0.0399\,(1.12 - 1.05)]^2$$
$$= \frac{3.01792}{3.0520} = 0.989$$

The critical value from Table A7 is 0.842. Since $W > 0.842$, we cannot reject the null hypothesis of a normal distribution. In Example 12.1 we could not reject the null hypothesis that the population is lognormal. Hence, the data are not sufficient to distinguish between normality and lognormality.

12.2 $n = 20$, $d = 5.757295$.

$$W = \frac{1}{5.757295}[0.4734\,(2.06 - 0.07) + 0.3211\,(1.89 - 0.13)$$
$$+ 0.2565\,(1.69 - 0.13) + 0.2085\,(1.39 - 0.45)$$
$$+ 0.1686\,(1.37 - 0.60) + 0.1334\,(1.26 - 0.76)$$
$$+ 0.1013\,(1.20 - 0.83) + 0.0711\,(1.12 - 0.89)$$
$$+ 0.0422\,(1.08 - 0.92) + 0.0140\,(1.05 - 1.00)]^2$$
$$= 0.968$$

The critical value from Table A7 is 0.905. Since $W > 0.905$, we cannot reject the null hypothesis of normality, the same test result as in Exercise 12.1.

13 Characterizing Lognormal Populations

The lognormal distribution is the most commonly used probability density model for environmental contaminant data. Therefore this chapter considers several estimation procedures for this distribution. More specifically, the chapter

- Gives optimal methods for estimating the mean and median
- Shows how to compute confidence limits about the mean and median
- Shows how to determine the number n of data needed to estimate the median
- Shows how to estimate quantiles
- Discusses the geometric mean and some problems with its use in evaluating compliance with environmental pollution limits.

13.1 ESTIMATING THE MEAN AND VARIANCE

We begin by giving four methods that can be used to estimate the mean μ and variance σ^2 of a lognormal distribution: (1) the sample mean $\bar{x}$, (2) the minimum variance unbiased (MVU) estimator $\hat{\mu}_1$, (3) an easily computed estimator $\hat{\mu}$, and (4) the probability-plotting estimator. Which of these is used in practice depends on circumstances, as discussed in what follows.

The arithmetic mean $\bar{x}$ is easy to compute. Furthermore, it is a statistically unbiased estimator of μ no matter what the underlying distribution may be (lognormal, normal, Weibull, etc.). If the underlying distribution is normal, it is also the MVU estimator of μ. Unfortunately, $\bar{x}$ does not have this MVU property when the underlying distribution is lognormal. Also, $\bar{x}$ is highly sensitive to the presence of one or more large data values. Nevertheless, even when the underlying distribution is lognormal, $\bar{x}$ is probably the preferred estimator if the coefficient of variation η is believed to be less than 1.2 (a rule suggested by Koch and Link, 1980).

If statistical tests support the hypothesis of a lognormal distribution, the MVU estimator $\hat{\mu}_1$ described in Section 13.1.1 may be used. As a general rule, $\hat{\mu}_1$ is preferred to $\bar{x}$ if $\eta > 1.2$, that is, if the lognormal distribution is highly skewed, assuming that one has a good estimate of σ_y^2, the variance of the transformed variable $Y = \ln X$. Finally, the easily computed estimator $\hat{\mu}$, described in Section 13.1.2, may be used to estimate μ if n is reasonably large and the distribution is lognormal.

13.1.1 Minimum Variance Unbiased Estimators

This section shows how to estimate μ and σ^2 from the MVU estimators $\hat{\mu}_1$ and $\hat{\sigma}_1^2$, developed independently by Finney (1941) and Sichel (1952, 1966). An MVU estimator of a parameter is one that is statistically unbiased and has the smallest sampling error variance of all unbiased estimators of the parameter. Hence, since $\hat{\mu}_1$ is an MVU estimator, it has a smaller variance than $\bar{x}$ or the alternative estimators of μ given in Sections 13.1.2 and 13.1.3. However, $\hat{\mu}_1$ is a biased estimator of μ if the distribution is not lognormal.

To obtain $\hat{\mu}_1$, we first estimate the parameters μ_y and σ_y^2 of the lognormal distribution by computing

$$\bar{y} = \frac{1}{n} \sum_{i=1}^{n} y_i \tag{13.1}$$

$$s_y^2 = \frac{1}{n-1} \sum_{i=1}^{n} (y_i - \bar{y})^2 \tag{13.2}$$

where $\bar{y}$ and s_y^2 are the arithmetic mean and variance of the n transformed values $y_i = \ln x_i$. Then compute

$$\hat{\mu}_1 = [\exp(\bar{y})]\Psi_n\left(\frac{s_y^2}{2}\right) \tag{13.3}$$

where $\exp(\bar{y})$ is the sample geometric mean, and $\Psi_n(t)$ (with $t = s_y^2/2$) is the infinite series

$$\Psi_n(t) = 1 + \frac{(n-1)t}{n} + \frac{(n-1)^3t^2}{2!n^2(n+1)} + \frac{(n-1)^5t^3}{3!n^3(n+1)(n+3)} + \frac{(n-1)^7t^4}{4!n^4(n+1)(n+3)(n+5)} + \cdots \tag{13.4}$$

This series can be programmed on a computer, or one may use tables of $\Psi_n(t)$ given by Aitchison and Brown (1969, Table A2), Koch and Link (1980, Table A7), or Sichel (1966). (Sichel's table is entered with $t = (n-1)s_y^2/n$ rather than $t = s_y^2/2$.) Portions of these tables are given here as Table A9. Also, Agterberg (1974, p. 235) gives a table from which $\Psi_n(t)$ can be obtained for t up to 20, and Thoni (1969) published tables for use when logarithms to base 10 are used.

An unbiased estimator of the variance of $\hat{\mu}_1$ is (from Bradu and Mundlak, 1970, Eq. 4.3)

$$s^2(\hat{\mu}_1) = \exp(2\bar{y})\left\{\left[\Psi_n\left(\frac{s_y^2}{2}\right)\right]^2 - \Psi_n\left[\frac{s_y^2(n-2)}{n-1}\right]\right\} \tag{13.5}$$

where $\bar{y}$ and s_y^2 are computed from Eqs. 13.1 and 13.2. To obtain the first and second $\Psi_n(t)$ terms in Eq. 13.5, enter Table A9 with $t = s_y^2/2$ and $t = s_y^2(n-2)/(n-1)$, respectively.

The MVU estimator of the variance σ^2 of a two-parameter lognormal distribution was found by Finney (1941) to be

$$\hat{\sigma}_1^2 = \exp(2\bar{y})\left\{\Psi_n(2s_y^2) - \Psi_n\left[\frac{s_y^2(n-2)}{n-1}\right]\right\} \quad \textbf{13.6}$$

EXAMPLE 13.1

Table 13.1 gives 10 data that were drawn at random (using a computer) from a 2-parameter lognormal distribution with parameters $\mu_y = 1.263$ and $\sigma_y^2 = 1.099$. We use $\hat{\mu}_1$ to estimate the mean $\mu = \exp(\mu_y + \sigma_y^2/2) = 6.126$ of this distribution, using the 10 data. We also estimate $\text{Var}(\hat{\mu}_1)$ using Eq. 13.5. Equations 13.1 and 13.2 give $\bar{y} = 1.48235$ and $s_y^2 = 0.56829$ (see Table 13.1). Using linear interpolation in Table A9, we find $\Psi_{10}(0.56829/2) = 1.2846$. Therefore, Eq. 13.3 gives

$$\hat{\mu}_1 = 4.403(1.2846) = 5.66$$

which is smaller than the true mean $\mu = 6.126$ of the distribution. Equation 13.5 gives

$$s^2(\hat{\mu}_1) = \exp(2.964)\,\{[\Psi_{10}(0.28414)]^2 - \Psi_{10}(0.5051)\}$$

$$= 1.97$$

or the standard error $s(\hat{\mu}_1) = 1.40$. Hence, our estimate of μ is $\hat{\mu}_1 = 5.66$, and its standard error is 1.40. These estimates may be compared with $\bar{x} = 5.89$ and $s(\bar{x}) = 1.80$. For this data set $\bar{x}$ is closer than $\hat{\mu}_1$ to $\mu = 6.126$.

Table 13.1 Ten Data Drawn at Random from a Two-Parameter Lognormal Distribution with Parameters $\mu_y = 1.263$ and $\sigma_y^2 = 1.099$

x_i	$y_i = \ln x_i$
3.161	1.1509
4.151	1.4233
3.756	1.3234
2.202	0.7894
1.535	0.4285
20.76	3.0330
8.42	2.1306
7.81	2.0554
2.72	1.0006
4.43	1.4884
$\bar{x} = 5.89$	$\bar{y} = 1.48235$
$s_x^2 = 32.331$	$s_y^2 = 0.56829$
$s_x = 5.69$	$s_y = 0.75385$
$s(\bar{x}) = 1.80$	$\exp(\bar{y}) = 4.40$ = sample geometric mean

As estimate of σ^2 is obtained from Eq. 13.6:

$$\hat{\sigma}_1^2 = 19.389[\Psi_{10}(1.1366) - \Psi_{10}(0.50515)]$$
$$= 19.8$$

which is considerably smaller than the true variance $\sigma^2 = (6.126)^2 \cdot [\exp(1.099) - 1] = 75.1$. This discrepancy shows the importance of obtaining precise estimates of μ_y and σ_y^2 when using $\hat{\mu}_1$ and $\hat{\sigma}_1^2$ to estimate μ and σ^2. Using more than $n = 10$ data is clearly desirable.

13.1.2 Less Efficient But Simpler Estimators

A simple method of estimating the mean μ and variance σ^2 of the two-parameter lognormal distribution is to replace μ_y and σ_y^2 by $\bar{y}$ and s_y^2 in the formulas for μ and σ^2. We get

$$\hat{\mu} = \exp\left(\bar{y} + \frac{s_y^2}{2}\right) \qquad \textbf{13.7}$$

and

$$\hat{\sigma}^2 = \hat{\mu}^2 [\exp(s_y^2) - 1] \qquad \textbf{13.8}$$

For example, using the data in Table 13.1, we obtain

$$\hat{\mu} = \exp\left(1.48235 + \frac{0.56829}{2}\right) = 5.85$$

and

$$\hat{\sigma}^2 = (5.85)^2[\exp(0.56829) - 1] = 26.2$$

The variance of $\hat{\mu}$ may be approximated as follows by using a result in Kendall and Stuart (1961, p. 69):

$$s^2(\hat{\mu}) \cong \exp\left(2\bar{y} + \frac{s_y^2}{n}\right)\left[\left(1 - \frac{2s_y^2}{n}\right)^{-(n-1)/2} \cdot \exp\left(\frac{s_y^2}{n}\right) - \left(1 - \frac{s_y^2}{n}\right)^{-(n-1)}\right] \qquad \textbf{13.8a}$$

Using the data in Table 13.1, we find $s_y^2/n = 0.056829$, so Eq. 13.8a gives $s^2(\hat{\mu}) = 2.6387$, or $s(\hat{\mu}) = 1.6$.

The mathematical expected value (over many repetitions of the experiment) of $\hat{\mu}$ is (from Kendall and Stuart, 1961, p. 68):

$$E\left[\exp\left(\bar{y} + \frac{s_y^2}{2}\right)\right] = \mu\left(1 - \frac{\sigma_y^2}{n}\right)^{-(n-1)/2} \exp\left(-\frac{n-1}{2n}\sigma_y^2\right)$$
$$= \text{(true mean) (bias factor)}$$

Hence, $\hat{\mu}$ is biased upward for μ, but the bias factor approaches zero as n becomes large. For example, if $n = 20$ and $\sigma_y^2 = 2$, then the bias factor = 1.0522, indicating a 5.22% positive bias on the average. If $n = 100$, the bias factor is only 1.010, a 1% bias.

Note that $\hat{\mu}$ will tend to decrease as n increases because the bias goes to zero for large n. This effect should be kept in mind if $\hat{\mu}$ is used to evaluate compliance with environmental pollution guidelines. For example, one facility emitting pollutants might be declared in compliance, whereas another is not, solely because the first took more samples, not because it was emitting lower levels of pollution. This problem does not occur if $\bar{x}$ or $\hat{\mu}_1$ are used to estimate μ. The same problem occurs if the geometric mean is used to estimate the true median of a lognormal distribution, as discussed in Section 13.3.3. Also see Landwehr (1978).

13.1.3 Probability Plotting

In Section 11.2 we used probability plotting to estimate the mean and variance of a normal distribution. A similar procedure may be used to estimate the parameters μ_y and σ_y^2 of the lognormal distribution, which can in turn be used to estimate the mean, μ, and variance, σ^2, of the distribution.

First, order the n untransformed data from smallest to largest to obtain the order statistics $x_{[1]} \leq x_{[2]} \leq \cdots \leq x_{[n]}$. Then plot $x_{[i]}$ versus $(i - 0.5)100/n$ on log-probability paper and fit a straight line by eye (or use the objective method of Mage, 1982) if the plotted points fall approximately on a straight line. Then the 0.16, 0.50, and 0.84 quantiles ($x_{0.16}$, $x_{0.50}$, and $x_{0.84}$, respectively) are read from the plot and are used as follows to estimate μ_y and σ_y^2 (from Aitchison and Brown, 1969, p. 32):

$$\hat{\mu}_y = \ln x_{0.50} \qquad \textbf{13.9}$$

$$\hat{\sigma}_y^2 = \left\{ \ln \left[\frac{1}{2} \left(\frac{x_{0.50}}{x_{0.16}} + \frac{x_{0.84}}{x_{0.50}} \right) \right] \right\}^2 \qquad \textbf{13.10}$$

The mean and standard deviation of the distribution are then estimated by computing

$$\hat{\mu} = \exp \left(\hat{\mu}_y + \frac{\hat{\sigma}_y^2}{2} \right) \qquad \textbf{13.11}$$

$$\hat{\sigma} = \hat{\mu}[\exp(\hat{\sigma}_y^2) - 1]^{1/2} \qquad \textbf{13.12}$$

Estimates of the geometric mean, $\exp(\mu_y)$, and the geometric standard deviation, $\exp(\sigma_y)$, are given by $x_{0.50}$ and $\frac{1}{2}(x_{0.50}/x_{0.16} + x_{0.84}/x_{0.50})$, respectively.

Probability plotting is a quick way to evaluate whether the data are likely to have come from a two-parameter lognormal distribution—that is, by checking whether a straight line fits the plotted points. If so, the foregoing procedure is used to estimate μ. If not, probability plotting techniques for other hypothesized distributions, such as the normal, Weibull, gamma, and exponential, can be tried by using methods given by, for example, Hahn and Shapiro (1967) and King (1971).

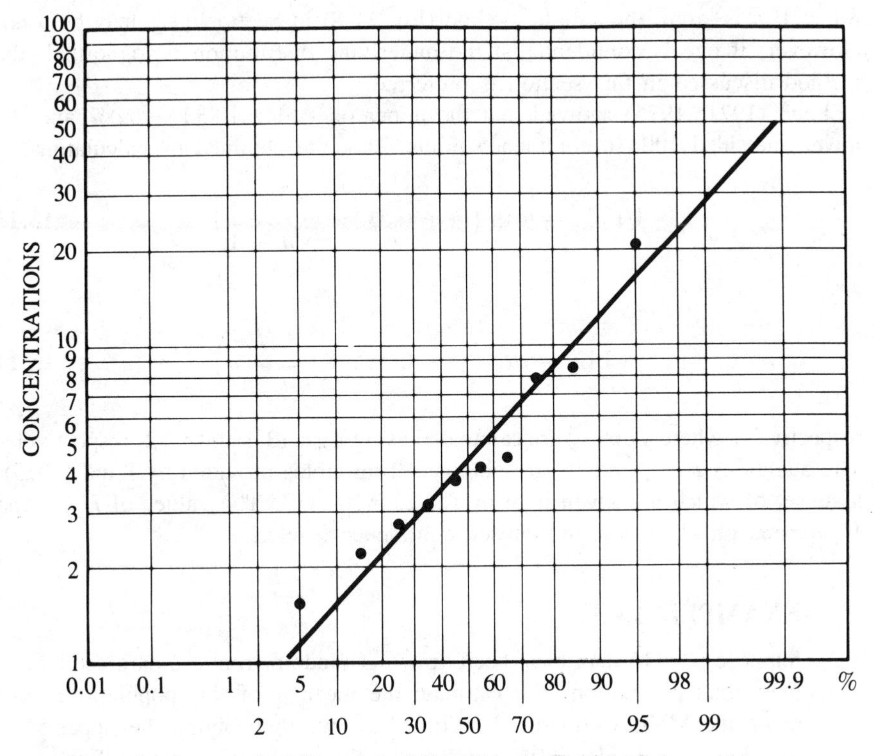

Figure 13.1 Log-probability plot of the data in Table 13.1.

EXAMPLE 13.2

Return to the $n = 10$ data in Table 13.1. Since these data are drawn from a 2-parameter lognormal distribution, the ordered data should plot as a straight line on log-probability paper. This plot is given in Figure 13.1. The eyeball fit straight line is used to obtain the percentiles $\hat{x}_{0.16} = 1.95$, $\hat{x}_{0.50} = 4.3$, and $\hat{x}_{0.84} = 9.5$. Using these in Eqs. 13.9 and 13.10 gives $\hat{\mu}_y = 1.459$ and $\hat{\sigma}_y^2 = 0.6268$. These values deviate from the true values $\mu_y = 1.263$ and $\sigma_y^2 = 1.099$ because of random sampling error and the subjective (eyeball) fit of the line to the points. Using $\hat{\mu}_y$ and $\hat{\sigma}_y^2$ in Eq. 13.11 estimates the mean to be 5.88 as compared to $\bar{x} = 5.89$, $\hat{\mu} = 5.85$, and $\hat{\mu}_1 = 5.66$.

13.2 CONFIDENCE LIMITS FOR THE MEAN

Thus far we have discussed methods for estimating the mean μ of a two-parameter lognormal distribution. We now see how to obtain confidence limits for μ. If n is large, the simple method (Eq. 11.8) in Section 11.7 may be used. However, if one is confident that the underlying distribution is lognormal, the method discussed in this section is preferred.

Land (1971, 1975) showed that the upper one-sided $100(1 - \alpha)\%$ and the lower one-sided $100\alpha\%$ confidence limits for μ are obtained by calculating

$$\mathrm{UL}_{1-\alpha} = \exp\left(\bar{y} + 0.5s_y^2 + \frac{s_y H_{1-\alpha}}{\sqrt{n-1}}\right) \tag{13.13}$$

and

$$\mathrm{LL}_{\alpha} = \exp\left(\bar{y} + 0.5s_y^2 + \frac{s_y H_{\alpha}}{\sqrt{n-1}}\right) \tag{13.14}$$

respectively, where $\bar{y}$ and s_y^2 are calculated using Eqs. 13.1 and 13.2, respectively. The quantities $H_{1-\alpha}$ and H_α are obtained from tables provided by Land (1975), a subset of which are given here in Tables A10–A13. The values of $H_{1-\alpha}$ and H_α depend on s_y, n, and the chosen confidence level α.

EXAMPLE 13.3

Suppose $n = 15$ data have been drawn at random from a 2-parameter lognormal population. We estimate the mean μ of this population, using the MVU estimator $\hat{\mu}_1$ (Eq. 13.3); we then obtain the upper and lower one-sided 90% confidence limits about μ, using Eqs. 13.13 and 13.14.

Suppose the $n = 15$ data give $\bar{y} = 1.8$ and $s_y^2 = 4.0$. Then Eq. 13.3 gives

$$\hat{\mu}_1 = \exp(1.8)\,\Psi_{15}(2) = 6.0496(5.439) = 33$$

where $\Psi_{15}(2)$ is obtained from Table A9. Entering Table A10 with $n = 15$ and $s_y = 2.0$, we obtain $H_{0.90} = 3.244$. Hence,

$$\mathrm{UL}_{0.90} = \exp\left(1.8 + 0.5(4) + \frac{2(3.244)}{\sqrt{14}}\right)$$

$$= 253$$

Entering Table A11 with $n = 15$ and $s_y = 2.0$, we find $H_{0.10} = -1.733$. Hence,

$$\mathrm{LL}_{0.10} = \exp\left(1.8 + 0.5(4) + \frac{2(-1.733)}{\sqrt{14}}\right)$$

$$= 17.7$$

In summary, μ is estimated to be 33 with lower and upper one-sided 90% limits of 18 and 250. The interval 18 to 250 is the two-sided 80% confidence interval about μ. To obtain one-sided upper and lower 95% limits (equivalent to a two-sided 90% confidence interval about μ) use Tables A12 and A13.

In practice, H may be required for values of s_y and n not given in Tables A10–A13. Land (1975) indicates that cubic interpolation (four-point Lagrangian interpolation; Abramowitz and Stegun, 1964, p. 879) appears to be adequate with these tables.

13.3 ESTIMATING THE MEDIAN

The true median of an underlying distribution is that value above which and below which half the distribution lies. If the distribution is symmetrical, then the median equals the mean μ. But for the two-parameter lognormal or other right-skewed distributions, the true median is less than μ. For left-skewed distributions the median exceeds μ.

13.3.1 Sample Median

The median of any distribution, no matter what its shape, can be estimated by the sample median. First, the data are ranked from smallest to largest. Then the sample median (median of the n data) is computed from the sample order statistics $x_{[1]} \leq x_{[2]} \cdots \leq x_{[n]}$ as follows:

$$\text{sample median} = x_{[(n+1)/2]} \qquad \text{if } n \text{ is odd} \qquad \textbf{13.15}$$

$$= \tfrac{1}{2}(x_{[n/2]} + x_{[(n+2)/2]}) \qquad \text{if } n \text{ is even} \qquad \textbf{13.16}$$

Uses of the sample median are discussed in Section 14.2.2.

EXAMPLE 13.4

We estimate the median of the population from which the $n = 10$ data in Table 13.1 were drawn. Since $n = 10$ is even, the sample median is

$$\tfrac{1}{2}(x_{[10/2]} + x_{[12/2]}) = \tfrac{1}{2}(x_{[5]} + x_{[6]}) = \tfrac{1}{2}(3.756 + 4.151) = 3.95$$

Hence, for this particular set of data, the sample median is larger than $\exp(\mu_y) = \exp(1.263) = 3.536$, the true median of the population.

13.3.2 Minimum Variance Unbiased Estimator

As we noted earlier, using the sample median to estimate the true median is appropriate no matter what the underlying distribution may be. However, if the distribution is known to be lognormal, other methods can also be used. One approach is to prepare a log-probability plot of the data, as discussed in Section 13.1.3. Then the true media is estimated by the 50th percentile, $x_{0.50}$, obtained from the straight line.

If the distribution is truly lognormal, and if the lognormal distribution parameter σ_y^2 (defined in Table 12.1) is known a priori, an unbiased estimator of the true median is

$$\hat{M}_1 = \exp(\bar{y}) \exp\left(-\frac{\sigma_y^2}{2}\right) \qquad \textbf{13.17}$$

If an estimate s_y^2 of σ_y^2 is available, s_y^2 could be used in place of σ_y^2 in Eq. 13.17. However, the preferred approach is to use the MVU estimator given by (from Bradu and Mundlak, 1970)

$$\hat{M}_2 = \exp(\bar{y})\, \Psi_n(t)$$

where $t = -s_y^2/[2(n-1)]$ and $\Psi_n(t)$ is the infinite series defined by Eq. 13.4. Using this value for t and the first five terms of Eq. 13.4, we obtain the approximate expression

$$\hat{M}_2 \cong \exp(\bar{y})\left[1 - \frac{s_y^2}{2n} + \frac{(n-1)(s_y^2)^2}{2!(2n)^2(n+1)} - \frac{(n-1)^2(s_y^2)^3}{3!(2n)^3(n+1)(n+3)} + \frac{(n-1)^3(s_y^2)^4}{4!(2n)^4(n+1)(n+3)(n+5)}\right] \qquad \textbf{13.18}$$

Additional terms in the $\Psi_n(t)$ series may be required before $\Psi_n(t)$ stabilizes if s_y^2 is very large and n is small. An unbiased estimator of $\text{Var}(\hat{M}_2)$ is (Bradu and Mundlak, 1970, Eq. 4.3)

$$s^2(\hat{M}_2) = \exp(2\bar{y})\left\{\left[\Psi_n\left(-\frac{s_y^2}{2(n-1)}\right)\right]^2 - \Psi_n\left(-\frac{2s_y^2}{n-1}\right)\right\}$$

13.3.3 Sample Geometric Mean

The sample geometric mean (GM) is computed as

$$\text{GM} = \prod_{i=1}^{n} x_i^{1/n} = \exp(\bar{y})$$

where

$$\bar{y} = \frac{1}{n}\sum_{i=1}^{n} y_i = \frac{1}{n}\sum_{i=1}^{n} \ln x_i$$

It is tempting to estimate the true median $\exp(\mu_y)$ of a lognormal distribution by computing the GM—that is, by simply replacing μ_y in this expression by the estimate $\bar{y}$. However, in the previous section we learned that the GM is a biased estimator of $\exp(\mu_y)$. The bias factor is $\exp(\sigma_y^2/2n)$, since (from Aitchison and Brown, 1969, p. 45)

$$E(\text{GM}) = E[\exp(\bar{y})] = \exp(\mu_y)\exp\left(\frac{\sigma_y^2}{2n}\right)$$

$$= \text{(true median) (bias factor)} \qquad \textbf{13.19}$$

The bias factor is positive and decreases as n increases and/or as σ_y^2 (or equivalently the skewness of the lognormal distribution) becomes small.

The GM is also a biased estimator of the *mean* of the two-parameter lognormal distribution. This fact can be seen by writing Eq. 13.19 in a different way (from Aitchison and Brown, 1969, p. 45):

$$E(\text{GM}) = \mu \exp\left(-\frac{n-1}{2n}\sigma_y^2\right)$$

$$= \text{(true mean) (bias factor)}$$

In this case the bias factor is less than 1, so the GM tends to underestimate the true mean. Furthermore, this bias does not go to zero as n increases.

In summary, if the underlying distribution is known to be lognormal, then the sample GM estimates the true GM if n is reasonably large, but the sample GM should not be used to estimate the mean μ unless σ_y^2 is very near zero.

Landwehr (1978) discusses some properties of the geometric mean and some problems inherent in its use for evaluating compliance with water quality standards for microbiological counts. The basic problem, which is generic to all compliance situations, is that the sample GM will tend, on the average, to be smaller if a larger number of samples is collected. Hence, one facility emitting pollutants might be declared in compliance, whereas another is not, solely because the first took more samples. We discussed in the last paragraph of Section 13.1.2 that this same problem occurs if $\hat{\mu}$ (Eq. 13.7) is used to estimate the lognormal mean. Landwehr (1978) shows that this problem also occurs for other distributions, such as the Weibull and gamma.

13.4 CONFIDENCE LIMITS FOR THE MEDIAN

An approximate two-sided $100(1 - \alpha)\%$ confidence interval for the true median of a lognormal distribution is obtained by computing

$$\exp(\bar{y}) \exp(-t_{1-\alpha/2,n-1} s_{\bar{y}}) \leq \exp(\mu_y) \leq \exp(\bar{y}) \exp(t_{1-\alpha/2,n-1} s_{\bar{y}}) \qquad \textbf{13.20}$$

or equivalently,

$$\frac{\exp(\bar{y})}{[\exp(s_{\bar{y}})]^{t_{1-\alpha/2,n-1}}} \leq \exp(\mu_y) \leq \exp(\bar{y}) [\exp(s_{\bar{y}})]^{t_{1-\alpha/2,n-1}}$$

where $\exp(s_{\bar{y}})$ is the sample geometric standard error, and $t_{1-\alpha/2,n-1}$ is obtained from Table A2. These lower and upper limits, given by the left and right sides of Eq. 13.20 are approximate when n is small and σ_y^2 is large since in this situation $\exp(\bar{y})$ is a biased estimator of $\exp(\mu_y)$. However, the bias factor $(\sigma_y^2/2n)$ (from Eq. 13.19) will be near zero even for rather small n unless σ_y^2 is very large.

The use of Eq. 13.20 requires the underlying distribution to be lognormal. However, a two-sided $100(1 - \alpha)\%$ confidence interval for the true median from *any* continuous underlying distribution can be easily obtained from Table A14 if the data are not correlated. This table gives values for integers l and u such that the order statistics $x_{[l]}$ and $x_{[u]}$ are the estimated lower and upper limits.

For example, suppose two-sided 95% confidence limits are desired, and a random sample of $n = 10$ independent measurements from some unknown underlying distribution has been obtained. From Table A14 we find for $\alpha = 0.05$ and $n = 10$ that $l = 2$ and $u = 9$. Hence, the lower and upper 95% confidence limits are given by the second and ninth largest data values, respectively, in the random sample of ten measurements. For the $n = 10$ data in Table 13.1, we find $x_{[2]} = 2.202$ and $x_{[9]} = 8.42$, which are the lower and upper 95% limits about the true median. Values of l and u for n as large as 499 and for $\alpha = 0.10, 0.05, 0.025, 0.01, 0.005$, and 0.001 are given by Geigy (1982, pp. 103–107).

If n is fairly large, say $n > 20$, approximate values for l and u may be obtained as follows:

$$l = \frac{n+1}{2} - \frac{Z_{1-\alpha/2}\sqrt{n}}{2} \qquad \textbf{13.21}$$

$$u = \frac{n+1}{2} + \frac{Z_{1-\alpha/2}\sqrt{n}}{2} \qquad \textbf{13.22}$$

where $Z_{1-\alpha/2}$ is obtained from Table A1. For example, if $n = 100$ independent measurements are drawn at random from a population and a 95% confidence interval about the true median is desired, then $Z_{0.975} = 1.96$, and Eqs. 13.21 and 13.22 give $l = 40.7$ and $u = 60.3$. The value of the lower confidence limit, $x_{[40.7]}$, is obtained by linear interpolation between the 40th and 41st sample order statistics. Similarly, the upper limit, $x_{[60.3]}$, is obtained by linear interpolation between the 60th and 61st sample order statistics.

13.5 CHOOSING *n* FOR ESTIMATING THE MEDIAN

Hale (1972) derived the following expression for approximating the number of independent observations, n, required for estimating the true median of a lognormal distribution:

$$n = \frac{Z^2_{1-\alpha/2} s^2_y}{[\ln (d + 1)]^2 + Z^2_{1-\alpha/2} s^2_y / N} \qquad \textbf{13.23}$$

where d is the prespecified relative error in the estimated median that can be tolerated, $100(1 - \alpha)\%$ is the percent confidence required that this error is not exceeded, and s^2_y is given by Eq. 13.2.

For example, suppose we choose $d = 0.10$ (10% relative error) and $\alpha = 0.05$, and that prior studies give $s^2_y = 2.0$. Also, suppose the size, N, of the population is very large. Then Eq. 13.23 gives

$$n = \frac{(1.96)^2(2)}{(\ln 1.1)^2} = 845.8 \cong 846$$

If the budget will not allow collecting this much data, we must either accept a larger percent error or smaller confidence (larger α). For example, if d is set at 0.50 (50% relative error) and $\alpha = 0.05$, we obtain $n = 47$.

13.6 ESTIMATING QUANTILES

Estimates of quantiles other than the median are frequently needed to evaluate compliance with standards or guidelines. For example, as discussed by Crager (1982), one way to evaluate whether air quality standards for ozone have been violated is to estimate the $1 - 1/365 = 0.99725$th quantile of the population of daily maximum ozone readings. By definition, the 0.99725th quantile is that daily maximum reading that exceeds $100(0.99726) = 99.726\%$ of the population and is exceeded by $100(0.00274) = 0.274\%$ of the population.

One method of approximating quantiles of a lognormal distribution is to read the quantile directly off a log-probability plot. For example, using the log-probability plot in Figure 13.1, the 0.75 and 0.99 quantiles are estimated to be 7.4 and 28, respectively. This method is subjective, since the line is drawn by eye and it is a matter of judgment whether the plotted data are adequately fit by a straight line. These problems can be overcome by using the objective fitting and testing procedures of Mage (1982). However, estimating extreme quantiles from a region of the line that extends beyond the range of the data is not recommended.

Since the data in Table 13.1 were drawn from a two-parameter lognormal distribution, the true pth quantile is $\exp(\mu_y + Z_p \sigma_y)$, where Z_p cuts off $100(1 - p)\%$ of the upper tail of the standard normal distribution. Since $\mu_y = 1.263$, $\sigma_y^2 = 1.099$, and $Z_{0.99} = 2.3263$ (from Table A1), the true 0.99 quantile is $\exp[1.263 + 2.3263(1.04833)] = 40.5$, as compared to the estimate of 28 obtained from the probability plot. The difference between 40.5 and 28 illustrates the difficulty of estimating extreme quantiles by probability plotting when only ten data from the population are available.

Rather than use probability plotting to estimate a quantile x_p of a two-parameter lognormal distribution, we may compute

$$\hat{x}_p = \exp(\bar{y} + Z_p s_y) = \exp(\bar{y}) \exp(Z_p s_y) \qquad \mathbf{13.24}$$

Using the data in Table 13.1, we obtain, using Eq. 13.24,

$$\hat{x}_{0.99} = \exp[1.48235 + 2.3263(0.75385)] = 25.4$$

which is similar to the estimated 0.99 quantile obtained earlier by probability plotting.

We note that Saltzman (1972) gives a nomograph for estimating quantiles, using Eq. 13.24, that eliminates the need for looking up Z_p in Table A1. The nomograph gives the value of $\exp(Z_p s_y)$ for given values of p and s_y. Then $\exp(Z_p s_y)$ is multiplied by the sample geometric mean $\exp(\bar{y})$ to obtain x_p.

13.7 SUMMARY

This chapter focused on the lognormal distribution, since it is so frequently used. Methods for estimating the mean and median of that distribution have been given, with the warning that they will result in biased results if the underlying distribution is not lognormal. Alternative methods are also given since they perform well under certain conditions and are easier to compute. We have also provided methods for computing confidence limits for the mean and median, for estimating quantiles other than the median, and for deciding how many samples, n, are required for estimating the median.

EXERCISES

13.1 In Example 13.3 find the two-tailed 90% confidence interval about μ, using $\bar{y} = 1.8$, $s_y^2 = 4.0$, and Land's method.

13.2 Suppose $n = 30$ observations from a lognormal distribution have been obtained and that Eqs. 13.1 and 13.2 yield $\bar{y} = 1.2$ and $s_y^2 = 4.0$, using those data. Estimate the median of the lognormal population using Eq. 13.18. Estimate the standard error of this estimate.

13.3 Suppose 30 observations are drawn at random from a lognormal distribution and the sample mean and variance of the logarithms of these data are $\bar{y} = 1.2$ and $s_y^2 = 4.0$, as in Exercise 13.2. Compute a 95% confidence interval for the true median of the underlying distribution.

13.4 In Exercise 13.3 determine the order statistics of the sample of size 30 that are the 95% limits on the true median (use Eqs. 13.21 and 13.22).

13.5 How many randomly drawn observations from a lognormal distribution should be taken to estimate the median with a relative error no larger than 20% with 90% confidence if preliminary data give $s_y^2 = 1.5$. Assume the size of the target population is very large. Repeat the calculations, using a relative error of 50%.

ANSWERS

13.1

$$\text{UL}_{0.95} = \exp\left(1.8 + 0.5(4) + \frac{2(4.564)}{\sqrt{14}}\right) = 513$$

$$\text{LL}_{0.05} = \exp\left(1.8 + 0.5(4) + \frac{2(-2.144)}{\sqrt{14}}\right) = 14$$

13.2

$$\begin{aligned}\hat{M}_2 &= 3.320[1 - 0.066667 + 0.002078853 \\ &\quad - 0.000040597 + 0.000000561] \\ &= 3.1 \\ s^2(\hat{M}_2) &= 11.02318\{[\Psi_{30}(-0.06897)]^2 \\ &\quad - \Psi_{30}(-0.27586)\} \\ \Psi_{30}(-0.06897) &= 0.935368 \\ \Psi_{30}(-0.27586) &= 0.76414\end{aligned}$$

Therefore, $s(\hat{M}_2) = 1.1$ = standard error of $\hat{M}_2$.

13.3 When $n = 30$, $t_{0.975,29} = 2.045$ (from Table A2). By Eq. 13.20, the lower and upper limits are 1.6 and 7.0.

13.4 Since $n = 30$ and $Z_{0.975} = 1.96$, we have $l = 10.13$ and $u = 20.87$. Therefore the lower limit is 13% between the 10th and 11th largest observations. The upper confidence limit is 87% between the 20th and 21st largest observations.

13.5 Using Eq. 13.23 with $d = 0.20$ gives $n = 123$, and with $d = 0.50$ gives $n = 25$.

14 Estimating the Mean and Variance from Censored Data Sets

In some environmental sampling situations the pollution measurements are considerably greater than zero, and measurement errors are small compared to variations in true concentrations over time and/or space. In other situations the true concentration of the sample being measured may be very near zero, in which case the measured value may be less than the measurement limit of detection (LOD). In this situation, analytical laboratories may report them as not detected (ND), zeros, or less-than (LT) values. Data sets containing these types of data are said to be "censored on the left" because data values below the LOD are not available.

These missing data make it difficult to summarize and compare data sets and can lead to biased estimates of means, variances, trends, and other population parameters. Also, some statistical tests cannot be computed, or they give misleading results. One problem, the topic of this chapter, has to do with how to estimate the mean μ and variance σ^2 of a population when only a censored data set is available. We begin by considering the several ways laboratories may report measurements, and the biased estimates of μ and σ^2 that can result when actual measurements are not available. We then discuss the median, trimmed mean, and Winsorized mean and standard deviation, methods that may be used on censored data sets. Then two methods (probability plotting and maximum likelihood) are given for using a censored data set to estimate μ and σ^2 of a population that has a normal or two-parameter lognormal distribution.

14.1 DATA NEAR DETECTION LIMITS

Keith et al. (1983) define the LOD as "the lowest concentration level that can be determined to be statistically different from a blank." When a measurement is less than the LOD (however it is defined), the analytical laboratory may: (1) report the datum as "below LOD," (2) report the datum as zero, (3) report an LT value—that is, a numerical value (usually the LOD) preceded by a "<" sign, (4) report some value between zero and the LOD, for example, one half the LOD, as suggested by Nehls and Akland (1973), or whenever possible (5) report the actual concentration (positive or negative) whether or not it is below the LOD.

The last option, the reporting of actual concentrations, is the best procedure from both practical and statistical analysis points of view, as discussed by Rhodes (1981), assuming the very small measurement values are not the result of a measurement bias in the laboratory. Environmental Protection Agency (1980, Chapter 6) discusses detection limits for radionuclides, emphasizing that these limits are estimated quantities that should not be used as a posteriori criterion for the presence of radioactivity. Reporting only ''below LOD'' or zero throws away information useful to the data analyst. Evidence of this loss is supplied by Gilliom, Hirsch, and Gilroy (1984), who showed, using computer Monte Carlo experiments, that linear trends in data near detection limits are more likely to be detected if data sets are not censored–that is, if the actual concentrations for all analyses are used rather than only those above the detection limit.

Keith et al. (1983) recommend that measurements below the LOD (as they define it) be reported as ND and that the LOD be given in parentheses. It is strongly recommended here that, whenever the measurement technique permits, report the actual measurement, whatever it may be, even if it is negative. Similar recommendations are also made by Environmental Protection Agency (1980, Chapter 6) and American Society of Testing Materials (1984).

14.2 ESTIMATORS OF THE MEAN AND VARIANCE

14.2.1 Biased Estimators

If only LT values are reported when a measurement is below the LOD, the mean μ and variance σ^2 of the population might be estimated by computing the sample mean $\bar{x}$ and variance s^2 in one of the following ways:

1. Compute $\bar{x}$ and s^2 using all the measurements, including the LT values.
2. Ignore LT values and compute $\bar{x}$ and s^2 using only the remaining ''detected'' values.
3. Replace LT values by zero and then compute $\bar{x}$ and s^2.
4. Replace LT values by some value between zero and the LOD, such as one half the LOD; then compute $\bar{x}$ and s^2.

The first three methods are biased for both μ and σ^2. The bias of the second method is illustrated in Environmental Protection Agency (1980, Chapter 6). The fourth method is unbiased for μ (but not for σ^2) if the analytical measurement technique cannot result in negative measurements, and if all measurements between zero and the LOD are equally likely to occur—that is, if they have a uniform distribution. Kushner (1976) studied this fourth method when aerometric data below the detection limit are lognormal. For his application (pollution data) he concluded that biases in using the midpoint would be overshadowed by measurement error.

If the reported data set consists almost entirely of LT values, one could use the first method (averaging all the data, including LT values) and report the resulting value of $\bar{x}$ preceded by a ''<'' sign. In this case the complete data set should be reported, if possible, including the LT values. As a minimum, the number of LT values used in computing $\bar{x}$ and s^2 should be indicated.

14.2.2 Sample Median

Instead of computing $\bar{x}$ when the data set is censored, one could compute the sample median. This approach is appropriate for estimating the mean if the underlying distribution is symmetric. The median can be estimated even if almost half the data set consists of NDs, LT values, or "trace." The reason is that the median is computed by using only the middle value of the ordered measurements if n is odd, or the average of the two middle values if n is even (see Eqs. 13.15, 13.16). The median is also not affected by erratic extreme values (errors or mistakes), that is, it is robust or resistant to outliers. If the distribution is asymmetric, then the sample median estimates the median of the population rather than the mean. Hence, the sample median will tend to be smaller than the true mean μ if the distribution is skewed to the right and larger than the true mean if the distribution is skewed to the left.

14.2.3 Trimmed Mean

An alternative to computing the median of n data values is to compute a $100p\%$ trimmed mean, where $0 < p < 0.50$, that is, to compute the arithmetic mean on the $n(1 - 2p)$ data values remaining after the largest np data values and the smallest np data values are eliminated (trimmed away). If the number of measurements reported as NDs, LT, or "trace" are no more than np, then the trimmed mean can be computed. The degree of trimming (p) that can be used will depend on the number of these values that are present. The number of data trimmed off both ends of the ordered data set is the integer part of the product pn. When n is even, the most extreme case is when all but the middle two data are trimmed away. In that situation the trimmed mean is just the sample median.

The trimmed mean is usually recommended as a method of estimating the true mean of a symmetric distribution to guard against outlier data (very large data that are mistakes or are unexplainable). Hence, it may be useful even if the data set does not contain NDs or LT values. When the underlying distribution is symmetric, Hoaglin, Mosteller, and Tukey (1983) suggest that a 25% trimmed mean (the midmean) is a good estimator of μ. Hill and Dixon (1982) considered asymmetric distributions and found that a 15% trimmed mean was a "safe" estimator to use, in the sense that its performance did not vary markedly from one situation to another. Mosteller and Rourke (1973) give an introductory discussion of trimmed means. David (1981) considers the statistical efficiency of trimmed means.

EXAMPLE 14.1

Suppose $n = 27$ data are collected from a symmetric distribution with true mean μ. If we want to estimate μ using a 25% trimmed mean, we first compute $0.25n = 0.25(27) = 6.75$. Hence, the 6 smallest and 6 largest data are discarded. The arithmetic mean of the remaining $27 - 12 = 15$ data is the estimate of μ.

14.2.4 Winsorized Mean and Standard Deviation

"Winsorization" can be used to estimate the mean, μ, and standard deviation, σ, of a symmetric distribution even though the data set has a few missing or unreliable values at either or both ends of the ordered data set. A detailed discussion of the method is given by Dixon and Tukey (1968).

Suppose n data are collected, and there are three ND values. The Winsorization procedure is as follows:

1. Replace the three ND values by the next largest datum.
2. Replace the three largest values by the next smallest datum.
3. Compute the sample mean, $\bar{x}_w$, and standard deviation, s, of the resulting set of n data.
4. Then $\bar{x}_w$, the Winsorized mean, is an unbiased estimator of μ. The Winsorized standard deviation is

$$s_w = \frac{s(n - 1)}{v - 1}$$

which is an approximately unbiased estimator for σ, where n is the total number of data values and v is the number of data not replaced during the Winsorization. (The quantity v equals $n - 6$ in this example because 3 ND values are present.)

5. If the data are from a normal distribution, the upper and lower limits of a two-sided $100(1 - \alpha)\%$ confidence interval about μ are

$$\bar{x}_w \pm t_{1-\alpha/2, v-1} \frac{s_w}{\sqrt{n}} \qquad \textbf{14.1}$$

where $t_{1-\alpha/2, v-1}$ is the value of the t variate (from Table A2) that cuts off $(100\alpha/2)\%$ of the upper tail of the t distribution with $v - 1$ degrees of freedom. (Note: Equation 14.1 is identical to the usual limits, Eq. 11.5, except the degrees of freedom are $v - 1$ instead of $n - 1$, and s_w replaces s.) One-sided limits on μ can be obtained from Eqs. 11.6 and 11.7 using $v - 1$ degrees of freedom and s_w.

Note the distinction between trimming and Winsorizing. Trimming discards data in both tails of the data set, and the trimmed mean is computed on the remaining data. Winsorizing replaces data in the tails with the next most extreme datum in each tail and then computes the mean on the new data set.

EXAMPLE 14.2

Suppose groundwater has been sampled monthly for 12 months from the same well, yielding the following concentrations for a hazardous chemical (ordered from smallest to largest):

trace trace 0.78 2.3 3.0
3.1 3.2 4.0 4.1 5.6 6.7 9.3

Replace the two trace concentrations by 0.78 and the two largest concentrations by 5.6. The sample mean and standard deviation of the new data set are $\bar{x}_w = 3.24$ and $s = 1.838$, respectively. This $\bar{x}_w$ is a statistically unbiased estimate of μ. The Winsorized standard deviation is

$$s_w = s\left(\frac{n-1}{\nu-1}\right)$$

$$= \frac{1.838(11)}{7} = 2.888$$

There are $\nu - 1 = 7$ degrees of freedom, and Table A2 gives $t_{0.975,7} = 2.365$. Therefore, assuming the population is normally distributed, the 95% upper and lower limits about μ are (using Eq. 14.1) $3.24 \pm 2.365(2.888)/\sqrt{12}$, or 1.27 and 5.21.

If the data set is skewed to the right, the logarithms of the data may be approximately symmetric. For example, if the untransformed data are from a lognormal distribution, the logarithms are from a normal distribution. If so, Winsorization could be used to estimate the mean and standard deviation of the log-transformed data. These Winsorized estimates, $\bar{x}_w$ and s_w, could then be used in Eqs. 13.7 and 13.8 to estimate the mean and variance of the underlying lognormal distribution.

14.3 TWO-PARAMETER LOGNORMAL DISTRIBUTION

This section gives two methods for estimating μ and σ^2 when only a censored data set from a two-parameter lognormal distribution is available. These methods were developed for normal distributions, but they may also be used to estimate the parameters μ_y and σ_y^2 of the two-parameter lognormal distribution, which can be used in turn to estimate μ and σ^2.

Two types of censoring can occur, depending on whether the number of measurements falling below the point of censorship is or is not specified before the measurements are made. If the number is not specified (i.e., if the number is a random variable) the censoring is called *Type I*. If the number *is* specified, we have *Type II* censoring. Type I censoring is perhaps more common for pollution data. Probability plotting can be used for either type. The maximum likelihood method differs slightly for the two types as described in Section 14.3.2.

Data sets may also be "censored on the right," meaning that all data values *above* some known point of censorship are not available. The two methods given here may be used with censoring either on the left or on the right.

14.3.1 Probability Plotting

In Section 13.1.3 log-probability plotting was used to estimate the parameters μ_y and σ_y^2 of a two-parameter lognormal distribution when a complete (uncensored) data set of size n was available. The same plotting procedure is used with a left-censored data set except that the data below the point of censorship, x_0 (which may be the LOD or some other value), cannot be plotted. If the n'

smallest of the n data are missing, then the $(n' + 1)$th ordered datum $x_{[n'+1]}$ (smallest reported datum) is plotted versus $[(n' + 1) - 0.5]100/n$ on log-probability paper, and similarly for the $(n' + 2)$th datum $x_{[n'+2]}$ versus $[(n'+2) - 0.5]100/n$, and so on. If the measurements above x_0 are from a lognormal distribution, then the resulting plot should be linear on log-probability paper. If so, a straight line is drawn through the points, and μ_y and σ_y^2 are estimated by Eqs. 13.9 and 13.10, respectively. Then the mean, μ, and the variance, σ^2, of the lognormal population are estimated by Eq. 13.11 and the square of Eq. 13.12, respectively. From Eq. 13.10 we see that if more than 16% of the data set is censored, $\hat{\sigma}_y^2$ cannot be determined by Eq. 13.10. If the distribution is known with assurance to be lognormal, the straight line might be extended down to the 0.16 quantile. However, extrapolation into a region where no data are available is always risky.

14.3.2 Maximum Likelihood Estimators

Cohen (1959, 1961) used the method of maximum likelihood to obtain estimates of the mean and variance of a normal distribution when the data set is either left or right censored. His procedure can also be applied to estimate the mean and the variance of the logarithms of left or right-censored lognormally distributed data, since the logarithms are normally distributed. Then the mean and the standard deviation of the lognormal distribution can be estimated by Eqs. 13.11 and 13.12.

We now describe Cohen's procedure for the lognormal case. Let n = total number of measurements x_i, k = number out of n that are above the LOD, $y_i = \ln x_i$, and $y_0 = \ln$ LOD. To estimate the mean μ_y and the variance σ_y^2 of the log-transformed population, we

1. Compute $h = (n - k)/n$ = proportion of measurements below the LOD.
2. Compute

$$\bar{y}_u = \frac{1}{k}\sum_{i=1}^{k} y_i \qquad \textbf{14.2}$$

and

$$s_u^2 = \frac{1}{k}\sum_{i=1}^{k} (y_i - \bar{y}_u)^2 \qquad \textbf{14.3}$$

the sample mean and variance of the k measurements above the LOD.

3. Compute

$$\hat{\gamma} = \frac{s_u^2}{(\bar{y}_u - y_0)^2} \qquad \textbf{14.4}$$

4. Obtain an estimate $\hat{\lambda}$ of the parameter λ from Table A15. Enter the table with h and $\hat{\gamma}$ and use linear interpolation in both horizontal and vertical planes if necessary.
5. Estimate the mean and the variance of the log-transformed data as follows:

$$\hat{\mu}_y = \bar{y}_u - \hat{\lambda}(\bar{y}_u - y_0) \qquad \textbf{14.5}$$

$$\hat{\sigma}_y^2 = s_u^2 + \hat{\lambda}(\bar{y}_u - y_0)^2 \qquad \textbf{14.6}$$

These estimates may then be used to estimate the mean and the variance of the lognormal distribution by computing

$$\hat{\mu} = \exp\left(\hat{\mu}_y + \frac{\hat{\sigma}_y^2}{2}\right) \qquad \textbf{14.7}$$

$$\hat{\sigma}^2 = \hat{\mu}^2[\exp(\hat{\sigma}_y^2) - 1] \qquad \textbf{14.8}$$

This procedure is appropriate for Type I censored samples. For Type II censoring the procedure is the same except that y_0 in Eqs. 14.5 and 14.6 is replaced by the logarithm of the smallest fully measured concentration—that is, of the smallest observed concentration above the LOD (Cohen, 1961, Eq. 3).

If the censored data set is from a normal distribution, Eqs. 14.2–14.6 are used to estimate μ and σ^2 of that distribution, where $y_i = \ln x_i$ and $y_0 = \ln \text{LOD}$ are replaced by x_i and the LOD, respectively.

EXAMPLE 14.3

We use the mercury data in Table 12.3. For the sake of illustration, suppose the LOD is 0.20 ppm, so the smallest observation in Table 12.3 is not available. The calculations are laid out in Table 14.1, yielding the following estimates of the mean and standard deviation of the lognormal population: $\hat{\mu} = 1.1$ and $\hat{\sigma} = 0.93$.

Table 14.1 Computations to Estimate the Mean and Variance of a Two-Parameter Lognormal Distribution Using a Left Censored Data Set (Example 14.3)

1. The nine log-transformed data above the LOD = 0.20 are (from Table 12.3)

$y_1 = -0.7985$	$y_6 = 0.1823$
$y_2 = -0.5108$	$y_7 = 0.3148$
$y_3 = -0.2744$	$y_8 = 0.5247$
$y_4 = 0.04879$	$y_9 = 0.7227$
$y_5 = 0.1133$	

2. $n = 10$, $k = 9$, $h = (10 - 9)/10 = 0.1$, $y_0 = \ln \text{LOD} = -1.6094$

3. Using Eqs. 14.2, 14.3, and 14.4, we obtain

$$\bar{y}_u = 0.03588 \quad s_u^2 = 0.21193$$
$$\hat{\gamma} = 0.21193/(0.03588 + 1.6094)^2 = 0.07829$$

Entering Table A15 with $h = 0.1$ and $\hat{\gamma} = 0.07829$, we find using linear interpolation, that $\hat{\lambda} = 0.1164$.

4. Therefore, Eqs. 14.5 and 14.6 give

$$\hat{\mu}_y = 0.03588 - 0.1164(0.03588 + 1.6094) = -0.1556$$
$$\hat{\sigma}_y^2 = 0.21193 + 0.1164(0.03588 + 1.6094)^2 = 0.5270$$

5. Therefore, using Eqs. 14.7 and 14.8,

$$\hat{\mu} = \exp\left(-0.1556 + \frac{0.5270}{2}\right) = 1.114 \text{ or } 1.1$$

$$\hat{\sigma}^2 = (1.114)^2\,[\exp(0.5270) - 1] = 0.8611 \text{ or } \hat{\sigma} = 0.93$$

14.4 THREE-PARAMETER LOGNORMAL DISTRIBUTION

Suppose a left-censored data set has been drawn from a three-parameter lognormal rather than from a two-parameter distribution. Recall from Chapter 12 that the third parameter is τ, which shifts the two-parameter distribution to the right or left, depending on whether τ is positive or negative, respectively. If τ is known a priori, then the procedures given in Sections 14.3.1 and 14.3.2 can be applied to the transformed data $x_i - \tau$ rather than to the x_i. But estimation procedures become more complicated if τ is not known a priori. In that case the simplest approach is to estimate τ, subtract this estimate from each x_i, and use the procedure in Sections 14.3.1 or 14.3.2.

The optimum method for estimating μ and σ^2 is to use maximum likelihood methods if n is reasonably large (see discussions by Harter and Moore, 1966; Tiku, 1968; Ott and Mage, 1976; and Mage and Ott, 1978). These require iterative solutions on a computer. Two simpler approaches, the method of quantiles and a graphical trial-and-error procedure are illustrated by Gilbert and Kinnison (1981) and Aitchison and Brown (1969).

14.5 SUMMARY

This chapter considered methods for estimating the mean and variance of populations when data less than some known point, frequently the limit of detection, are censored—that is, they are not available to the data analyst. Simple procedures such as treating these missing values as if they were zero can lead to biased estimates. Three alternative methods that are unbiased for estimating the mean when the population distribution is symmetric are the median, trimmed mean, and Winsorized mean. Probability plotting and maximum likelihood methods are illustrated when the two-parameter lognormal distribution applies. The methods given here may also be used when the data set is censored on the right (above some point) rather than on the left.

EXERCISES

14.1 Compute the median and the 15% trimmed mean on the following data set:

34 18 22 32 48 35 5
22 21 8 10 12 2 80 95

14.2 Assume the two smallest data in Exercise 14.1 were reported as "not detected." Compute a Winsorized mean and standard deviation for this censored data set.

14.3 Suppose the smallest mercury datum in Table 12.3 was less than the limit of detection (LOD), where LOD = 0.20. Assume the remaining (censored) data set was drawn from a normal distribution and use Cohen's (1961) procedure to estimate μ and σ^2 of that normal distribution.

ANSWERS

14.1 The median is 22. The trimmed mean is 23.8.

14.2 $\bar{x}_w = 24.9$ and $s_w = s(n - 1)/(v - 1) = 15.026(14/10) = 21$.

14.3 $h = (10 - 9)/10 = 0.10$. Using the untransformed data in Eqs. 14.2, 14.3, and 14.4, we obtain $\bar{x}_u = 1.144$, $s_u^2 = 0.23620$, $\hat{\gamma} = 0.23620/(1.144 - 0.20)^2 = 0.26505$. From Table A15, $\hat{\lambda} = 0.1286$. Therefore, by Eqs. 14.5 and 14.6, $\hat{\mu} = 1.0$, $\hat{\sigma}^2 = 0.351$ or $\hat{\sigma} = 0.59$.

15 Outlier Detection and Control Charts

An unavoidable problem in the statistical analysis of environmental pollution data is dealing with outliers. Hunt et al. (1981) define an *outlier* to be "an observation that does not conform to the pattern established by other observations." Outliers may arise from mistakes such as transcription, keypunch or data-coding errors. They may also arise as a result of instrument breakdowns, calibration problems and power failures. In addition, outliers may be manifestations of a greater amount of inherent spatial or temporal variability than expected for the pollutant. They could also be an indication of unsuspected factors of practical importance such as malfunctioning pollutant effluent controls, spills, and plant shutdowns.

This chapter briefly discusses data-screening and validation procedures, followed by recommendations on how to handle outliers in practice. The chapter then illustrates Rosner's procedure for detecting up to k outliers and a method for detecting outliers in correlated variables. The remainder of the chapter is concerned with how to use Shewhart control charts to look for consistent data over time or shifts in the mean or standard deviation of a time process.

Many methods for detecting outliers are discussed by Beckman and Cook (1983), Hawkins (1980), and Barnett and Lewis (1978). Burr (1976) and Vardeman and David (1984) provide many references on control chart techniques. Kinnison (1985) discusses extreme value statistics, which are closely connected with the ideas of outlier detection.

15.1 DATA SCREENING AND VALIDATION

Statistical tests for outliers are one part of the data validation process wherein data are screened and examined in various ways before being placed in a data bank and used for estimating population parameters or making decisions. Nelson, Armentrout, and Johnson (1980) and Curran (1978) discuss data screening and validation procedures for air quality data.

Nelson and co-workers identify four categories of data validation procedures.

1. Routine checks made during the processing of data. Examples include looking for errors in identification codes (those indicating time, location of sampler,

method of sampling, etc.), in computer processing procedures, or in data transmission.

2. Tests for the internal consistency of a data set. These include plotting data for visual examination by an experienced analyst and testing for outliers.
3. Comparing the current data set with historical data to check for consistency over time. Examples are visually comparing data sets against gross upper limits obtained from historical data sets, or testing for historical consistency using the Shewhart control chart test. The Shewhart test was recommended by Hunt, Clark, and Goranson (1978) for screening 24-h air pollution measurements (one measurement per day) and is discussed in Section 15.6.
4. Tests to check for consistency with parallel data sets, that is, data sets obtained presumably from the same population (e.g., from the same time period, region of the aquifer, air mass, or volume of soil). Three tests for doing so are the sign test, the Wilcoxon signed-ranks test, and the Wilcoxon rank sum test. These tests are discussed in Chapter 18.

15.2 TREATMENT OF OUTLIERS

After an outlier has been identified, one must decide what to do with it. Outliers that are obvious mistakes are corrected when possible, and the correct value is inserted. If the correct value is not known and cannot be obtained, the datum might be excluded, and statistical methods that were developed specifically for missing-value situations could be used. Examples of such methods are general analysis of variance and covariance (available in most commercial packages of statistical computer codes), estimating the mean and the variance from censored data sets by methods given in Chapter 14, and testing for trend by the Mann-Kendall nonparametric test (discussed in Chapter 16). Alternatively, the outlier could be retained, and a robust method of statistical analysis could be used, that is, a method that is not seriously affected by the presence of a few outliers. Examples of robust methods are the sample median, trimmed mean, and Winsorized mean (discussed in Chapter 14); the trend estimation and testing techniques given in Chapters 16 and 17; and the nonparametric tests for comparing populations in Chapter 18.

It is important that no datum be discarded solely on the basis of a statistical test. Indeed, there is always a small chance (the α level of the test) that the test incorrectly declares the suspect datum to be an outlier. Also, multiple outliers should not be automatically discarded since the presence of two or more outliers may indicate that a different model should be adopted for the frequency distribution of the population. For example, several unusually large measurements may be an indication that the data set should be modeled by a skewed distribution such as the lognormal. There should always be some plausible explanation other than the test result that warrants the exclusion or replacement of outliers. The use of robust methods that have the effect of eliminating or giving less weight to extreme values should also be justified as being appropriate.

If no plausible explanation for an outlier can be found, the outlier might be excluded, accompanied by a note to that effect in the data base and in the report. In addition, one could examine the effect on final analysis procedures applied to the data set when the outlier was both included and excluded. A description of major effects should be included in the report. In some cases it may be feasible to take another sample for comparison with the old.

15.3 ROSNER'S TEST FOR DETECTING UP TO k OUTLIERS

This section gives Rosner's (1983) "many-outlier" sequential procedure for identifying up to $k = 10$ outliers. The procedure is an improved version of Rosner's (1975) "extreme studentized deviate" outlier test. Simonoff (1982) found that this earlier test performed well compared to other outlier tests, although Rosner (1983) points out that it tends to detect more outliers than are actually present. This problem does not exist with the improved version discussed here. Rosner's (1983) method assumes that the main body of data is from a normal distribution. If the assumption of a lognormal distribution is more plausible, all computations should be performed on the logarithms of the data.

Rosner's approach is designed to avoid masking of one outlier by another. *Masking* occurs when an outlier goes undetected because it is very close in value to another outlier. In Figure 15.1 datum B could mask datum A if an inappropriate outlier test for A is performed.

To use Rosner's approach, we need to specify an upper limit k on the number of potential outliers present. Then we repeatedly delete the datum (large or small) farthest from the mean and recompute the test statistic after each deletion. Table A16 (from Rosner, 1983) is used to evaluate the test statistic when $n \geq 25$. This table is also used when the null hypothesis specifies a lognormal distribution. Linear interpolation may be used to obtain critical values not given in the tables for n between 50 and 500. A formula for obtaining approximate critical values when $n > 500$ is given in the footnote to Table A16. If $n < 25$, Rosner's test cannot be used. In that situation, a test for a single outlier, such as that by Dixon (1953), may be used, but the problem of masking may occur.

Rosner's tests are two-tailed since the procedure identifies either suspiciously large or suspiciously small data. When a one-tailed test is needed, that is, when there is interest in detecting only large values or only small values, then the skewness test for outliers discussed by Barnett and Lewis (1978) is suitable.

Some notation is needed to illustrate Rosner's procedure. Let $\bar{x}^{(i)}$ and $s^{(i)}$ be the arithmetic mean and the standard deviation, respectively, of the $n - i$ observations in the data set that remain after the i most extreme observations have been deleted. That is,

$$\bar{x}^{(i)} = \frac{1}{n - i} \sum_{j=1}^{n-i} x_j \qquad \textbf{15.1}$$

$$s^{(i)} = \left[\frac{1}{n - i} \sum_{j=1}^{n-i} (x_j - \bar{x}^{(i)})^2 \right]^{1/2} \qquad \textbf{15.2}$$

where i ranges from zero to k. For example, $\bar{x}^{(1)}$ and $s^{(1)}$ are the sample mean

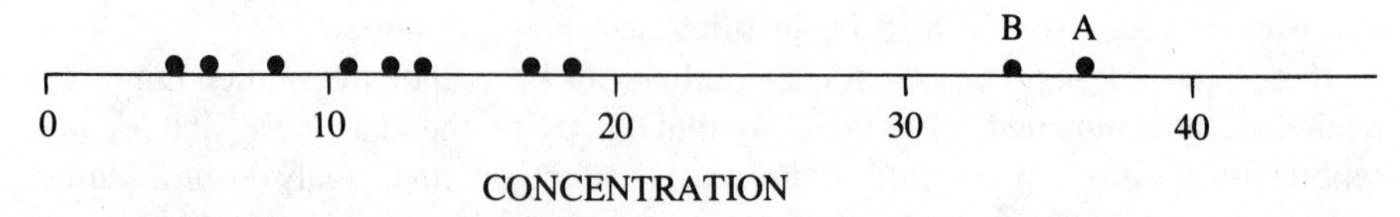

Figure 15.1 How datum B can "mask" datum A if an inappropriate outlier test is conducted for the suspected outlier A.

and the standard deviation of the $n - 1$ data remaining after the most extreme datum from $\bar{x}^{(0)}$ has been removed. Let $x^{(i)}$ denote the most outlying observation (farthest from the mean $\bar{x}^{(i)}$) remaining in the data set after i more extreme data (large or small) have been removed. Then

$$R_{i+1} = \frac{|x^{(i)} - \bar{x}^{(i)}|}{s^{(i)}}$$

= test statistic for deciding whether the $i + 1$ most extreme values in the complete data set are outliers from a normal distribution **15.3**

λ_{i+1} = tabled critical value (Table A16) for comparison with R_{i+1}

Rosner provides a FORTRAN IV computer code for computing the values of R_{i+1} for an arbitrary unordered data set. This code is given here in Table 15.1.

The null hypothesis being tested by Rosner's method is

H_0: The entire data set is from a normal distribution

There are a series of alternative hypotheses:

$H_{A,k}$: There are k outliers.
$H_{A,k-1}$: There are $k - 1$ outliers.
⋮
$H_{A,1}$: There is one outlier.

The first test is H_0 versus $H_{A,k}$, which is made by comparing R_k with λ_k. If H_0 is not rejected, we test H_0 versus $H_{A,k-1}$ by comparing R_{k-1} with λ_{k-1}, and so on, until one of the tests is statistically significant or all tests are nonsignificant. If the test of H_0 versus $H_{A,k-1}$ is significant, we conclude that $k - 1$ outliers from the assumed normal distribution are present. When one of the tests is significant, then no more tests are made.

If the null hypothesis is

H_0: The entire data set is from a lognormal distribution

then Eqs. 15.1, 15.2, and 15.3 are computed on the logarithms of the data, in which case we use the notation $y^{(i)}$, $\bar{y}^{(i)}$, $s_y^{(i)}$, and $R_{y,i+1}$ instead of $x^{(i)}$, $\bar{x}^{(i)}$, $s^{(i)}$, and R_{i+1}, respectively.

EXAMPLE 15.1

Table 15.2 gives the logarithms of $n = 55$ total suspended particulate (TSP) air data that were collected every sixth day at a monitoring site (from Nelson, Armentrout, and Johnson, 1980, Table 3.11). The logarithms are ordered from smallest to largest. Since TSP data are frequently approximately lognormally distributed, we use Rosner's procedure to test the null hypothesis H_0: The entire data set is from a lognormal distribution. We are interested in testing whether outliers from this assumed lognormal distribution are present. We use $\alpha = 0.05$.

Suppose that prior to seeing the data set, we had set $k = 3$. Then H_0 is tested versus the following series of $k = 3$ alternative hypotheses:

Table 15.1 FORTRAN IV Program to Compute Values of R_{i+1} for Rosner's (1983) Test for up to k Outliers

```
      SUBROUTINE WT(X,N,NSAM,
      NOUT,WTVEC,N1,Q)
C COMPUTE ESD OUTLIER STATISTICS
      DIMENSION X(N), WTVEC(N1),Q(N)
      II=1
      SUM=0.0
      SUMSQ=0.0
      FN=0.0
      DO 2 I=1,NSAM
      Q(I)=0.0
      SUM=SUM+X(I)
      SUMSQ=SUMSQ+X(I)**2
      FN=FN+1.0
    2 CONTINUE
    1 SS=SUMSQ-(SUM**2)/FN
      S=(SS/(FN-1.0))**.5
      XBAR=SUM/FN
      BIG=0.0
      IBIG=0
      DO 3 I=1,NSAM
      IF(Q(I).EQ.1.0) GO TO 3
      A=ABS(X(I)-XBAR)
      IF(A.LE.BIG) GO TO 3
      BIG=A
      IBIG=I
    3 CONTINUE
      WTVEC(II)=BIG/S
      Q(IBIG)=1.0
      II=II+1
      IF (II.GT.NOUT) GO TO 999
      SUM=SUM-X(IBIG)
      SUMSQ=SUMSQ-X(IBIG)**2
      FN=FN-1.0
      GO TO 1
  999 RETURN
      END
```

where

NSAM = sample size of data set

NOUT = maximum number of outliers to be detected

N = maximum sample size for all data sets used with this subroutine on a given computer run

N1 = maximum number of outliers to be detected for all data sets used with this subroutine on a given computer run

X = single-precision input vector of dimension N, whose first NSAM elements represent the unordered data set to which this subroutine is applied

WTVEC = single-precision output vector of dimension N1 whose first NOUT elements are R_1, . . . , R_{NOUT}

Q = single-precision input vector of dimension N used internally in the subroutine

Source: After Rosner, 1983.

Table 15.2 Total Suspended Particulate (TSP) Data (Units of $\log_e$ $\mu g/m^3$) Ordered from Smallest to Largest

2.56	3.58	3.83	4.06	4.32
3.18	3.58	3.91	4.08	4.33
3.33	3.64	3.91	4.09	4.33
3.33	3.64	3.95	4.17	4.34
3.40	3.69	3.97	4.17	4.44
3.43	3.69	3.99	4.17	4.47
3.43	3.71	4.03	4.23	4.48
3.43	3.74	4.04	4.23	4.48
3.50	3.76	4.04	4.26	4.62
3.50	3.76	4.04	4.29	4.68
3.50	3.81	4.04	4.32	5.16

Source: Data from Nelson, Armentrout, and Johnson, 1980.

Table 15.3 Computations for Using Rosner's (1983) Test for up to Three Outliers from a Lognormal Distribution

i	$n - i$	$\bar{y}^{(i)}$	$s_y^{(i)}$	$y^{(i)}$	$R_{y,i+1} = \lvert y^{(i)} - \bar{y}^{(i)} \rvert / s_y^{(i)}$	λ_{i+1} ($\alpha = 0.05$)
0	55	3.94	0.444	2.56	3.11	3.165
1	54	3.96	0.406	5.16	2.96	3.155
2	53	3.94	0.374	4.68	1.98	3.150

$H_{A,3}$: There are three outliers.
$H_{A,2}$: There are two outliers.
$H_{A,1}$: There is one outlier.

Values of $\bar{y}^{(i)}$, $s_y^{(i)}$, and $R_{y,i+1}$ for $i = 0$, 1, and 2 were computed from the data in Table 15.2. These values are summarized in Table 15.3. The terms $\bar{y}^{(0)}$ and $s_y^{(0)}$ are the mean and the standard deviation for all the data, whereas $\bar{y}^{(1)}$ and $s_y^{(1)}$ were computed after deleting 2.56, and $\bar{y}^{(2)}$ and $s_y^{(2)}$ after deleting 2.56 and 5.16. The most extreme datum, $y^{(i)}$, at each stage is also shown. The critical values in the last column of Table 15.3 were obtained by linear interpolation between the $\alpha = 0.05$ entries for $n = 50$ and $n = 60$ in Table A16. That is, since there are 55 measurements in the data set, each of the λ_{i+1} in Table 15.3 is halfway between the tabled values for $n = 50$ and $n = 60$.

We first test H_0 versus $H_{A,3}$ by comparing $R_{y,3}$ with λ_3. Since $R_{y,3} = 1.98$ is less than $\lambda_3 = 3.150$, we cannot reject H_0 in favor of $H_{A,3}$. Next, we test H_0 against $H_{A,2}$ by comparing $R_{y,2}$ with λ_2. Since $R_{y,2} = 2.96$ is less than $\lambda_2 = 3.155$, we cannot reject H_0 in favor of $H_{A,2}$. Finally, we test H_0 against $H_{A,1}$ by comparing $R_{y,1}$ with λ_1. Since $R_{y,1} = 3.11$ is less than $\lambda_1 = 3.165$, we cannot reject H_0 in favor of $H_{A,1}$. We conclude that there are no outliers from the assumed lognormal distribution.

15.4 DETECTING OUTLIERS IN CORRELATED VARIABLES

Suppose the following n paired data on two correlated variables x and z are obtained: $(x_1, z_1), (x_2, z_2), \ldots, (x_n, z_n)$. We wish to determine whether any of the $2n$ observations are outliers. Rosner's test could be applied independently on each variable. However, if two variables are correlated, this additional information can be used to uncover outliers that would not be found by using a univariate procedure applied separately on each variable.

An indispensable tool for outlier detection with bivariate data is the simple scatter plot of x_i against z_i for the n pairs of data. Visual inspection of these plots will identify points that seem too far removed from the main cloud of points. That is, these points may not be from the same bivariate distribution as the remaining points. An approximate probability plotting technique suggested by Healy (1968) and discussed by Barnett and Lewis (1978, p. 212) may be used to supplement scatter plots. The method consists of computing a "distance,"

D, from the cloud of points for each of the n pairs of observations. The distance D_i for the ith pair is calculated as follows:

$$D_i = \left[\left(\frac{x_i - \bar{x}}{s_x}\right)^2 - 2r\left(\frac{x_i - \bar{x}}{s_x}\right)\left(\frac{z_i - \bar{z}}{s_z}\right) + \left(\frac{z_i - \bar{z}}{s_z}\right)^2\right]^{1/2} \quad \textbf{15.4}$$

where $\bar{x}$, $\bar{z}$, s_x, and s_z are the sample means and the standard deviations of the two variables, and r is the estimated correlation between x and z, computed as

$$r = \frac{\sum_{i=1}^{n} (x_i - \bar{x})(z_i - \bar{z})}{\sqrt{\sum_{i=1}^{n} (x_i - \bar{x})^2 \sum_{i=1}^{n} (z_i - \bar{z})^2}} \quad \textbf{15.5}$$

If the pairs are assumed to be from a bivariate normal distribution except for possible outliers, then the ordered D values are plotted on normal probability paper. An obvious deviation from a straight line suggests either that the assumption of a bivariate normal distribution is erroneous or that one or more outliers are present. When a nonlinear plot occurs, the scatter plot should be examined to identify possible outliers. If these suspect points are deleted and the calculations and plotting redone, a resulting linear plot would suggest that the remaining points are from a bivariate normal distribution. However, if a nonlinear plot is still obtained, this would suggest that the "nonoutliers" are not from a bivariate normal distribution.

Healy (1968) and Barnett and Lewis (1978, p. 212) illustrate the technique on a data set of 39 pairs of data. If the population distribution is assumed to be bivariate lognormal rather than bivariate normal, then the D_i would be calculated using the logarithms of the data, and would be plotted on normal probability paper.

Nelson, Armentrout, and Johnson (1980) give a graphical procedure for identifying outliers from a bivariate normal or bivariate lognormal distribution that is closely related to Healy's method. They illustrate the technique, using TSP air quality data obtained at the same times at two measurement sites.

15.5 OTHER OUTLIER TESTS

Methods for detecting multivariate outliers are discussed by Everitt (1978, pp. 67–73), Beckman and Cook (1983), Barnett and Lewis (1978), Rohlf (1975), Gnanadesikan and Kettenring (1972), and Hawkins (1974). Outliers in designed experiments are difficult to detect, but methods are discussed by Stefansky (1972), Snedecor and Cochran (1980, p. 280), and Barnett and Lewis (1978, pp. 238–249). Outliers from linear regression may be detected with methods discussed by Snedecor and Cochran (1980, pp. 167–169), Barnett and Lewis (1978, pp. 252–256), and Marasinghe (1985).

Rather than identifying outliers and discarding them before doing least squares regression, one could do robust regression, as discussed and illustrated by Mosteller and Tukey (1977) and Reckhow and Chapra (1983). The objective of robust regression is to reduce the impact of data far removed from the regression line. Standard least squares methods are highly sensitive to divergent data points, whereas robust methods assign less weight to these points. However,

Reckhow and Chapra (1983) caution that robust regression should be applied only after the investigator is satisfied that less weight should be applied to the divergent data. Nonparametric regression discussed by Hollander and Wolfe (1973, p. 201) and Reckhow and Chapra (1983) is an alternative to either standard least squares regression or robust regression.

15.6 CONTROL CHARTS

Outlier tests discussed in Sections 15.2 and 15.3 check for the internal consistency of a data set. This section illustrates the use of Shewhart control charts for checking whether current data are consistent with past data. A lack of consistency over time may be an indication of data outliers. However, it could also indicate shifts or trends in mean concentrations or in levels of variability. Control charts are useful graphical tools because they provide a basis for action, that is, they indicate when changing data patterns over time should be examined to determine causes.

The essential features of the chart for means is illustrated in Figure 15.2. The features of charts for ranges and standard deviations are similar, as will be illustrated. The control chart for means can detect outliers and shifts in average concentrations, whereas charts for ranges and standard deviations check for shifts in variability. For completeness, the control chart for means should be accompanied by a control chart for standard deviations or ranges.

The general idea underlying the control chart for means is first to select k historical data sets and to compute the mean $\bar{x}_i$, range R_i, and standard deviation s_i for each, where the ith data set contains n_i data. This information is used to construct the center line and the upper and lower control limits. Then if the k subgroup means all fall between the control limits, the time process being

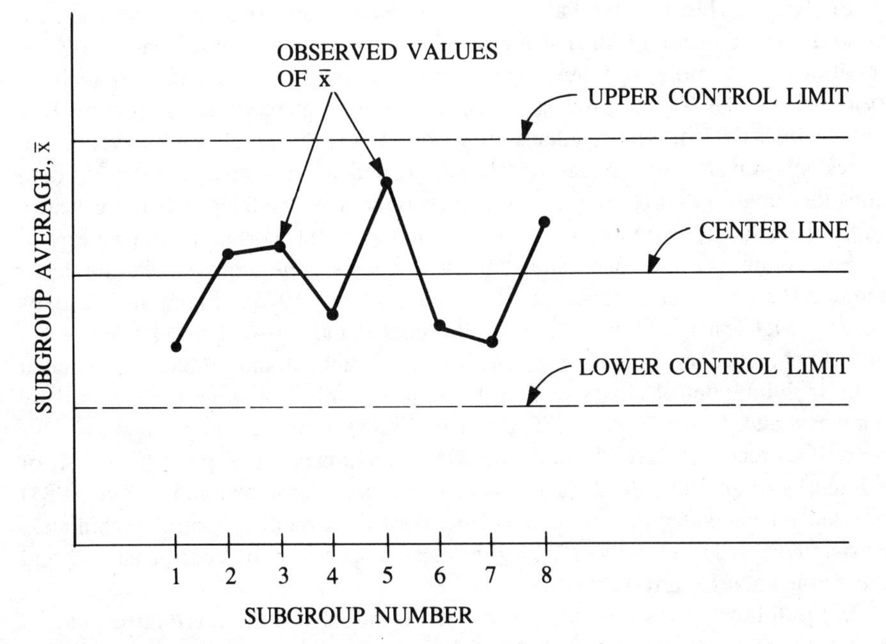

Figure 15.2 Essential features of a control chart for means.

measured is said to be "in control"—that is, operating consistently over time. In that situation we may think of each new subgroup of measurements as simply a new set of data from the same population. If a process is in control, then it makes sense to use the control limits to judge whether future subgroups of data are from this same population. If one or more sufficiently large outliers are present in a future subgroup, they will inflate the sample mean $\bar{x}_i$ enough so that it exceeds the upper control chart limit. Also, if the population mean has changed greatly from what it was during the historical period, the sample mean $\bar{x}_i$ is likely to exceed the upper or lower control limits.

15.6.1 Assumptions Underlying Control Charts

The assumptions underlying control charts are that the data are independent and normally distributed with constant mean μ and constant variance σ^2. These assumptions can be expressed as the simple stochastic model

$$x_t = \mu + e_t \qquad t = 1, 2, \cdots, n \qquad \textbf{15.6}$$

where x_t is the datum at time t, μ is the mean, n is the number of data values, and e_t are random, independent disturbances that are normally distributed with mean zero and variance σ^2.

For many environmental pollutants these assumptions are not realistic even for a process with constant mean. As discussed several times in this book, environmental data are commonly correlated and nonnormally distributed. Also, the process variance may change over time. Berthouex et al. (1978) examine whether Shewhart control charts are appropriate for data from processes with these characteristics. Based on their experience with modeling sewage treatment plants, they conclude that, given some care, standard quality control charts can provide useful qualitative information for purposes of process control.

Berthouex, Hunter, and Pallesen (1978) show how to modify the standard control chart procedures so that the usual assumptions are more nearly satisfied to allow for a more rigorous quantitative analysis. They develop a realistic stochastic model for the process, using time series methods developed by Box and Jenkins (1976). The residuals from this model (residual = observed value − value estimated using the model) are treated as raw data and are used to construct control charts. If the model is correct, the residuals will more nearly fulfill the standard assumptions of normal, independent, constant variance errors.

For comprehensive discussions of Box-Jenkins time series modeling techniques, the reader is referred to Box and Jenkins (1976), Fuller and Tsokos (1971), McClearly and Hay (1980), Berthouex and co-workers (1975, 1978), and Hipel and co-workers (1977*a*, 1977*b*). The potential and problems associated with applying control charts to detect pollution limit violations is discussed by Vaughan and Russell (1983). They also discuss other graphical methods for detecting trends (CUSUM and MOSUM techniques) and provide a list of references to environmental quality control papers. Vardeman and David (1984) provide an annotated list of papers and books on quality control techniques. Burr (1976) and Wetherill (1977) give clear discussions of control charts, and the former gives many references.

In the following material we present the standard control charting techniques based on the usual assumptions (Eq. 15.6). If the data are normally distributed and serial correlation is not large, these methods should work well. If the data are lognormally distributed, the standard procedures should be applied to the logarithms of the data.

15.6.2 Historical Data Sets

Before a control chart can be constructed, it is necessary to define meaningful historical data sets. These data sets were called *rational subgroups* by Shewhart (1931). When sampling over time, a rational subgroup may be the n data collected at a particular point in space by a given instrument during a specified time period, such as a day, week, month or quarter. For example, Hunt, Clark, and Goranson (1978) and Nelson, Armentrout, and Johnson (1980) used the five air quality (24-h) measurements made during a month at a particular sampling site as a rational data set. If groundwater samples are being collected monthly at a particular well, the three samples within each calendar quarter might be used as a subgroup.

Rational subgroups must be chosen with care because the variability within subgroups is used to construct control chart limits. Each subgroup should consist of data that are homogeneous—that is, for which there are only nonassignable (chance) causes of variability present. Then the control chart will be more sensitive to changes over time that may be due to assignable causes such as a new source of pollution, sampling or measurement biases, seasonal cycles, trends in mean concentrations, or outliers. If subgroups consist of data collected over several months, the variability within subgroups may be large due to seasonal effects. This seasonal variation will cause the control limits to be more widely spaced, so only very extreme outliers or very large shifts in mean or variance may be detected.

The definition of a rational subgroup is somewhat subjective, since it depends on one's concept of the model representing a controlled situation. Also, a series of data may consist of segments that are considered to represent an in control situation and other segments that are out of control. The latter segments are not used when estimating the control limits on the control chart.

The number of data in each rational subgroup should be as large as possible while still maintaining nonassignable variability within subgroups. The number of subgroups, k, should also be as large as possible with the warnings that there should be no outliers within subgroups and no trends over time in subgroup means or within subgroup variances. Tests for outliers may be applied to historical subgroups, and trends may be identified by simple time plots of data or by statistical tests. Techniques for detecting trends and estimating their magnitude are discussed in Chapters 16 and 17.

For 24-h air quality data, Nelson, Armentrout, and Johnson (1980) suggest that each data set contain between 4 and 15 data and that at least 10 to 15 such data sets be available. They also suggest that the chosen time period (e.g., week or month) should relate to the National Ambient Air Quality Standards (NAAQS) of interest. They cite the example of using months or quarters for 24-h TSP, SO_2, and NO_2 data collected at 6- or 12-day intervals.

15.6.3 Construction of Control Charts

The information needed from historical data sets to construct control charts for means, ranges and standard deviations is

Data Set (Subgroup) i	Number of Data in the Data Set n_i	Sample Mean $\bar{x}_i$	Sample Range R_i	Sample Standard Deviation s_i
1	n_1	$\bar{x}_1$	R_1	s_1
2	n_2	$\bar{x}_2$	R_2	s_2
⋮	⋮	⋮	⋮	⋮
k	n_k	$\bar{x}_k$	R_k	s_k

The formulas for computing the center line and control limits for means control charts are given in Table 15.4, while those for range and standard deviation charts are in Table 15.5. Table 15.6 defines the quantities used in Tables 15.4 and 15.5.

The value of the constants d_2, d_3, and c_4 depend on n_i and are given in Table A17 for n_i from 2 to 25. The derivation and meaning of these constants is discussed by Burr (1976, Chapter 5) and American Society for Testing and Materials (1976). They were derived under the assumption that the data are normally distributed.

The quantity Z_p in the limit formulas is usually set equal to 2 or 3, although other values can be used. If the data are normally distributed, we find from Table A1 that, for a process in control, the probability of a plotted point falling outside the $Z_p = 2$ limit lines is 0.0456 (about 1 chance in 20). This probability is 0.0026 when $Z_p = 3$ is used (about three changes in 1000). The $Z_p = 2$ line might be used as a "warning" line, whereas the $Z_p = 3$ line could be used to indicate action of some kind.

Although environmental data are frequently nonnormally distributed, control charts constructed under the assumption of normal data are still useful for indicating when the measurements are not likely to be from the same distribution

Table 15.4 Formulas for Control Charts for Means

	Equal n_i	Unequal n_i
Center line	$\bar{\bar{x}} = \frac{1}{k} \sum_{i=1}^{k} \bar{x}_i$	$\bar{\bar{x}} = \frac{\sum_{i=1}^{k} n_i \bar{x}_i}{\sum_{i=1}^{k} n_i}$
Control limits[a]		
$n_i \leq 10$	$\bar{\bar{x}} \pm \frac{Z_p \bar{R}_1}{\sqrt{n}}$	$\bar{\bar{x}} \pm \frac{Z_p \bar{R}_1}{\sqrt{n_i}}$
$n_i \geq 2$	$\bar{\bar{x}} \pm \frac{Z_p \bar{s}_1}{\sqrt{n}}$	$\bar{\bar{x}} \pm \frac{Z_p \bar{s}_1}{\sqrt{n_i}}$

Source: Formulas from American Society for Testing and Materials, 1976. Formulas are defined in Table 15.6.

[a] Control limits computed using $\bar{s}_1$ are appropriate for any subgroup size $n_i \geq 2$. Control limits using $\bar{R}_1$ are easier to compute but are less efficient than those computed with $\bar{s}_1$ if most $n_i > 10$.

$$\bar{R}_1 = \frac{1}{k} \sum_{i=1}^{k} \frac{R_i}{d_{2i}}, \quad \bar{s}_1 = \frac{1}{k} \sum_{i=1}^{k} \frac{s_i}{c_{4i}}.$$

d_{2i} and d_{3i} are constants from Table A17 that depend on n_i. $c_{4i} = 1$ if $n_i > 25$.

Table 15.5 Formulas for Range and Standard Deviation Control Charts

	Equal n_i	*Unequal* n_i^a
Center line		
Range: Use when $n_i \le 10$	$\bar{R} = \frac{1}{k} \sum_{i=1}^{k} R_i$	$\bar{R}_{2i} = d_{2i}\bar{R}_1$
Standard deviation: Use when $n_i > 10$	$\bar{s} = \frac{1}{k} \sum_{i=1}^{k} s_i$	$\bar{s}_{2i} = c_{4i}\bar{s}_1$
Control limits[b]		
Range: Use when $n_i \le 10$	$\bar{R}\left(1 \pm \frac{Z_p d_3}{d_2}\right)$	$\bar{R}_{2i}\left(1 \pm \frac{Z_p d_{3i}}{d_{2i}}\right)$
Standard deviation: Use when $n_i > 10$	$\bar{s}\left[1 \pm Z_p \sqrt{\frac{1 - c_4^2}{c_4^2}}\right]$	$\bar{s}_{2i}\left[1 \pm Z_p \sqrt{\frac{1 - c_{4i}^2}{c_{4i}^2}}\right]$

Source: Formulas from American Society for Testing and Materials, 1976.

[a] Formulas for $\bar{R}_1$ and $\bar{s}_1$ are given in Table 15.4. d_2, d_3, and c_4 are constants from Table A17 that depend on n_i.

[b] Control limits computed using $\bar{s}$ or $\bar{s}_{2i}$ are appropriate for any subgroup size $n_i \ge 2$. Limits computed using $\bar{R}$ or $\bar{R}_{2i}$ are simpler to compute but are less efficient than those computed with $\bar{s}$ or $\bar{s}_{2i}$ if most $n_i > 10$.

Table 15.6 Definition of Quantities Used in the Control Chart Formulas in Tables 15.4 and 15.5

k = number of historical data sets (subgroups)

n_i = number of data in the ith subgroup

Z_p = 2 if 2-sigma control limit lines are desired

Z_p = 3 if 3-sigma control limit lines are desired

$\bar{\bar{x}}$ = grand average of all data over the k subgroups

$\bar{R}$ = average range for the k subgroups

$\bar{R}_1$ = estimator of the population standard deviation within subgroups when all n_i are not equal; $\bar{R}_1$ reduces to $\bar{R}/d_2$ when all n_i are equal

$\bar{R}_{2i}$ = approximate expression for the average range at time i when all n_i are not equal. $\bar{R}_{2i}$ reduces to $\bar{R}$ when all n_i are equal

$\bar{s}$ = average standard deviation for the k subgroups

$\bar{s}_1$ = estimator of the population standard deviation within subgroups when all n_i are not equal; $\bar{s}_1$ reduces to $\bar{s}/c_4$ when all n_i are equal

$\bar{s}_{2i}$ = approximate expression for the average standard deviation at time i when all n_i are not equal. $\bar{s}_{2i}$ reduces to $\bar{s}$ when $n_i > 25$ for each of the k subgroups or when all n_i are equal

d_2, d_3, c_4 = correction factors to improve the accuracy of the estimators; these factors (in Table A17) are appropriate when the data are normally distributed as in the past. Control charts are not constructed for the purpose of making precise probability statements. They are constructed as a guide for when investigative action is needed. If the data are known or suspected of being lognormally distributed, then control charts should be constructed with logarithms of the data in the formulas in Tables 15.4 and 15.5. Alternatively, control chart limits in the original (untransformed) scale can be constructed for lognormal data by the methods of Morrison (1958) or Ferrell (1958).

EXAMPLE 15.2

Suppose total suspended particulate (TSP) air measurements have been taken for several months at a site and that control charts are desired to look for outliers and changes in the mean and the standard deviation over time. The first step is to see if historical data sets indicate that the population is homogeneous, that is, to see if the TSP time process is in control. This step is done by constructing a control chart with the historical data.

For illustration purposes we use TSP data listed by Nelson, Armentrout, and Johnson (1980, Table 3-11, Site 14) that were collected every sixth day (with occasional missing values) for 12 months at a particular site. These data are listed in Table 15.7. The observations taken within a month are considered to be a rational subgroup. For illustration purposes we use the first 7 months of data to construct control charts for the mean $\bar{x}_i$ and range R_i. The reader is asked to construct the standard deviation control chart in Exercise 15.2.

The values of n_i, $\bar{x}_i$, R_i, and s_i for the subgroups are given in Table 15.7. Using the first 7 months of data, we compute in Table 15.8 the center line and control limits for the mean and range control charts, using the formulas in Tables 15.4 and 15.5. The $Z_p = 2$ and 3 limit lines are drawn in Figures 15.3 and 15.4 and the means and ranges plotted for the 12-month period.

We see from Figure 15.3 that $\bar{x}_6$ falls just outside the 2-sigma limit, and $\bar{x}_1$ and $\bar{x}_4$ are very close to the limit. However, no $\bar{x}_i$

Table 15.7 TSP Measurements (μg/m^3) Taken at Six-Day Intervals at a Site

	Subgroup (month)	n_i	*TSP Concentrations ($\mu g/m^3$)*	*Mean* $\bar{x}_i$	*Range* R_i	*Standard Deviation* s_i
Historical data	1	4	31, 34, 13, 40	29.5	27	11.62
	2	5	49, 19, 79, 39, 51	47.4	60	21.74
	3	5	46, 72, 49, 72, 33	54.4	39	17.16
	4	4	18, 24, 24, 47	28.2	29	12.82
	5	5	32, 66, 28, 68, 74	53.6	46	21.79
	6	5	83, 80, 37, 69, 55	64.8	46	19.03
	7	2	28, 51	39.5	23	16.26
	8	5	42, 53, 56, 129, 64	68.8	87	34.56
	9	5	46, 46, 57, 26, 41	43.2	31	11.26
	10	5	41, 90, 31, 63, 37	52.4	59	24.24
	11	5	69, 108, 37, 47, 43	60.8	71	29.02
	12	5	26, 25, 45, 39, 23	31.6	22	9.79

Source: After Nelson, Armentrout, and Johnson, 1980, Table 3-11, Site 14.

Table 15.8 Computations for Means and Range Control Charts Using the First Seven Subgroups of Data in Table 15.7

A. Means Control Chart

$$\bar{\bar{x}} = \frac{1}{30}\,[4(29.5) + 5(47.4) + \cdots + 2(39.5)] = 47$$

$$\bar{R}_1 = \frac{1}{7}\left(\frac{27}{2.059} + \frac{60}{2.326} + \cdots + \frac{23}{1.128}\right) = 18.5$$

				Limit Lines $\bar{\bar{x}} \pm Z_p\bar{R}_1/\sqrt{n_i}$			
			Center Line	$Z_p = 2$		$Z_p = 3$	
Months	n_i	R_i	$\bar{\bar{x}}$	*Lower*	*Upper*	*Lower*	*Upper*
1,4	4	18.5	47	28	66	19	75
2,3,5,6	5	18.5	47	30	64	22	72
7	2	18.5	47	21	73	7.7	86

B. Range Control Chart

					Limit Lines $\bar{R}_{2i}\ (1 \pm Z_pd_{3i}/d_{2i})$			
				Center Line	$Z_p = 2$		$Z_p = 3$	
Months	n_i	d_{2i}	d_{3i}	$\bar{R}_{2i} = d_{2i}\bar{R}_1$	*Lower*	*Upper*	Lower	Upper
1,4	4	2.059	0.880	38.2	5.5	71	0	87
2,3,5,6	5	2.326	0.864	43.1	11	75	0	91
7	2	1.128	0.853	20.9	0	52	0	68

exceeds the 3-sigma limit. These results show that the average monthly TSP concentrations can deviate substantially from the 7-month average, but no wild swings were present during that time period. From Figure 15.4 we see that none of the R_i during the first 7 months fall outside the 2-sigma limits. Taken together, these results suggest that the control charts constructed during the first 7 months of data will be useful for evaluating future TSP data sets.

Plotting the next 5 months of $\bar{x}_i$ and R_i on the 2 control charts, we see that both $\bar{x}_8$ and R_8 exceed their 2-sigma limits and $\bar{x}_{11}$ and R_{11} are very close to their 2-sigma limits. Hence the data for month 8 and possibly month 11 may contain outliers. Considering the full 12 months of data in the 2 control charts, we see that there appears to be no evidence for a permanent shift in either mean or range over the year.

Control charts should be updated periodically when there is no evidence of a trend or a permanent shift in mean or range. In this example it would be appropriate to treat the full 12 months of data as a new historical data set and to recompute the center line and limits for the 2 control charts. These values could then be used during the coming year to look for outliers and changes.

15.6.4 Seasonality

When mean pollution levels fluctuate or cycle over time, this component of variation must be properly handled when constructing control charts. Since the

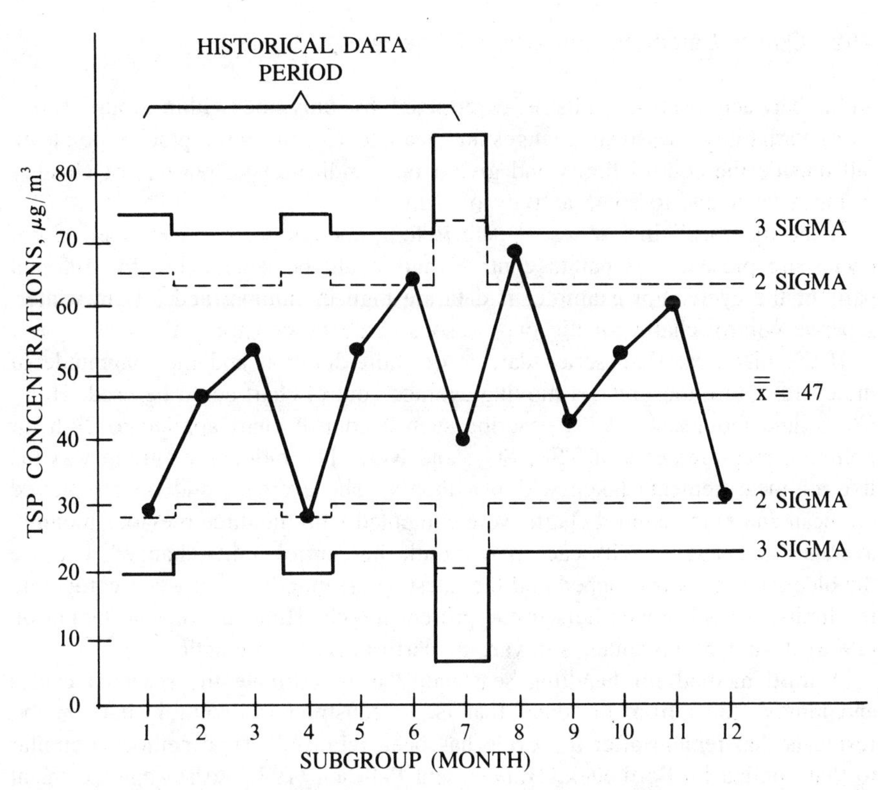

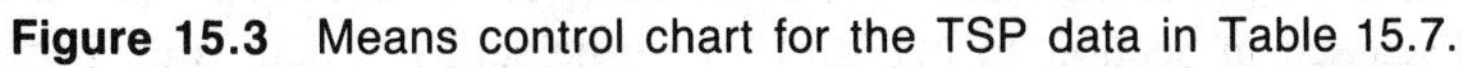
Figure 15.3 Means control chart for the TSP data in Table 15.7.

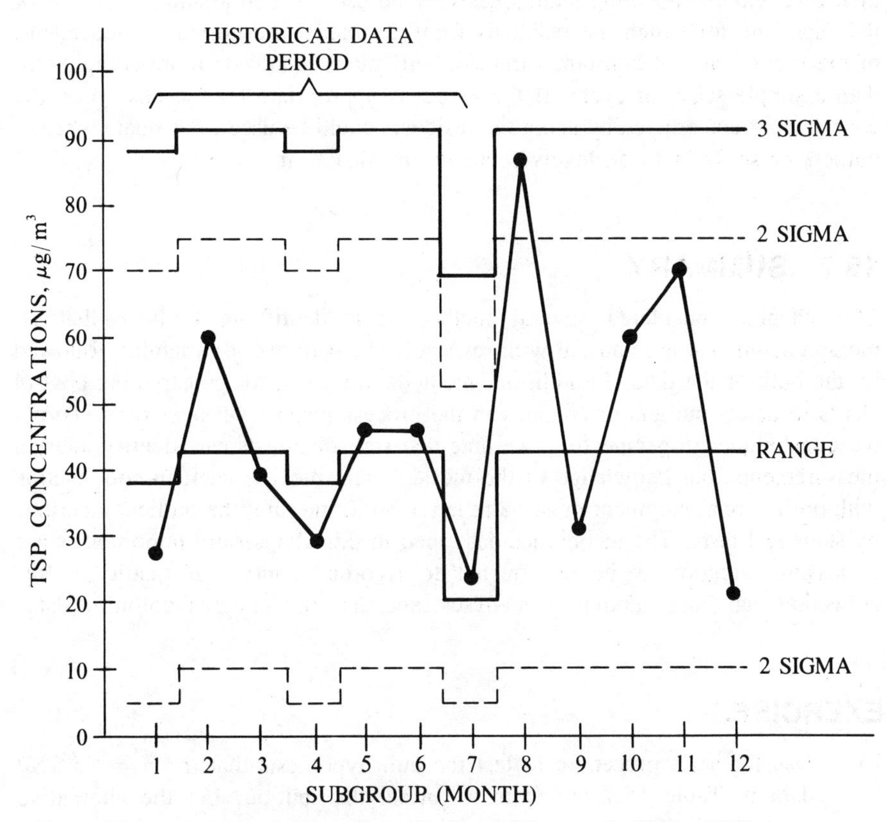

Figure 15.4 Range control chart for the TSP data in Table 15.7.

width between control limits is determined by only the within-group (short-term) variability, seasonal changes in mean level can cause plotted points to fall outside the control limits and give a false indication of outliers or changes in mean level due to some activity of man.

If the historical time series of data is long enough so that a number of full cycles are present, a separate control chart could be constructed for different parts of the cycle. For example, if data are high in summer and low in winter, separate control charts for the two seasons could be constructed.

If the historical time series data is of short duration and the magnitude of the cycles is not too great, a moving-average control chart could be used. Hunt, Clark, and Goranson (1978) describe such a control chart applied to 24-h air pollution measurements of TSP, SO_2, and NO_2. The rational subgroup was the five air measurements taken within a month. The averages and ranges plotted on mean and range control charts were computed with the three previous monthly averages and ranges. With each new month the control limits changed, because the oldest month was dropped and the latest (past) month added when computing the limits, to look for outliers in the present month. Hunt, Clark, and Goranson (1978) developed computer software to perform these calculations.

A third method for handling seasonality is to estimate the seasonal cycles and remove them from the data, that is, to construct the control chart on the residuals that remain after the cycle has been removed. This method is similar to that applied by Berthouex, Hunter, and Pallesen (1978) to sewage treatment plant data discussed in Section 15.6.1. If a long time series of data is available, then Box-Jenkins modeling techniques may be used to find a suitable model for the data, one for which the residuals from the model are normal, independent, of mean zero, and of constant variance. This model could be more complicated than a simple seasonal cycle. If the model is appropriate for the next year, the control chart constructed by using the residuals could be used to evaluate whether outliers or shifts in mean levels occur during that year.

15.7 SUMMARY

This chapter considered several methods for identifying outliers—that is, measurements that are unusual with respect to the patterns of variability followed by the bulk of the data. In addition, methods are given for constructing control charts to detect outliers or changes in the process mean level over time. To put these techniques in perspective, we note that statistical tests can identify unusual measurements, but knowledge of the measurement process itself in combination with professional judgment must be relied upon to interpret the outliers identified by statistical tests. The techniques described in this chapter are important, since increasing attention is being directed to rigorous control of quality in the collection, handling, laboratory analyses, and data reduction of pollution data.

EXERCISES

15.1 Use Rosner's procedure to test the null hypothesis that the $n = 55$ TSP data in Table 15.7 are from a normal distribution. Let the alternative hypothesis be that there are at most three outliers present. Test at the $\alpha = 0.05$ significance level.

15.2 Using the TSP data for the first 7 months given in Table 15.7, construct a standard deviation control chart and plot $s_1, s_2, \cdots, s_{12}$ on the chart. What do you conclude?

ANSWERS

15.1 Using the data in Table 15.7, we compute

i	$n - i$	$\bar{x}^{(i)}$	$s_x^{(i)}$	$x^{(i)}$	$R_{x,\,i+1} = \lvert x^{(i)} - \bar{x}^{(i)}\rvert / s_x^{(i)}$	λ_{i+1} ($\alpha = 0.05$)
0	55	49.0	22.9	129	3.49	3.165
1	54	47.5	20.3	108	2.98	3.155
2	53	46.4	18.6	90	2.34	3.150

Since $R_{x,3} < 3.150$ and $R_{x,2} < 3.155$, we cannot reject the null hypothesis of no outliers versus the alternative hypothesis of 2 or 1 outlier, respectively. Since $R_{x,1} > 3.165$, we reject the null hypothesis and conclude that datum 129 is an outlier from the assumed normal distribution.

15.2

$$\bar{s}_1 = \frac{1}{7}\left[\frac{11.62}{0.9213} + \frac{21.74}{0.94} + \frac{17.16}{0.94} + \frac{12.82}{0.9213} + \frac{21.79}{0.94} + \frac{19.03}{0.94} + \frac{16.26}{0.7979}\right] = 18.8164$$

Month	n_i	c_{4i}	$\sqrt{(1 - c_{4i}^2)/c_{4i}^2}$	Center Line $\bar{s}_{2i} = c_{4i}\bar{s}_1$	$\bar{s}_{2i}\left[1 + Z_p\sqrt{(1 - c_{4i}^2)/c_{4i}^2}\right]$ $Z_p = 2$ Lower	Upper	$Z_p = 3$ Lower	Upper
1,4	4	0.9213	0.4221	17.3	2.7	32	0	39
2,3,5,6	5	0.94	0.3630	17.7	4.8	31	0	37
7	2	0.7979	0.7555	15.0	0	38	0	49

Conclusion: S_8 exceeds the upper 2-sigma limit, and s_{11} is slightly less than that limit. These same results were obtained by using the range control chart (Example 15.2).

16 Detecting and Estimating Trends

An important objective of many environmental monitoring programs is to detect changes or trends in pollution levels over time. The purpose may be to look for increased environmental pollution resulting from changing land use practices such as the growth of cities, increased erosion from farmland into rivers, or the startup of a hazardous waste storage facility. Or the purpose may be to determine if pollution levels have declined following the initiation of pollution control programs.

The first sections of this chapter discuss types of trends, statistical complexities in trend detection, graphical and regression methods for detecting and estimating trends, and Box-Jenkins time series methods for modeling pollution processes. The remainder of the chapter describes the Mann-Kendall test for detecting monotonic trends at single or multiple stations and Sen's (1968b) nonparametric estimator of trend (slope). Extensions of the techniques in this chapter to handle seasonal effects are given in Chapter 17. Appendix B lists a computer code that computes the tests and trend estimates discussed in Chapters 16 and 17.

16.1 TYPES OF TRENDS

Figure 16.1 shows some common types of trends. A sequence of measurements with no trend is shown in Figure 16.1(a). The fluctuations along the sequence are due to random (unassignable) causes. Figure 16.1(b) illustrates a cyclical pattern wih no long-term trend, and Figure 16.1(c) shows random fluctuations about a rising linear trend line. Cycles may be caused by many factors including seasonal climatic changes, tides, changes in vehicle traffic patterns during the day, production schedules of industry, and so on. Such cycles are not "trends" because they do not indicate long-term change. Figure 16.1(d) shows a cycle with a rising long-term trend with random fluctuation about the cycle.

Frequently, pollution measurements taken close together in time or space are positively correlated, that is, high (low) values are likely to be followed by other high (low) values. This distribution is illustrated in Figure 16.1(e) for a rising trend but can occur for the other situations in Figure 16.1.

Figure 16.1(f) depicts a random sequence with a short-lived impulse of pollution. A permanent step change is illustrated in Figure 16.1(g). This latter type could be due to a new pollution abatement program, such as a water

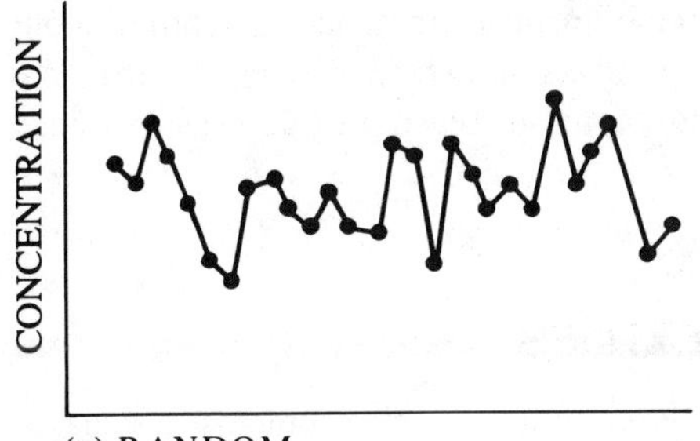

(a) RANDOM

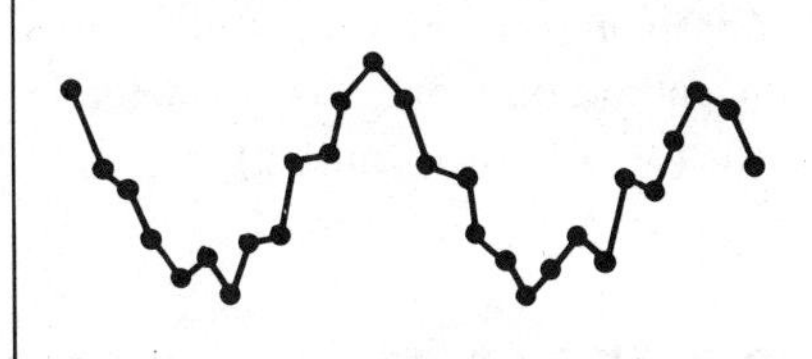

(b) CYCLE + RANDOM

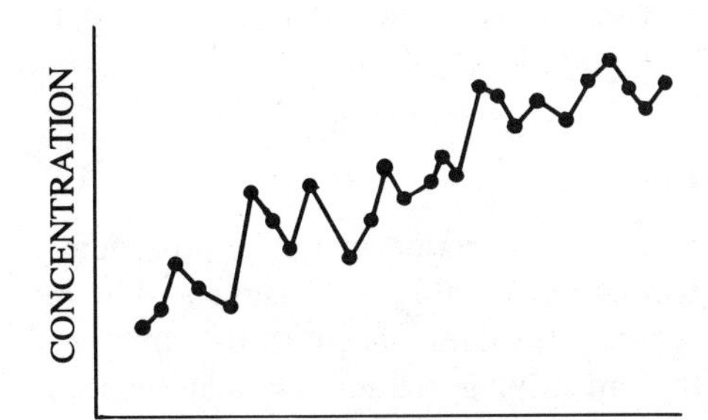

(c) TREND + RANDOM

(d) TREND + CYCLE + RANDOM

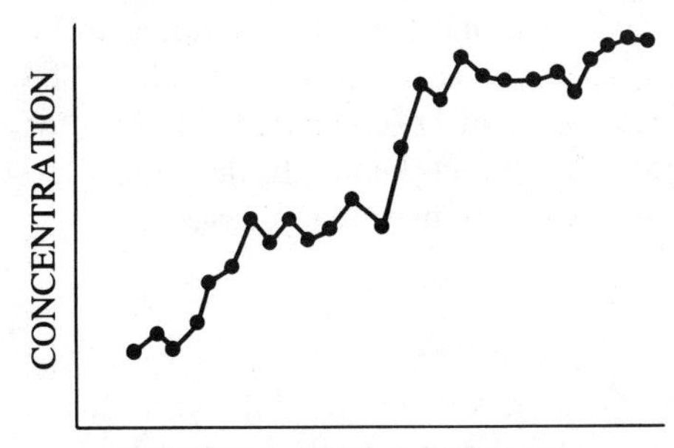

(e) TREND + NON-RANDOM

(f) RANDOM WITH IMPULSE

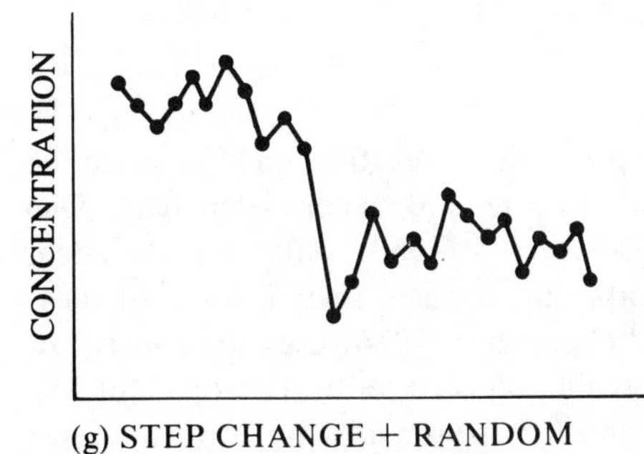

(g) STEP CHANGE + RANDOM

(h) RANDOM FOLLOWED BY TREND

Figure 16.1 Types of time series.

treatment plant. Finally, a sequence of random measurements fluctuating about a constant level may be followed by a trend as shown in Figure 16.1(h). We concentrate here on tests for detecting monotonic increasing or decreasing trends as in (c), (d), (e), and (h).

16.2 STATISTICAL COMPLEXITIES

The detection and estimation of trends is complicated by problems associated with characteristics of pollution data. In this section we review these problems, suggest approaches for their alleviation, and reference pertinent literature for additional information. Harned et al. (1981) review the literature dealing with statistical design and analysis aspects of detecting trends in water quality. Munn (1981) reviews methods for detecting trends in air quality data.

16.2.1 Changes in Procedures

A change of analytical laboratories or of sampling and/or analytical procedures may occur during a long-term study. Unfortunately, this may cause a shift in the mean or in the variance of the measured values. Such shifts could be incorrectly attributed to changes in the underlying natural or man-induced processes generating the pollution.

When changes in procedures or laboratories occur abruptly, there may not be time to conduct comparative studies to estimate the magnitude of shifts due to these changes. This problem can sometimes be avoided by preparing duplicate samples at the time of sampling: one is analyzed and the other is stored to be analyzed if a change in laboratories or procedures is introduced later. The paired, old-new data on duplicate samples can then be compared for shifts or other inconsistencies. This method assumes that the pollutants in the sample do not change while in storage, an unrealistic assumption in many cases.

16.2.2 Seasonality

The variation added by seasonal or other cycles makes it more difficult to detect long-term trends. This problem can be alleviated by removing the cycle before applying tests or by using tests unaffected by cycles. A simple nonparametric test for trend using the first approach was developed by Sen (1968a). The seasonal Kendall test, discussed in Chapter 17, uses the latter approach.

16.2.3 Correlated Data

Pollution measurements taken in close proximity over time are likely to be positively correlated, but most statistical tests require uncorrelated data. One approach is to use test statistics developed by Sen (1963, 1965) for dependent data. However, Lettenmaier (1975) reports that perhaps several hundred measurements are needed for their validity. Lettenmaier (1976) uses the concept of effective independent sample size to obtain adjusted critical values for the Wilcoxon rank sum test for a step trend and for Spearman's rho correlation test for a linear trend. Montgomery and Reckhow (1984) illustrate his procedure and provide tables of adjusted critical values for the Wilcoxon rank sum and Spearman tests. Their paper summarizes the latest statistical techniques for trend detection.

16.2.4 Corrections for Flow

The detection of trends in stream water quality is more difficult when concentrations are related to stream flow, the usual situation. Smith, Hirsch, and Slack (1982) obtain flow-adjusted concentrations by fitting a regression equation to the concentration-flow relationship. Then the residuals from regression are tested for trend by the seasonal Kendall test discussed in Chapter 17. Harned, Daniel, and Crawford (1981) illustrate two alternative methods, discharge compensation and discharge-frequency weighting. Methods for adjusting ambient air quality levels for meteorological effects are discussed by Zeldin and Meisel (1978).

16.3 METHODS

16.3.1 Graphical

Graphical methods are very useful aids to formal tests for trends. The first step is to plot the data against time of collection. Velleman and Hoaglin (1981) provide a computer code for this purpose, which is designed for interactive use on a computer terminal. They also provide a computer code for ''smoothing'' time series to point out cycles and/or long-term trends that may otherwise be obscured by variability in the data.

Cumulative sum (CUSUM) charts are also an effective graphical tool. With this method changes in the mean are detected by keeping a cumulative total of deviations from a reference value or of residuals from a realistic stochastic model of the process. Page (1961, 1963), Ewan (1963), Gibra (1975), Wetherill (1977), Berthouex, Hunter, and Pallesen (1978), and Vardeman and David (1984) provide details on the method and additional references.

16.3.2 Regression

If plots of data versus time suggest a simple linear increase or decrease over time, a linear regression of the variable against time may be fit to the data. A t test may be used to test that the true slope is not different from zero; see, for example, Snedecor and Cochran (1980, p. 155). This t test can be misleading if seasonal cycles are present, the data are not normally distributed, and/or the data are serially correlated. Hirsch, Slack, and Smith (1982) show that in these situations, the t test may indicate a significant slope when the true slope actually is zero. They also examine the performance of linear regression applied to deseasonalized data. This procedure (called *seasonal regression*) gave a t test that performed well when seasonality was present, the data were normally distributed, and serial correlation was absent. Their results suggest that the seasonal Kendall test (Chapter 17) is preferred to the standard or seasonal regression t tests when data are skewed, cyclic, and serially correlated.

16.3.3 Intervention Analysis and Box- Jenkins Models

If a long time sequence of equally spaced data is available, intervention analysis may be used to detect changes in average level resulting from a natural or man-induced intervention in the process. This approach, developed by Box and Tiao (1975), is a generalization of the autoregressive integrated moving-average (ARIMA) time series models described by Box and Jenkins (1976). Lettenmaier and Murray (1977) and Lettenmaier (1978) study the power of the method to detect trends. They emphasize the design of sampling plans to detect impacts from polluting facilities. Examples of its use are in Hipel et al. (1975) and Roy and Pellerin (1982).

Box-Jenkins modeling techniques are powerful tools for the analysis of time series data. McMichael and Hunter (1972) give a good introduction to Box-Jenkins modeling of environmental data, using both deterministic and stochastic components to forecast temperature flow in the Ohio River. Fuller and Tsokos (1971) develop models to forecast dissolved oxygen in a stream. Carlson, MacCormick, and Watts (1970) and McKerchar and Delleur (1974) fit Box-Jenkins models to monthly river flows. Hsu and Hunter (1976) analyze annual series of air pollution SO_2 concentrations. McCollister and Wilson (1975) forecast daily maximum and hourly average total oxidant and carbon monoxide concentrations in the Los Angeles Basin. Hipel, McLeod, and Lennox (1977*a*, 1977*b*) illustrate improved Box-Jenkins techniques to simplify model construction. Reinsel et al. (1981*a*, 1981*b*) use Box-Jenkins models to detect trends in stratospheric ozone data. Two introductory textbooks are McCleary and Hay (1980) and Chatfield (1984). Box and Jenkins (1976) is recommended reading for all users of the method.

Disadvantages of Box-Jenkins methods are discussed by Montgomery and Johnson (1976). At least 50 and preferably 100 or more data collected at equal (or approximately equal) time intervals are needed. When the purpose is forecasting, we must assume the developed model applies to the future. Missing data or data reported as trace or less-than values can prevent the use of Box-Jenkins methods. Finally, the modeling process is often nontrivial, with a considerable investment in time and resources required to build a satisfactory model. Fortunately, there are several packages of statistical programs that contain codes for developing time series models, including Minitab (Ryan, Joiner, and Ryan 1982), SPSS (1985), BMDP (1983), and SAS (1985). Codes for personal computers are also becoming available.

16.4 MANN-KENDALL TEST

In this section we discuss the nonparametric Mann-Kendall test for trend (Mann, 1945; Kendall, 1975). This procedure is particularly useful since missing values are allowed and the data need not conform to any particular distribution. Also, data reported as trace or less than the detection limit can be used (if it is acceptable in the context of the population being sampled) by assigning them a common value that is smaller than the smallest measured value in the data set. This approach can be used because the Mann-Kendall test (and the seasonal Kendall test in Chapter 17) use only the relative magnitudes of the data rather than their measured values. We note that the Mann-Kendall test can be viewed as a nonparametric test for zero slope of the linear regression of time-ordered data versus time, as illustrated by Hollander and Wolfe (1973, p. 201).

16.4.1 Number of Data 40 or Less

If n is 40 or less, the procedure in this section may be used. When n exceeds 40, use the normal approximation test in Section 16.4.2. We begin by considering the case where only one datum per time period is taken, where a time period may be a day, week, month, and so on. The case of multiple data values per time period is discussed in Section 16.4.3.

The first step is to list the data in the order in which they were collected over time: $x_1, x_2, \ldots, x_n$, where x_i is the datum at time i. Then determine the sign of all $n(n - 1)/2$ possible differences $x_j - x_k$, where $j > k$. These differences are $x_2 - x_1, x_3 - x_1, \ldots, x_n - x_1, x_3 - x_2, x_4 - x_2, \ldots, x_n - x_{n-2}, x_n - x_{n-1}$. A convenient way of arranging the calculations is shown in Table 16.1.

Let $\text{sgn}(x_j - x_k)$ be an indicator function that takes on the values 1, 0, or -1 according to the sign of $x_j - x_k$:

$$\begin{aligned} \text{sgn}(x_j - x_k) &= 1 \quad \text{if } x_j - x_k > 0 \\ &= 0 \quad \text{if } x_j - x_k = 0 \\ &= -1 \quad \text{if } x_j - x_k < 0 \end{aligned} \qquad \mathbf{16.1}$$

Then compute the Mann-Kendall statistic

$$S = \sum_{k=1}^{n-1} \sum_{j=k+1}^{n} \text{sgn}(x_j - x_k) \qquad \mathbf{16.2}$$

which is the number of positive differences minus the number of negative differences. These differences are easily obtained from the last two columns of Table 16.1. If S is a large positive number, measurements taken later in time tend to be larger than those taken earlier. Similarly, if S is a large negative number, measurements taken later in time tend to be smaller. If n is large, the computer code in Appendix B may be used to compute S. This code also computes the tests for trend discussed in Chapter 17.

Suppose we want to test the null hypothesis, H_0, of no trend against the alternative hypothesis, H_A, of an upward trend. Then H_0 is rejected in favor of H_A if S is positive and if the probability value in Table A18 corresponding to the computed S is less than the a priori specified α significance level of the test. Similarly, to test H_0 against the alternative hypothesis H_A of a downward trend, reject H_0 and accept H_A if S is negative and if the probability value in the table corresponding to the absolute value of S is less than the a priori specified α value. If a two-tailed test is desired, that is, if we want to detect either an upward or downward trend, the tabled probability level corresponding to the absolute value of S is doubled and H_0 is rejected if that doubled value is less than the a priori α level.

EXAMPLE 16.1

We wish to test the null hypothesis H_0, of no trend versus the alternative hypothesis, H_A, of an upward trend at the $\alpha = 0.10$

Table 16.1 Differences in Data Values Needed for Computing the Mann-Kendall Statistic S to Test for Trend

Data Values Listed in the Order Collected Over Time								
x_1	x_2	x_3	x_4	$\cdots$	x_{n-1}	x_n	*No. of + Signs*	*No. of − Signs*
	$x_2 - x_1$	$x_3 - x_1$	$x_4 - x_1$	$\cdots$	$x_{n-1} - x_1$	$x_n - x_1$		
		$x_3 - x_2$	$x_4 - x_2$	$\cdots$	$x_{n-1} - x_2$	$x_n - x_2$		
			$x_4 - x_3$	$\cdots$	$x_{n-1} - x_3$	$x_n - x_3$		
				$\ddots$	$\vdots$	$\vdots$		
					$x_{n-1} - x_{n-2}$	$x_n - x_{n-2}$		
						$x_n - x_{n-1}$		
						$S =$	$\overline{\begin{pmatrix}\text{sum of}\\ +\text{ signs}\end{pmatrix}}$	$+\ \overline{\begin{pmatrix}\text{sum of}\\ -\text{ signs}\end{pmatrix}}$

Table 16.2 Computation of the Mann-Kendall Trend Statistic S for the Time Ordered Data Sequence 10, 15, 14, 20

Time *Data*	*1* *10*	*2* *15*	*3* *14*	*4* *20*		*No. of +* *Signs*		*No. of −* *Signs*
		15 − 10	14 − 10	20 − 10		3		0
			14 − 15	20 − 15		1		1
				20 − 14		1		0
				S	=	5	−	1 = 4

significance level. For ease of illustration suppose only 4 measurements are collected in the following order over time or along a line in space: 10, 15, 14, and 20. There are 6 differences to consider: 15 − 10, 14 − 10, 20 − 10, 14 − 15, 20 − 15, and 20 − 14. Using Eqs. 16.1 and 16.2, we obtain $S = +1 + 1 + 1 - 1 + 1 + 1 = +4$, as illustrated in Table 16.2. (Note that the sign, not the magnitude of the difference is used.) From Table A18 we find for $n = 4$ that the tabled probability for $S = +4$ is 0.167. This number is the probability of obtaining a value of S equal to +4 or larger when $n = 4$ and when no upward trend is present. Since this value is greater than 0.10, we cannot reject H_0.

If the data sequence had been 18, 20, 23, 35, then $S = +6$, and the tabled probability is 0.042. Since this value is less than 0.10, we reject H_0 and accept the alternative hypothesis of an upward trend.

Table A18 gives probability values only for $n \leq 10$. An extension of this table up to $n = 40$ is given in Table A.21 in Hollander and Wolfe (1973).

16.4.2 Number of Data Greater Than 40

When n is greater than 40, the normal approximation test described in this section is used. Actually, Kendall (1975, p. 55) indicates that this method may be used for n as small as 10 unless there are many tied data values. The test procedure is to first compute S using Eq. 16.2 as described before. Then compute the variance of S by the following equation, which takes into account that ties may be present:

$$\mathrm{VAR}(S) = \frac{1}{18}\left[n(n-1)(2n+5) - \sum_{p=1}^{q} t_p(t_p - 1)(2t_p + 5)\right] \qquad \textbf{16.3}$$

where g is the number of tied groups and t_p is the number of data in the pth group. For example, in the sequence {23, 24, trace, 6, trace, 24, 24, trace, 23} we have $g = 3$, $t_1 = 2$ for the tied value 23, $t_2 = 3$ for the tied value 24, and $t_3 = 3$ for the three trace values (considered to be of equal but unknown value less than 6).

Then S and VAR(S) are used to compute the test statistic Z as follows:

$$Z = \frac{S-1}{[\mathrm{VAR}(S)]^{1/2}} \quad \text{if } S > 0$$

$$= 0 \quad \text{if } S = 0$$

$$= \frac{S+1}{[\mathrm{VAR}(S)]^{1/2}} \quad \text{if } S < 0 \qquad \textbf{16.4}$$

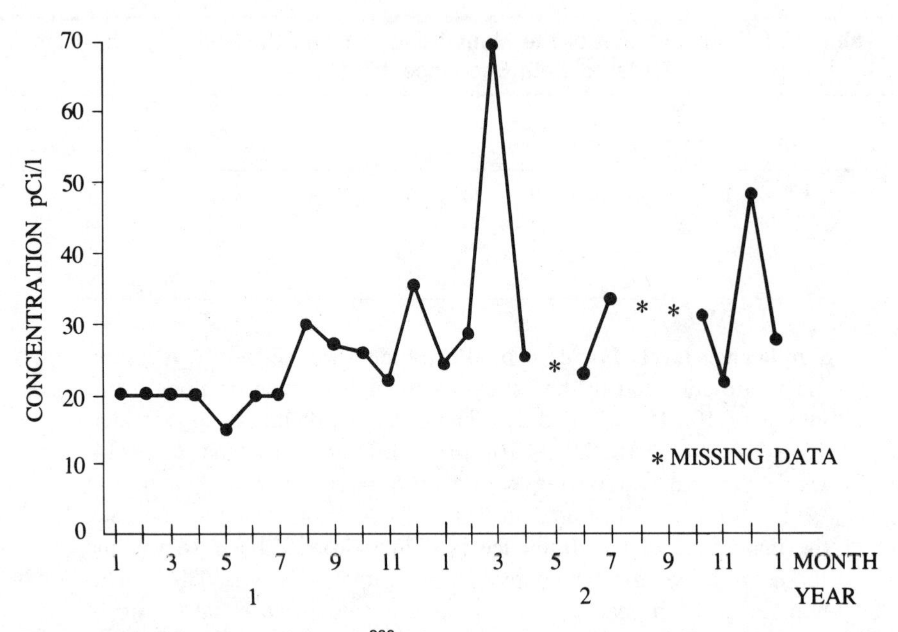

Figure 16.2 Concentrations of ^{238}U in ground water in well E at the former St. Louis Airport storage site for January 1981 through January 1983 (after Clark and Berven, 1984).

A positive (negative) value of Z indicates an upward (downward) trend. If the null hypothesis, H_0, of no trend is true, the statistic Z has a standard normal distribution, and hence we use Table A1 to decide whether to reject H_0. To test for either upward or downward trend (a two-tailed test) at the α level of significance, H_0 is rejected if the absolute value of Z is greater than $Z_{1-\alpha/2}$, where $Z_{1-\alpha/2}$ is obtained from Table A1. If the alternative hypothesis is for an upward trend (a one-tailed test), H_0 is rejected if Z (Eq. 16.4) is greater than $Z_{1-\alpha}$. We reject H_0 in favor of the alternative hypothesis of a downward trend if Z is negative and the absolute value of Z is greater than $Z_{1-\alpha/2}$. Kendall (1975) indicates that using the standard normal tables (Table A1) to judge the statistical significance of the Z test will probably introduce little error as long as $n \geq 10$ unless there are many groups of ties and many ties within groups.

EXAMPLE 16.2

Figure 16.2 is a plot of $n = 22$ monthly ^{238}U concentrations x_1, x_2, $x_3, \ldots, x_{22}$ obtained from a groundwater monitoring well from January 1981 through January 1983 (reported in Clark and Berven, 1984). We use the Mann-Kendall procedure to test the null hypothesis at the $\alpha = 0.05$ level that there is no trend in ^{238}U groundwater concentrations at this well over this 2-year period. The alternative hypothesis is that an upward trend is present.

There are $n(n - 1)/2 = 22(21)/2 = 231$ differences to examine for their sign. The computer code in Appendix B was used to obtain S and Z (Eqs. 16.2 and 16.4). We find that $S = +108$. Since there are 6 occurrences of the value 20 and 2 occurrences of both 23 and 30, we have $g = 3$, $t_1 = 6$, and $t_2 = t_3 = 2$. Hence, Eq. 16.3 gives

$$\text{VAR}(S) = \tfrac{1}{18}[22(21)(44 + 5) - 6(5)(12 + 5) - 2(1)(4 + 5) - 2(1)(4 + 5)] = 1227.33$$

or $[\text{VAR}(S)]^{1/2} = 35.0$. Therefore, since $S > 0$, Eq. 16.4 gives $Z = (108 - 1)/35.0 = 3.1$. From Table A1 we find $Z_{0.95} = 1.645$. Since Z exceeds 1.645, we reject H_0 and accept the alternative hypothesis of an upward trend. We note that the three missing values in Figure 16.2 do not enter into the calculations in any way. They are simply ignored and constitute a regrettable loss of information for evaluating the presence of trend.

16.4.3 Multiple Observations per Time Period

When there are multiple observations per time period, there are two ways to proceed. First, we could compute a summary statistic, such as the median, for each time period and apply the Mann-Kendall test to the medians. An alternative approach is to consider the $n_i \geq 1$ multiple observations at time i (or time period i) as ties in the time index. For this latter case the statistic S is still computed by Eq. 16.2, where n is now the sum of the n_i, that is, the total number of observations rather than the number of time periods. The differences between data obtained at the same time are given the score 0 no matter what the data values may be, since they are tied in the time index.

When there are multiple observations per time period, the variance of S is computed by the following equation, which accounts for ties in the time index:

$$\text{VAR}(S) = \frac{1}{18}\left[n(n-1)(2n+5) - \sum_{p=1}^{g} t_p(t_p - 1)(2t_p + 5) - \sum_{q=1}^{h} u_q(u_q - 1)(2u_q + 5)\right] + \frac{\sum_{p=1}^{g} t_p(t_p - 1)(t_p - 2) \sum_{q=1}^{h} u_q(u_q - 1)(u_q - 2)}{9n(n-1)(n-2)} + \frac{\sum_{p=1}^{g} t_p(t_p - 1) \sum_{q=1}^{h} u_q(u_q - 1)}{2n(n-1)} \qquad \textbf{16.5}$$

where g and t_p are as defined following Eq. 16.3, h is the number of time periods that contain multiple data, and u_q is the number of multiple data in the qth time period. Equation 16.5 reduces to Eq. 16.3 when there is one observation per time period.

Equations 16.3 and 16.5 assume all data are independent and, hence, uncorrelated. If observations taken during the same time period are highly correlated, it may be preferable to apply the Mann-Kendall test to the medians of the data in each time period rather than use Eq. 16.5 in Eq. 16.4.

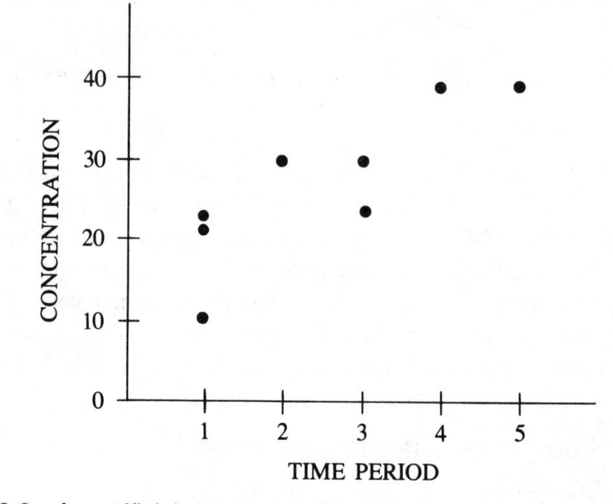

Figure 16.3 An artificial data set to illustrate the Mann-Kendall test for trend when ties in both the data and time are present.

EXAMPLE 16.3

To illustrate the computation of S and VAR(S), consider the following artificial data set:

(concentration, time period)

= (10, 1), (22, 1), (21, 1), (30, 2), (22, 3), (30, 3), (40, 4), (40, 5)

as plotted in Figure 16.3. There are 5 time periods and $n = 8$ data. To illustrate computing S, we lay out the data as follows:

Time Period :	1	1	1	2	3	3	4	5
Data :	10	22	21	30	22	30	40	40

We shall test at the $\alpha = 0.05$ level the null hypothesis, H_0, of no trend versus the alternative hypothesis, H_A, of an upward trend, a one-tailed test.

Now, look at all 8(7)/2 = 28 possible data pairs, remembering to give a score of 0 to the 4 pairs within the same time index. The differences are shown in Table 16.3. Ignore the magnitudes of the differences, and sum the number of positive and negative signs to obtain $S = 19$. It is clear from Figure 16.3 that there are $g = 3$ tied data groups (22, 30, and 40) with $t_1 = t_2 = t_3 = 2$. Also, there are $h = 2$ time index ties (times 1 and 3) with $u_1 = 3$ and $u_2 = 2$. Hence, Eq. 16.5 gives

$$\text{VAR}(S) = \frac{1}{18}[8(7)(16 + 5) - 3(2)(1)(4 + 5) - 3(2)(6 + 5)$$

$$- 2(1)(4 + 5)] + 0 + \frac{[3(2)(1)][3(2) + 2(1)]}{2(8)(7)}$$

$$= 58.1$$

or $[\text{VAR}(S)]^{1/2} = 7.6$. Hence, Eq. 16.4 gives $Z = (19 - 1)/7.6$

Table 16.3 Illustration of Computing S for Example 16.3

Time Period	*1*	*1*	*1*	*2*	*3*	*3*	*4*	*5*	*Sum of +*	*Sum of −*
Data	*10*	*22*	*21*	*30*	*22*	*30*	*40*	*40*	*Signs*	*Signs*
		NC	NC	+20	+12	+20	+30	+30	5	0
			NC	+8	0	+8	+18	+18	4	0
				+9	+1	+9	+19	+19	5	0
					−8	0	+10	+10	2	1
						NC	+18	+18	2	0
							+10	+10	2	0
								0	0	0
								S	= 20	− 1
									= 19	

NC = Not computed since both data values are within the same time period.

= 2.4. Referring to Table A1, we find $Z_{0.95} = 1.645$. Since $Z > 1.645$, reject H_0 and accept the alternative hypothesis of an upward trend.

16.4.4 Homogeneity of Stations

Thus far only one station has been considered. If data over time have been collected at $M > 1$ stations, we have data as displayed in Table 16.4 (assuming one datum per sampling period). The Mann-Kendall test may be computed for each station. Also, an estimate of the magnitude of the trend at each station can be obtained using Sen's (1968*b*) procedure, as described in Section 16.5.

When data are collected at several stations within a region or basin, there may be interest in making a basin-wide statement about trends. A general statement about the presence or absence of monotonic trends will be meaningful if the trends at all stations are in the same direction—that is, all upward or all downward. Time plots of the data at each station, preferably on the same graph to make visual comparison easier, may indicate when basin-wide statements are possible. In many situations an objective testing method will be needed to help make this decision. In this section we discuss a method for doing this that

Table 16.4 Data Collected over Time at Multiple Stations

		Station 1						*Station M*			
		Sampling Time				. . .		*Sampling Time*			
		1	*2*		*K*			*1*	*2*		*K*
Year	1	x_{111}	x_{211}	. . .	x_{K11}	. . .	1	x_{11M}	x_{21M}	. . .	x_{K1M}
	2	x_{121}	x_{221}	. . .	x_{K21}	. . .	2	x_{12M}	x_{22M}	. . .	x_{k2M}
	⋮						⋮				
	L	x_{1L1}	x_{2L1}	. . .	x_{KL1}	. . .	L	x_{1LM}	x_{2LM}	. . .	x_{KLM}
Mann-Kendall Test			S_1			. . .		S_M			
			Z_1			. . .		Z_M			

M = number of stations
K = number of sampling times per year
L = number of years
x_{ilj} = datum for the ith sampling time in the lth year at the jth station

makes use of the Mann-Kendall statistic computed for each station. This procedure was originally proposed by van Belle and Hughes (1984) to test for homogeneity of trends between seasons (a test discussed in Chapter 17).

To test for homogeneity of trend direction at multiple stations, compute the homogeneity chi-square statistic, χ^2_{homog}, where

$$\chi^2_{\text{homog}} = \chi^2_{\text{total}} - \chi^2_{\text{trend}} = \sum_{j=1}^{M} Z_j^2 - M\bar{Z}^2 \qquad \textbf{16.6}$$

$$Z_j = \frac{S_j}{[\text{VAR}(S_j)]^{1/2}} \qquad \textbf{16.7}$$

S_j is the Mann-Kendall trend statistic for the jth station,

and $\quad \bar{Z} = \frac{1}{M}\sum_{j=1}^{M} Z_j$

If the trend at each station is in the same direction, then χ^2_{homog} has a chi-square distribution with $M - 1$ degrees of freedom (df). This distribution is given in Table A19. To test for trend homogeneity between stations at the α significance level, we refer our calculated value of χ^2_{homog} to the α critical value in Table A19 in the row with $M - 1$ df. If χ^2_{homog} exceeds this critical value, we reject the H_0 of homogeneous station trends. In that case no regional-wide statements should be made about trend direction. However, a Mann-Kendall test for trend at each station may be used. If χ^2_{homog} does not exceed the α critical level in Table A19, then the statistic $\chi^2_{\text{trend}} = M\bar{Z}^2$ is referred to the chi-square distribution with 1 df to test the null hypothesis H_0 that the (common) trend direction is significantly different from zero.

The validity of these chi-square tests depends on each of the Z_j values (Eq. 16.7) having a standard normal distribution. Based on results in Kendall (1975), this implies that the number of data (over time) for each station should exceed 10. Also, the validity of the tests requires that the Z_j be independent. This requirement means that the data from different stations must be uncorrelated. We note that the Mann-Kendall test and the chi-square tests given in this section may be computed even when the number of sampling times, K, varies from year to year and when there are multiple data collected per sampling time at one or more times.

EXAMPLE 16.4

We consider a simple case to illustrate computations. Suppose the following data are obtained:

	Time				
	1	*2*	*3*	*4*	*5*
Station 1	10	12	11	15	18
Station 2	10	9	10	8	9

We wish to test for homogeneous trend direction at the $M = 2$ stations at the $\alpha = 0.05$ significance level. Equation 16.2 gives $S_1 = 1 + 1 + 1 + 1 - 1 + 1 + 1 + 1 + 1 + 1 = +9 - 1 =$

8 and $S_2 = -1 + 0 - 1 - 1 + 1 - 1 + 0 - 1 - 1 + 1 = 2 - 6 = -4$. Equation 16.3 gives

$$\text{VAR}(S_1) = \frac{5(4)(15)}{18} = 16.667 \quad \text{and} \quad \text{VAR}(S_2)$$

$$= \frac{[5(4)(15) - 2(1)(9) - 2(1)(9)]}{18} = 14.667$$

Therefore Eq. 16.4 gives

$$Z_1 = \frac{7}{(16.667)^{1/2}} = 1.71 \quad \text{and} \quad Z_2 = \frac{-3}{(14.667)^{1/2}} = -0.783$$

Thus

$$\chi^2_{\text{homog}} = 1.71^2 + (-0.783)^2 - 2\left(\frac{1.71 - 0.783}{2}\right)^2 = 3.1$$

Referring to the chi-square tables with $M - 1 = 1$ df, we find the $\alpha = 0.05$ level critical value is 3.84. Since $\chi^2_{\text{homog}} < 3.84$, we cannot reject the null hypothesis of homogeneous trend direction over time at the 2 stations. Hence, an overall test of trend using the statistic χ^2_{trend} can be made. [Note that the critical value 3.84 is only approximate (somewhat too small), since the number of data at both stations is less than 10.] $\chi^2_{\text{trend}} = M\bar{Z}^2 = 2(0.2148) = 0.43$. Since $0.43 < 3.84$, we cannot reject the null hypothesis of no trend at the 2 stations.

We may test for trend at each station using the Mann-Kendall test by referring $S_1 = 8$ and $S_2 = -4$ to Table A18. The tabled value for $S_1 = 8$ when $n = 5$ is 0.042. Doubling this value to give a two-tailed test gives 0.084, which is greater than our prespecified $\alpha = 0.05$. Hence, we cannot reject H_0 of no trend for station 1 at the $\alpha = 0.05$ level. The tabled value for $S_2 = -4$ when $n = 5$ is 0.242. Since $0.484 > 0.05$, we cannot reject H_0 of no trend for station 2. These results are consistent with the χ^2_{trend} test before. Note, however, that station 1 still appears to be increasing over time, and the reader may confirm it is significant at the $\alpha = 0.10$ level. This result suggests that this station be carefully watched in the future.

16.5 SEN'S NONPARAMETRIC ESTIMATOR OF SLOPE

As noted in Section 16.3.2, if a linear trend is present, the true slope (change per unit time) may be estimated by computing the least squares estimate of the slope, b, by linear regression methods. However, b computed in this way can deviate greatly from the true slope if there are gross errors or outliers in the data. This section shows how to estimate the true slope at a sampling station by using a simple nonparametric procedure developed by Sen (1968b). His procedure is an extension of a test by Theil (1950), which is illustrated by Hollander and Wolfe (1973, p. 205). Sen's method is not greatly affected by gross data errors or outliers, and it can be computed when data are missing. Sen's estimator is closely related to the Mann-Kendall test, as illustrated in the following paragraphs. The computer code in Appendix B computes Sen's estimator.

First, compute the N' slope estimates, Q, for each station:

$$Q = \frac{x_{i'} - x_i}{i' - i} \qquad \textbf{16.8}$$

where $x_{i'}$ and x_i are data values at times (or during time periods) i' and i, respectively, and where $i' > i$; N' is the number of data pairs for which $i' > i$. The median of these N' values of Q is Sen's estimator of slope. If there is only one datum in each time period, then $N' = n(n - 1)/2$, where n is the number of time periods. If there are multiple observations in one or more time periods, then $N' < n(n - 1)/2$, where n is now the total number of observations, not time periods, since Eq. 16.8 cannot be computed with two data from the same time period, that is, when $i' = i$. If an x_i is below the detection limit, one half the detection limit may be used for x_i.

The median of the N' slope estimates is obtained in the usual way, as discussed in Section 13.3.1. That is, the N' values of Q are ranked from smallest to largest (denote the ranked values by $Q_{[1]} \leq Q_{[2]} \leq \cdots \leq Q_{[N'-1]} \leq Q_{[N']}$) and we compute

$$\begin{aligned} \text{Sen's estimator} &= \text{median slope} \\ &= Q_{[(N'+1)/2]} && \text{if } N' \text{ is odd} \\ &= \tfrac{1}{2}\left(Q_{[N'/2]} + Q_{[(N'+2)/2]}\right) && \text{if } N' \text{ is even} \end{aligned} \qquad \textbf{16.9}$$

A $100(1 - \alpha)\%$ two-sided confidence interval about the true slope may be obtained by the nonparametric technique given by Sen (1968b). We give here a simpler procedure, based on the normal distribution, that is valid for n as small as 10 unless there are many ties. This procedure is a generalization of that given by Hollander and Wolfe (1973, p. 207) when ties and/or multiple observations per time period are present.

1. Choose the desired confidence coefficient α and find $Z_{1-\alpha/2}$ in Table A1.
2. Compute $C_\alpha = Z_{1-\alpha/2}[\text{VAR}(S)]^{1/2}$, where VAR($S$) is computed from Eqs. 16.3 or 16.5. The latter equation is used if there are multiple observations per time period.
3. Compute $M_1 = (N' - C_\alpha)/2$ and $M_2 = (N' + C_\alpha)/2$.
4. The lower and upper limits of the confidence interval are the M_1th largest and $(M_2 + 1)$th largest of the N' ordered slope estimates, respectively.

EXAMPLE 16.5

We use the data set in Example 16.3 to illustrate Sen's procedure. Recall that the data are

Time Period	1	1	1	2	3	3	4	5
Data	10	22	21	30	22	30	40	40

There are $N' = 24$ pairs for which $i' > i$. The values of individual

Table 16.5 Illustration of Computing an Estimate of Trend Slope Using Sen's (1968*b*) Nonparametric Procedure (for Example 16.5). Tabled Values Are Individual Slope Estimates, *Q*

Time Period	*1*	*1*	*1*	*2*	*3*	*3*	*4*	*5*
Data	*10*	*22*	*21*	*30*	*22*	*30*	*40*	*40*
		NC	NC	+20	+6	+10	+10	+7.5
			NC	+8	0	+4	+6	+4.5
				+9	+0.5	+4.5	+6.33	+4.75
					−8	0	+5	+3.33
						NC	+18	+9
							+10	+5
								0

NC = Cannot be computed since both data values are within the same time period.

slope estimates Q for these pairs are obtained by dividing the differences in Table 16.3 by $i' - i$. The 24 Q values are given in Table 16.5.

Ranking these Q values from smallest to largest gives

−8, 0, 0, 0, 0.5, 3.33, 4, 4.5, 4.5, 4.75, 5, 5, 6, 6, 6.33, 7.5, 8, 9, 9, 10, 10, 10, 18, 20

Since $N' = 24$ is even, the median of these Q values is the average of the 12th and 13th largest values (by Eq. 16.8), which is 5.5, the Sen estimate of the true slope. That is, the average (median) change is estimated to be 5.5 units per time period.

A 90% confidence interval about the true slope is obtaied as follows. From Table A1 we find $Z_{0.95} = 1.645$. Hence,

$$C_\alpha = 1.645[\mathrm{VAR}(S)]^{1/2} = 1.645[58.1]^{1/2} = 12.54$$

where the value for VAR(S) was obtained from Example 16.3. Since $N' = 24$, we have $M_1 = (24 - 12.54)/2 = 5.73$ and $M_2 + 1 = (24 + 12.54)/2 + 1 = 19.27$. From the list of 24 ordered slopes given earlier, the lower limit is found to be 2.6 by interpolating between the 5th and 6th largest values. The upper limit is similarly found to be 9.3 by interpolating between the 19th and 20th largest values.

16.6 CASE STUDY

This section illustrates the procedures presented in this chapter for evaluating trends. The computer program in Appendix B is used on the hypothetical data listed in Table 16.6 and plotted in Figure 16.4. These data, generated on a computer, represent measurements collected monthly at two stations for 48 consecutive months. The model for station 1 is $x_{il1} = \exp[0.83 e_{il} - 0.35] - 1.0$, where x_{il1} is the datum for month i in year l at station 1. The model used at station 2 was $x_{il2} = \exp[0.83 e_{il} - 0.35] - 1.0 + 0.40(i/12 + l)$. For both stations the measurement errors e_{il} were generated to have mean 0 and variance 1. The data for station 1 are lognormally distributed with no trend,

Table 16.6 Simulated Monthly Data at Two Stations over a Four-Year Period

NUMBER OF YEARS = 4
NUMBER OF STATIONS = 2

STATION 1	NUMBER OF DATA POINTS 48		STATION 2	NUMBER OF DATA POINTS 48	
YEAR	MONTH	STATION 1	YEAR	MONTH	STATION 2
1	1	6.00	1	1	5.09
1	2	5.41	1	2	5.07
1	3	4.58	1	3	4.93
1	4	4.34	1	4	4.94
1	5	4.77	1	5	5.15
1	6	4.54	1	6	11.82
1	7	4.50	1	7	5.48
1	8	5.02	1	8	5.18
1	9	4.38	1	9	5.79
1	10	4.27	1	10	5.11
1	11	4.33	1	11	5.10
1	12	4.33	1	12	5.94
2	13	5.00	2	13	6.91
2	14	5.02	2	14	7.11
2	15	4.14	2	15	5.40
2	16	5.16	2	16	6.77
2	17	6.33	2	17	5.35
2	18	5.49	2	18	6.04
2	19	4.54	2	19	5.45
2	20	6.62	2	20	6.95
2	21	4.64	2	21	5.54
2	22	4.45	2	22	5.71
2	23	4.57	2	23	6.14
2	24	4.09	2	24	7.13
3	25	5.06	3	25	5.80
3	26	4.83	3	26	5.91
3	27	4.92	3	27	5.88
3	28	6.02	3	28	7.21
3	29	4.77	3	29	8.29
3	30	5.03	3	30	6.00
3	31	7.15	3	31	6.28
3	32	4.30	3	32	5.69
3	33	4.15	3	33	6.52
3	34	5.13	3	34	6.27
3	35	5.28	3	35	6.46
3	36	4.31	3	36	6.94
4	37	6.53	4	37	6.28
4	38	5.11	4	38	6.74
4	39	4.31	4	39	6.91
4	40	4.64	4	40	7.81
4	41	4.87	4	41	6.53
4	42	4.89	4	42	6.26
4	43	4.92	4	43	7.01
4	44	4.94	4	44	7.42
4	45	4.69	4	45	8.35
4	46	4.50	4	46	6.27
4	47	4.80	4	47	6.69
4	48	4.80	4	48	6.99

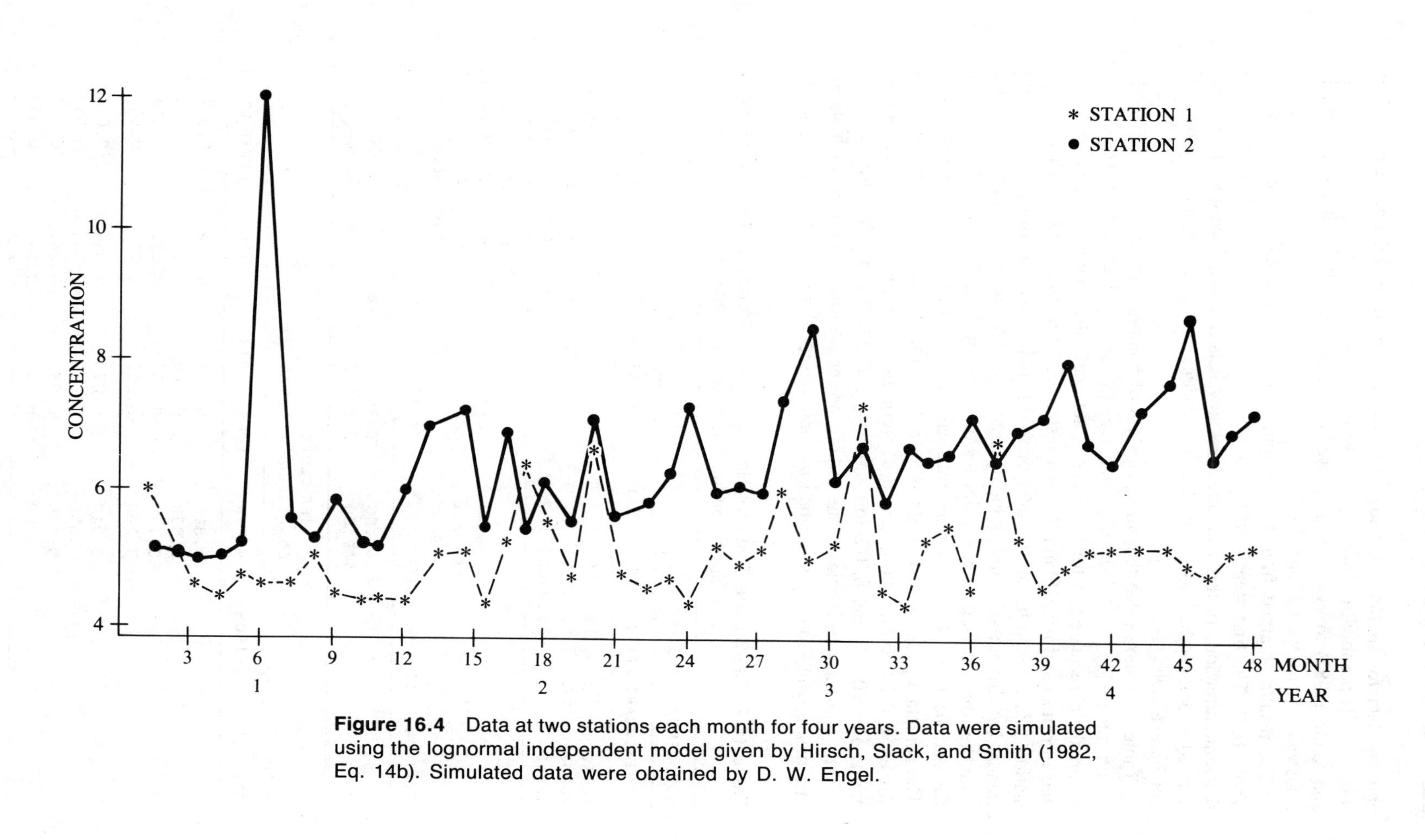

Figure 16.4 Data at two stations each month for four years. Data were simulated using the lognormal independent model given by Hirsch, Slack, and Smith (1982, Eq. 14b). Simulated data were obtained by D. W. Engel.

and the data for station 2 are lognormal with a trend of 0.4 units per year or 0.0333 units per month. These models were among those used by Hirsch, Slack, and Smith (1982) to evaluate the power of the seasonal Kendall test for trend, a test we discuss in Chapter 17.

The results obtained from the computer code in Appendix B are shown in Table 16.7. The first step is to decide whether the two stations have trends in the same direction. In this example we know it is not so, since one station has a trend and the other does not. But in practice this a priori information will not be available.

Table 16.7 shows that the chi-square test of homogeneity (Eq. 16.6) is highly significant ($\chi^2_{\text{homog}} = 10.0$; computed significance level of 0.002). Hence, we ignore the chi-square test for trend that is automatically computed by the program and turn instead to the Mann-Kendall test results for each station. This test for station 1 is nonsignificant (*P* value of 0.70), indicating no strong evidence for trends, but that for station 2 is highly significant. All of these test results agree with the true situation. Sen's estimates of slope are 0.002 and 0.041 per month for stations 1 and 2, whereas the true values are 0.0 and 0.0333, respectively. The computer code computes $100(1 - \alpha)\%$ confidence limits for the true slope for $\alpha = 0.20$, 0.10, 0.05, and 0.01. For this example the 95% confidence limits are −0.009 and 0.012 for station 1, and 0.030 and 0.050 for station 2.

The computer code allows one to split up the 48 observations at each station into meaningful groups that contain multiple observations. For instance, suppose

Table 16.7 Chi-Square Tests for Homogeneity of Trends at the Two Stations, and Mann-Kendall Tests for Each Station

HOMOGENEITY TEST RESULTS

CHI-SQUARE STATISTICS		df	PROB. OF A LARGER VALUE	
TOTAL	23.97558	2	0.000	
HOMOGENEITY	10.03524	1	0.002 ←	Trend not equal at the 2 stations
TREND	13.94034	1	0.000 ←	Not meaningful

STATION	SEASON	MANN-KENDALL S STATISTIC	Z STATISTIC	n	PROB. OF EXCEEDING THE ABSOLUTE VALUE OF THE Z STATISTIC (TWO-TAILED TEST) IF n > 10
1	1	45.00	0.39121	48	0.696
2	1	549.00	4.87122	48	0.000

SEN SLOPE CONFIDENCE INTERVALS

STATION	SEASON	ALPHA	LOWER LIMIT	SLOPE	UPPER LIMIT
1	1	0.010	-0.013	0.002	0.016
		0.050	-0.009	0.002	0.012
		0.100	-0.007	0.002	0.011
		0.200	-0.005	0.002	0.009
2	1	0.010	0.026	0.041	0.054
		0.050	0.030	0.041	0.050
		0.100	0.032	0.041	0.048
		0.200	0.034	0.041	0.046

Table 16.8 Analyses of the Data in Table 16.6 Considering the Data as Twelve Multiple Observations in Each of Four Years

```
NUMBER OF YEARS    = 4
NUMBER OF SEASONS  = 1
NUMBER OF STATIONS = 2
```

HOMOGENEITY TEST RESULTS

SOURCE	CHI-SQUARE	df	PROB. OF A LARGER VALUE
TOTAL	21.45468	2	0.00
HOMOGENEITY	5.79732	1	0.016
TREND	15.65736	1	0.000

STATION	SEASON	MANN-KENDALL S STATISTIC	Z STATISTIC	n	PROB. OF EXCEEDING THE ABSOLUTE VALUE OF THE Z STATISTIC (TWO-TAILED TEST) IF n > 10
1	1	119.00	1.08623	48	0.277
2	1	489.00	4.49132	48	0.000

		SEN SLOPE CONFIDENCE INTERVALS			
STATION	SEASON	ALPHA	LOWER LIMIT	SLOPE	UPPER LIMIT
1	1	0.010	-0.120	0.080	0.225
		0.050	-0.065	0.080	0.190
		0.100	-0.037	0.080	0.176
		0.200	-0.014	0.080	0.153
2	1	0.010	0.290	0.467	0.670
		0.050	0.353	0.467	0.620
		0.100	0.370	0.467	0.600
		0.200	0.390	0.467	0.575

we regard the data in this example as 12 multiple data points in each of four years. Applying the code using this interpretation gives the results in Table 16.8.

The conclusions of the tests are the same as obtained in Table 16.7 when the data were considered as one time series of 48 single observations. However, this may not be the case with other data sets or groupings of multiple observations. Indeed, the Mann-Kendall test statistic Z for station 1 is larger in Table 16.8 than in Table 16.7, so that the test is closer to (falsely) indicating a significant trend when the data are grouped into years. For station 2 the Mann-Kendall test in Table 16.8 is smaller than in Table 16.7, indicating the test has less power to detect the trend actually present. The best strategy appears to be to not group data unnecessarily. The estimates of slope are now 0.080 and 0.467 per year, whereas the true values are 0.0 and 0.40, respectively.

16.7 SUMMARY

This chapter began by identifying types of trends and some of the complexities that arise when testing for trend. It also discussed graphical methods for detecting and estimating trends, intervention analysis, and problems that arise when using regression methods to detect and estimate trends.

Next, the Mann-Kendall test for trend was described and illustrated in detail, including how to handle multiple observations per sampling time (or period). A chi-square test to test for homogenous trends at different stations within a basin was also illustrated. Finally, methods for estimating and placing confidence limits on the slope of a linear trend by Sen's nonparameter procedure were given and the Mann-Kendall test on a simulated data set was illustrated.

EXERCISES

16.1 Use the Mann-Kendall test to test for a rising trend over time, using the following data obtained sequentially over time.

Time	1	2	3	4	5	6	7
Data	ND	1	ND	3	1.5	1.2	4

Use $\alpha = 0.05$. What problem is encountered in using Table A18? Use the normal approximate test statistic Z.

16.2 Use the data in Exercise 16.1 to estimate the magnitude of the trend in the population. Handle NDs in two ways: (a) treat them as missing values, and (b) set them equal to one half the detection limit. Assume the detection limit is 0.5. What method do you prefer? Why?

16.3 Compute a 90% confidence interval about the true slope, using the data in part (b) of Exercise 16.2.

ANSWERS

16.1 $n = 7$. The 2 NDs are treated as tied at a value less than 1.1. $S = 16 - 4 = 12$. Since there is a tie, there is no probability value in Table A18 for $S = +12$, but the probability lies between 0.035 and 0.068. Using the large sample approximation gives $\text{Var}(S) = 43.3$ and $Z = 1.67$. Since $1.67 > 1.645$, we reject H_0 of no trend.

16.2 (a) The median of the 10 estimates of slope is 0.23. (b) The median of the 21 estimates of slope is 0.33.

Pros and Cons: Using one half of the detection limit assumes the actual measurements of ND values are equally likely to fall anywhere between zero and the detection limit. One half of the detection limit is the mean of that distribution. This method, though approximate, is preferred to treating NDs as missing values.

16.3 From Eq. 16.3, $\text{VAR}(S) = 44.3$. (The correction for ties in Eq. 16.3 is not used because the 2 tied values were originally ND values and were assigned to be equal.) $C_\alpha = 10.95$, $M_1 \cong 5$, $M_2 + 1 \cong 17$. Therefore, the limits are 0 and 0.94.

17 Trends and Seasonality

Chapter 16 discussed trend detection and estimation methods that may be used when there are no cycles or seasonal effects in the data. Hirsch, Slack, and Smith (1982) proposed the seasonal Kendall test when seasonality is present. This chapter describes the seasonal Kendall test as well as the extention to multiple stations developed by van Belle and Hughes (1984). It also shows how to estimate the magnitude of a trend by using the nonparametric seasonal Kendall slope estimator, which is appropriate when seasonality is present. All these techniques are included in the computer code listed in Appendix B. A computer code that computes only the seasonal Kendall test and slope estimator is given in Smith, Hirsch, and Slack (1982).

17.1 SEASONAL KENDALL TEST

If seasonal cycles are present in the data, tests for trend that remove these cycles or are not affected by them should be used. This section discusses such a test: the seasonal Kendall test developed by Hirsch, Slack, and Smith (1982) and discussed further by Smith, Hirsch, and Slack (1982) and by van Belle and Hughes (1984). This test may be used even though there are missing, tied, or ND values. Furthermore, the validity of the test does not depend on the data being normally distributed.

The seasonal Kendall test is a generalization of the Mann-Kendall test. It was proposed by Hirsch and colleagues for use with 12 seasons (months). In brief, the test consists of computing the Mann-Kendall test statistic S and its variance, VAR(S), separately for each month (season) with data collected over years. These seasonal statistics are then summed, and a Z statistic is computed. If the number of seasons and years is sufficiently large, this Z value may be referred to the standard normal tables (Table A1) to test for a statistically significant trend. If there are 12 seasons (e.g., 12 months of data per year), Hirsch, Slack, and Smith (1982) show that Table A1 may be used as long as there are at least three years of data for each of the 12 seasons.

Conceptually, the seasonal Kendall test may also be used for other "seasons" (e.g., four quarters of the year or the three 8-h periods of the day). However, the degree of approximation of Table A1 when there are fewer than 12 seasons has not, apparently, been given in the literature. For applications where an

Table 17.1 Data for the Seasonal Kendall Test at One Sampling Station

		Season			
		1	2		*K*
Year	1	x_{11}	x_{21}	...	x_{K1}
	2	x_{12}	x_{22}	...	x_{K2}
	⋮				
	L	x_{1L}	x_{2L}	...	x_{KL}
		S_1	S_2	...	S_K
		$S' = \sum_{i=1}^{K} S_i$	$\text{Var}(S') = \sum_{i=1}^{K} \text{Var}(S_i)$		

exact test is important, the exact distribution of the seasonal Kendall test statistic can be obtained on a computer for any combination of seasons and years by the technique discussed by Hirsch, Slack, and Smith (1982).

Let x_{il} be the datum for the ith season of the lth year, K the number of seasons, and L the number of years. The data for a given site (sampling station) are shown in Table 17.1. The null hypothesis, H_0, is that the x_{il} are independent of the time (season and year) they were collected. The hypothesis H_0 is tested against the alternative hypothesis, H_A, that for one or more seasons the data are not independent of time.

For each season we use data collected over years to compute the Mann-Kendall statistic S. Let S_i be this statistic computed for season i, that is,

$$S_i = \sum_{k=1}^{n_i - 1} \sum_{l=k+1}^{n_i} \text{sgn}(x_{il} - x_{ik}) \qquad \textbf{17.1}$$

where $l > k$, n_i is the number of data (over years) for season i, and

$$\begin{aligned} \text{sgn}(x_{il} - x_{ik}) &= 1 \qquad \text{if } x_{il} - x_{ik} > 0 \\ &= 0 \qquad \text{if } x_{il} - x_{ik} = 0 \\ &= -1 \qquad \text{if } x_{il} - x_{ik} < 0 \end{aligned}$$

VAR(S_i) is computed as follows:

$$\begin{aligned} \text{VAR}(S_i) = \frac{1}{18}\Bigg[& n_i(n_i - 1)(2n_i + 5) - \sum_{p=1}^{g_i} t_{ip}(t_{ip} - 1)(2t_{ip} + 5) \\ & - \sum_{q=1}^{h_i} u_{iq}(u_{iq} - 1)(2u_{iq} + 5)\Bigg] \\ & + \frac{\sum_{p=1}^{g_i} t_{ip}(t_{ip} - 1)(t_{ip} - 2) \sum_{q=1}^{h_i} u_{iq}(u_{iq} - 1)(u_{iq} - 2)}{9n_i(n_i - 1)(n_i - 2)} \\ & + \frac{\sum_{p=1}^{g_i} t_{ip}(t_{ip} - 1) \sum_{q=1}^{h_i} u_{iq}(u_{iq} - 1)}{2n_i(n_i - 1)} \end{aligned} \qquad \textbf{17.2}$$

where g_i is the number of groups of tied (equal-valued) data in season i, t_{ip} is the number of tied data in the pth group for season i, h_i is the number of sampling times (or time periods) in season i that contain multiple data, and u_{iq} is the number of multiple data in the qth time period in season i. These quantities are illustrated in Example 17.1.

After the S_i and Var(S_i) are computed, we pool across the K seasons:

$$S' = \sum_{i=1}^{K} S_i \qquad 17.3$$

and

$$\mathrm{VAR}(S') = \sum_{i=1}^{K} \mathrm{VAR}(S_i) \qquad 17.4$$

Next, compute

$$\begin{aligned} Z &= \frac{(S' - 1)}{[\mathrm{VAR}(S')]^{1/2}} && \text{if } S' > 0 \\ &= 0 && \text{if } S' = 0 \\ &= \frac{(S' + 1)}{[\mathrm{VAR}(S')]^{1/2}} && \text{if } S < 0 \end{aligned} \qquad 17.5$$

To test the null hypothesis, H_0, of no trend versus the alternative hypothesis, H_A, of either an upward or downward trend (a two-tailed test), we reject H_0 if the absolute value of Z is greater than $Z_{1-\alpha/2}$, where $Z_{1-\alpha/2}$ is from Table A1. If the alternative hypothesis is for an upward trend at the α level (a one-tailed test), we reject H_0 if Z (Eq. 17.5) is greater than $Z_{1-\alpha}$. Reject H_0 in favor of a downward trend (one-tailed test) if Z is negative and the absolute value of Z is greater than $Z_{1-\alpha}$. The computer code in Appendix B computes the seasonal Kendall test for multiple or single observations per time period. Example 17.1 in the next section illustrates this test. The +1 added to the S' in Eq. 17.5 is a correction factor that makes Table A1 more exact for testing the null hypothesis. This correction is not necessary if there are ten or more data for each season ($n_i \geq 10$).

17.2 SEASONAL KENDALL SLOPE ESTIMATOR

The seasonal Kendall slope estimator is a generalization of Sen's estimator of slope discussed in Section 16.5. First, compute the individual N_i slope estimates for the ith season:

$$Q_i = \frac{x_{il} - x_{ik}}{l - k}$$

where, as before, x_{il} is the datum for the ith season of the lth year, and x_{ik} is the datum for the ith season of the kth year, where $l > k$. Do this for each of the K seasons. Then rank the $N'_1 + N'_2 + \cdots + N'_K = N'$ individual slope estimates and find their median. This median is the seasonal Kendall slope estimator.

A $100(1 - \alpha)\%$ confidence interval about the true slope is obtained in the same manner as in Section 16.5:

1. Choose the desired confidence level α and find $Z_{1-\alpha/2}$ in Table A1.
2. Compute $C_\alpha = Z_{1-\alpha/2}[\text{VAR}(S')]^{1/2}$.
3. Computer $M_1 = (N' - C_\alpha)/2$ and $M_2 = (N' + C_\alpha)/2$.
4. The lower and upper confidence limits are the M_1th largest and the $(M_2 + 1)$th largest of the N' ordered slope estimates, respectively.

EXAMPLE 17.1

We use a simple data set to illustrate the seasonal Kendall test and slope estimator. Since the number of data are small, the tests and confidence limits are only approximations. All computations are given in Table 17.2. Suppose data are collected twice a year (e.g., December and June) for 3 years at a given location. The data are listed below and plotted in Figure 17.1.

	Year							
	1			*2*			*3*	
Season	1	1	2	1	2	2	1	2
Data	8	10	15	12	20	18	15	20

Note that two observations were made in season 1 of year 1 and in season 2 of year 2. Also, there is 1 tied data value, 20, in season 2.

Table 17.2, Part A, gives the $N_1' + N_2' = 5 + 5 = 10$ individual slope estimates for the 2 seasons and their ranking from smallest to largest. The seasonal Kendall slope estimate, 2.75, is the median of these 10 values. In Table 17.2, Part B, the seasonal Kendall Z statistic is calculated to be 2.1 by Eqs. 17.3–17.5. To test for an upward trend (one-tailed test) at the $\alpha = 0.05$ level, we reject the null hypothesis, H_0, of no trend if $Z > Z_{0.95}$, that is, if $Z > 1.645$. Since $Z = 2.10$, we reject H_0 and accept that an upward trend is present.

A 90% confidence interval on the true slope is obtained by computing $C_\alpha = 1.645[\text{VAR}(S')]^{1/2} = 1.645(3.808) = 6.264$, $M_1 = (10 - 6.264)/2 = 1.868$, and $M_2 + 1 = (10 + 6.264)/2 + 1 = 9.132$. Hence, the lower limit is found by interpolating between the first and second largest values to obtain 1.7. The upper limit is similarly found to be 4.1.

17.3 HOMOGENEITY OF TRENDS IN DIFFERENT SEASONS

Section 16.4.4 showed how to test for homogeneity of trend direction at different stations when no seasonal cycles are present. That test is closely related to the

Table 17.2 Illustration of the Seasonal Kendall Test and Slope Estimator. Tabled Values Are Individual Slope Estimates Obtained from Eq. 17.6

Part A. Computing the Seasonal Kendall Slope Estimate

	Season 1							*Season 2*					
Year	*1*	*1*	*2*	*3*	*Sum of*	*Sum of*		*1*	*2*	*2*	*3*	*Sum of*	*Sum of*
Data	*8*	*10*	*12*	*15*	*+ Signs*	*− Signs*		*15*	*20*	*18*	*20*	*+ Signs*	*− Signs*
		a	+4	+3.5	2	0			+5	+3	+2.5	3	0
			+2	+2.5	2	0				*a*	0	0	0
				+3	1	0					+2	1	0
				S_1 =	5 +	0	= 5				S_2 =	4 +	0 = 4

Ordered values of individual slope estimates:

0, 2, 2, 2.5, 2.5, 3, 3, 3.5, 4, 5

Median: Seasonal Kendall slope estimate = 2.75

80% Limits: 0.936 and 4.53

Part B. Computing the Seasonal Kendall Test

$n_1 = 4$ $\quad$ $n_2 = 4$

$g_1 = 0$ $\quad$ $g_2 = 1,\ t_{21} = 2$

$h_1 = 1,\ u_{11} = 2$ $\quad$ $h_2 = 1,\ u_{21} = 2$

$N_1' = 5$ $\quad$ $N_2' = 5$

$\text{Var}(S_1) = \frac{1}{18}[4(3)(13) - 2(1)(9)] + 0 + 0 = 7.667$

$\text{Var}(S_2) = \frac{1}{18}[4(3)(13) - 2(1)(9) - 2(1)(9)] + 0 + [2(1)][2(1)]/8(3)$

$= 6.667 + 0.1667 = 6.834$

$[\text{Var}(S_1)]^{1/2} = 2.8 \quad [\text{VAR}(S_2)]^{1/2} = 2.6$

$S' = S_1 + S_2 = 5 + 4 = 9$

$\text{VAR}(S') = \text{VAR}(S_1) + \text{VAR}(S_2) = 7.667 + 6.834 = 14.5$

$[\text{VAR}(S')]^{1/2} = 3.808 \quad Z = \dfrac{(9-1)}{3.808} = 2.1^b$

[a] Cannot be computed since both data values are within the same time period.

[b] Referring this value to Table A1 is only an approximate test for this example, since n_1 and n_2 are small and there are only two seasons.

procedure developed by van Belle and Hughes (1984) to test for homogeneity of trend direction in different seasons at a given station. This latter test is important, since if the trend is upward in one season and downward in another, the seasonal Kendall test and slope estimator will be misleading.

The procedure is to compute

$$\chi^2_{\text{homog}} = \chi^2_{\text{total}} - \chi^2_{\text{trend}} = \sum_{i=1}^{K} Z_i^2 - K\bar{Z}^2$$

where

$$Z_i = \frac{S_i}{[\text{VAR}(S_i)]^{1/2}}$$

S_i is the Mann-Kendall statistic, computed with data collected over years, during the ith season, and

$$\bar{Z} = \frac{1}{K}\sum_{i=1}^{K} Z_i$$

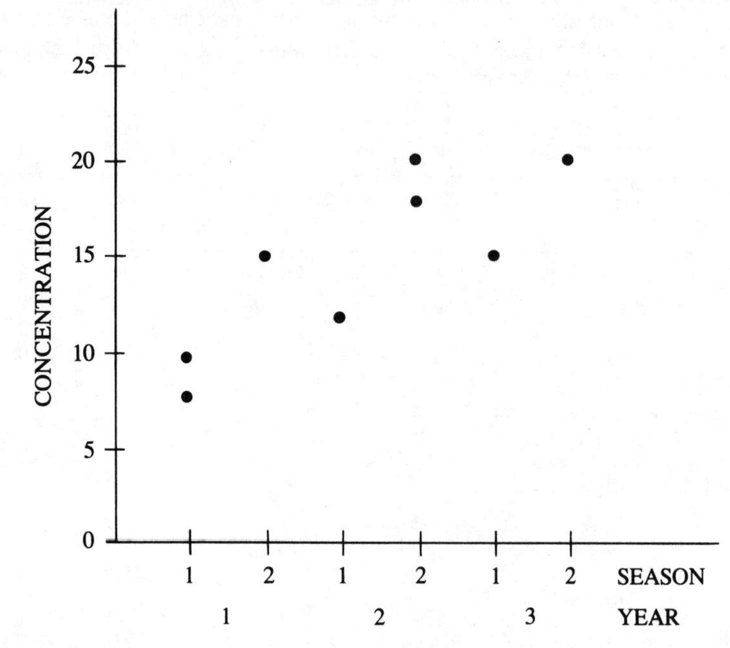

Figure 17.1 Artificial data set to illustrate computation of the seasonal Kendall slope estimator.

If χ^2_{homog} exceeds the α critical value for the chi-square distribution with $K - 1$ df, we reject the null hypothesis, H_0, of homogeneous seasonal trends over time (trends in the same direction and of the same magnitude). In that case the seasonal Kendall test and slope estimate are not meaningful, and it is best to compute the Mann-Kendall test and Sen's slope estimator for each individual season. If χ^2_{homog} does not exceed the critical value in the chi-square tables (Table A19), our calculated value of $\chi^2_{\text{trend}} = K\bar{Z}^2$ is referred to the chi-square distribution with 1 df to test for a common trend in all seasons.

The critical value obtained from the chi-square tables will tend to be too small unless (1) the number of data used to compute each Z_i is 10 or more, and (2) the data are spaced far enough apart in time so that the data in different seasons are not correlated. For some water quality variables Lettenmaier (1978) found that this implies that sampling should be at least two weeks apart.

Van Belle and Hughes (1984) show how to test whether there is a pattern to the trend heterogeneity when χ^2_{homog} is significantly large. They illustrate by showing how to test whether trends in summer and winter months are significantly different.

17.4 SEN'S TEST FOR TREND

The seasonal Kendall and chi-square tests are versatile and easy to use with the computer code in Appendix B. However, if seasonal cycles are present, van Belle and Hughes (1984) show that a nonparametric aligned rank test, used by Farrell (1980) (proposed by Sen, 1968a), is more likely to detect monotonic trends. It is especially true when only a few years of data are available. However, Sen's test is more difficult to compute than the seasonal Kendall test when there are missing values, and the test is inexact in that case. Given these facts, Sen's test is preferred to the seasonal Kendall test when no data are missing. The computer code in Appendix B also computes Sen's test. Computational procedures are given in van Belle and Hughes (1984).

17.5 TESTING FOR GLOBAL TRENDS

In Section 17.3 the χ^2_{homog} statistic was used to test for homogeneity of trend direction in different seasons at a given sampling station. This test is a special case of that developed by van Belle and Hughes (1984) for $M > 1$ stations. Their procedures allow one to test for homogeneity of trend direction at different stations when seasonality is present. The test for homogeneity given in Section 16.4.4 is a special case of this test. Van Belle and Hughes illustrate the tests, using temperature and biological oxygen demand data at two stations on the Willamette River.

The required data are illustrated in Table 17.3. The first step is to compute the Mann-Kendall statistic for each season at each station by Eq. 17.1. Let S_{im} denote this statistic for the ith season at the mth station. Then compute

$$Z_{im} = \frac{S_{im}}{[\text{VAR}(S_{im})]^{1/2}}, \qquad i = 1, 2, \cdots, K, \quad m = 1, 2, \cdots, M \qquad \textbf{17.6}$$

where $\text{VAR}(S_{im})$ is obtained by using Eq. 17.2. (For this application all quantities in Eq. 17.2 relate to the data set for the ith season and mth station.) Note that missing values, NDs, or multiple observations per time period are allowed, as discussed in Section 17.1. Also, note that the correction for continuity (± 1 added to S in Eq. 16.5 and S' in Eq. 17.5) is not used in Eq. 17.6 for reasons discussed by van Belle and Hughes (1984).

Next, compute

$$\bar{Z}_{i\cdot} = \frac{1}{M} \sum_{m=1}^{M} Z_{im}, \qquad i = 1, 2, \cdots, K$$

$$= \text{mean over } M \text{ stations for the } i\text{th season}$$

Table 17.3 Data to Test for Trends Using the Procedure of van Belle and Hughes (1984)

		Station 1				$\cdots$		*Station M*			
Season		*1*	*2*	$\cdots$	*K*	$\cdots$		*1*	*2*	$\cdots$	*K*
	1	x_{111}	x_{211}	$\cdots$	x_{K11}	$\cdots$	1	x_{11M}	x_{21M}	$\cdots$	x_{K1M}
	2	x_{121}	x_{221}	$\cdots$	x_{K21}	$\cdots$	2	x_{12M}	x_{22M}	$\cdots$	x_{K2M}
Year	$\vdots$						$\vdots$				
	L	x_{1L1}	x_{2L1}	$\cdots$	x_{KL1}	$\cdots$	L	x_{1LM}	x_{2LM}	$\cdots$	x_{KLM}
		S_{11}	S_{21}	$\cdots$	S_{K1}	$\cdots$		S_{1M}	S_{2M}	$\cdots$	S_{KM}
		Z_{11}	Z_{21}	$\cdots$	Z_{K1}	$\cdots$		Z_{1M}	Z_{2M}	$\cdots$	Z_{KM}

K = number of seasons; M = number of stations; L = number of years.
x_{ilj} = datum for the ith sampling time in the lth year at the jth station.
Multiple observations per year for one or more seasons are allowed but not shown here.

$$\bar{Z}_{\cdot m} = \frac{1}{K}\sum_{i=1}^{K} Z_{im}, \qquad m = 1, 2, \cdots, M$$

= mean over K seasons for the mth station

$$\bar{Z}_{\cdot\cdot} = \frac{1}{KM}\sum_{i=1}^{K}\sum_{m=1}^{M} Z_{im}$$

= grand mean over all KM stations and seasons

Now, compute the chi-square statistics in Table 17.4 in the following order: χ^2_{total}, χ^2_{trend}, χ^2_{station}, and χ^2_{season}. Then compute

$$\chi^2_{\text{homog}} = \chi^2_{\text{total}} - \chi^2_{\text{trend}}$$

and

$$\chi^2_{\text{station-season}} = \chi^2_{\text{homog}} - \chi^2_{\text{station}} - \chi^2_{\text{season}}$$

Refer χ^2_{station}, χ^2_{season}, and $\chi^2_{\text{station-season}}$ to the α level critical values in the chi-square tables with $M - 1$, $K - 1$, and $(M - 1)(K - 1)$ df, respectively.

If all three tests are nonsignificant, refer χ^2_{trend} to the chi-square distribution with 1 df to test for global trend. If χ^2_{season} is significant, but χ^2_{station} is not, that is, if trends have significantly different directions in different seasons but not at different stations, then test for a different trend direction in each season by computing the K seasonal statistics

$$M\bar{Z}^2_{i\cdot} \qquad i = 1, 2, \ldots, K \text{ seasons} \tag{17.7}$$

and referring each to the α-level critical value of the chi-square distribution with 1 df.

If χ^2_{station} is significant, but χ^2_{season} is not, that is, if trends have significantly different directions at different stations but not in different seasons, then test for a significant trend at each station by computing the M station statistics

Table 17.4. Testing for Trends Using the Procedure of van Belle and Hughes (1984)

Chi-Square Statistics	*Degrees of Freedom*	*Remarks*
$\chi^2_{\text{total}} = \sum_{i=1}^{K}\sum_{m=1}^{M} Z^2_{im}$	KM	
$\chi^2_{\text{homog}} = \sum_{i=1}^{K}\sum_{m=1}^{M} Z^2_{im} - KM\bar{Z}^2_{\cdot\cdot}$	$KM - 1$	Obtained by subtraction
$\chi^2_{\text{season}} = M\sum_{i=1}^{K}\bar{Z}^2_{i\cdot} - KM\bar{Z}^2_{\cdot\cdot}$	$K - 1$	Test for seasonal heterogeneity
$\chi^2_{\text{station}} = K\sum_{m=1}^{M}\bar{Z}^2_{\cdot m} - KM\bar{Z}^2_{\cdot\cdot}$	$M - 1$	Test for station heterogeneity
$\chi^2_{\text{station-season}} = \sum_{i=1}^{K}\sum_{m=1}^{M} Z^2_{im} - M\sum_{m=1}^{K}\bar{Z}^2_{i\cdot} - K\sum_{m=1}^{M}\bar{Z}^2_{\cdot j} + KM\bar{Z}^2_{\cdot\cdot}$	$(M - 1)(K - 1)$	Test for interaction Obtained by subtraction
$\chi^2_{\text{trend}} = KM\bar{Z}^2_{\cdot\cdot}$	1	Test for overall trend

M = number of stations; K = number of seasons; Z_{im} = Mann-Kendall statistic for the ith season–mth station data set (see Table 17.3).

$$K\bar{Z}^2_{.m} \qquad m = 1, 2, \ldots, M \text{ stations}$$

and refer to the α-level critical value of the chi-square distribution with 1 df.

If both χ^2_{station} and χ^2_{season} are significant or if $\chi^2_{\text{station-season}}$ is significant, then the χ^2 trend test should not be done. The only meaningful trend tests in that case are those for individual station-seasons. These tests are made by referring each Z_{im} statistic (see Table 17.3) to the α-level critical value of the standard normal table (Table A1), as discussed in Section 16.4.2 (or Section 16.4.3 if multiple observations per season have been collected). For these individual Mann-Kendall tests, the Z_{im} should be recomputed so as to include the correction for continuity (± 1) as given in Eq. 16.4.

The computer code listed in Appendix B computes all the tests we have described as well as Sen's estimator of slope for each station-season combination. In addition, it computes the seasonal Kendall test, Sen's aligned test for trends, the seasonal Kendall slope estimator for each station, the equivalent slope estimator (the "station Kendall slope estimator") for each season, and confidence limits on the slope.

The code will compute and print the K seasonal statistics (Eq. 17.7) to test for equal trends at different sites for each season only if (1) the computed P value of the χ^2_{season} test *is less than* α', and (2) the computed P value of the χ^2_{station} *exceeds* α', where α' is an a priori specified significance level, say $\alpha' = 0.01$, 0.05, or 0.10, chosen by the investigator. Similarly, the M station statistics (Eq. 17.8) are computed only if the computed P value of χ^2_{station} is less than α' and that for χ^2_{season} is greater than α'. The user of the code can specify the desired value of α'. A default value of $\alpha' = 0.05$ is used if no value is specified.

EXAMPLE 17.2

Table 17.5 gives a set of data collected monthly at 2 stations for 4 years (plotted in Fig. 17.2). These data were simulated on a computer using the lognormal, autoregressive, seasonal cycle model given in Hirsch, Slack, and Smith (1982, p. 112). The data at station 1 have no long-term trend (i.e., they have a slope of zero), whereas station 2 has an upward trend of 0.4 units per year for each season. Hence, seasonal trend directions are homogeneous, but the station trend directions are not.

The chi-square tests are given in Table 17.6. We obtain that $\chi^2_{\text{station}} = 8.16$ has a P value of 0.004. That is, the probability is only 0.004 of obtaining a χ^2_{station} value this large when trends over time at the 2 stations are in the same direction. Hence, the data suggest trend directions are different at the 2 stations, which is the true situation. Both χ^2_{season} and $\chi^2_{\text{station-season}}$ statistics (8.48 and 2.63) are small enough to be nonsignificant. This result is also expected, since trend direction does not change with season.

We chose $\alpha' = 0.05$. Since χ^2_{season} was not significant (computed P level exceeded $\alpha' = 0.05$), the K seasonal statistics (Eq. 17.7) were not computed. However, since χ^2_{station} was significant (P value less than $\alpha' = 0.05$) and χ^2_{season} was not, the 2 station statistics $12\bar{Z}^2_{.1}$ and $12\bar{Z}^2_{.2}$ were computed by Eq. 17.8 and found to equal 2.46 and 31.45, respectively (see Tble 17.6). These tests indicate some

Table 17.5 Simulated Water Quality Using a Lognormal Autoregressive, Seasonal Cycle Model Given by Hirsch, Slack, and Smith (1982, Eq. 14f)

NUMBER OF YEARS = 4
NUMBER OF SEASONS = 12
NUMBER OF STATIONS = 2

STATION 1	NUMBER OF DATA POINTS n = 48		STATION 2	NUMBER OF DATA POINTS n = 48	
YEAR	SEASON	STATION 1	YEAR	SEASON	STATION 2
1	1	6.32	1	1	6.29
1	2	6.08	1	2	6.11
1	3	5.16	1	3	5.66
1	4	4.47	1	4	5.16
1	5	4.13	1	5	4.75
1	6	3.65	1	6	6.79
1	7	3.48	1	7	4.51
1	8	3.78	1	8	4.37
1	9	3.94	1	9	4.95
1	10	4.40	1	10	5.22
1	11	4.94	1	11	5.73
1	12	5.32	1	12	6.72
2	1	5.82	2	1	7.42
2	2	5.76	2	2	7.56
2	3	4.88	2	3	6.13
2	4	4.84	2	4	6.24
2	5	4.87	2	5	5.07
2	6	4.13	2	6	4.95
2	7	3.51	2	7	4.59
2	8	4.32	2	8	5.22
2	9	4.06	2	9	5.13
2	10	4.47	2	10	5.69
2	11	5.05	2	11	6.41
2	12	5.20	2	12	7.53
3	1	5.83	3	1	7.02
3	2	5.65	3	2	6.93
3	3	5.32	3	3	6.55
3	4	5.33	3	4	6.66
3	5	4.20	3	5	6.69
3	6	3.85	3	6	5.23
3	7	4.45	3	7	5.14
3	8	3.56	3	8	5.06
3	9	3.85	3	9	5.71
3	10	4.72	3	10	6.17
3	11	5.38	3	11	6.78
3	12	5.33	3	12	7.64
4	1	6.59	4	1	7.46
4	2	5.93	4	2	7.56
4	3	4.98	4	3	7.30
4	4	4.61	4	4	7.22
4	5	4.18	4	5	6.07
4	6	3.79	4	6	5.53
4	7	3.64	4	7	5.65
4	8	3.77	4	8	5.94
4	9	4.05	4	9	6.68
4	10	4.50	4	10	6.42
4	11	5.15	4	11	7.10
4	12	5.57	4	12	7.86

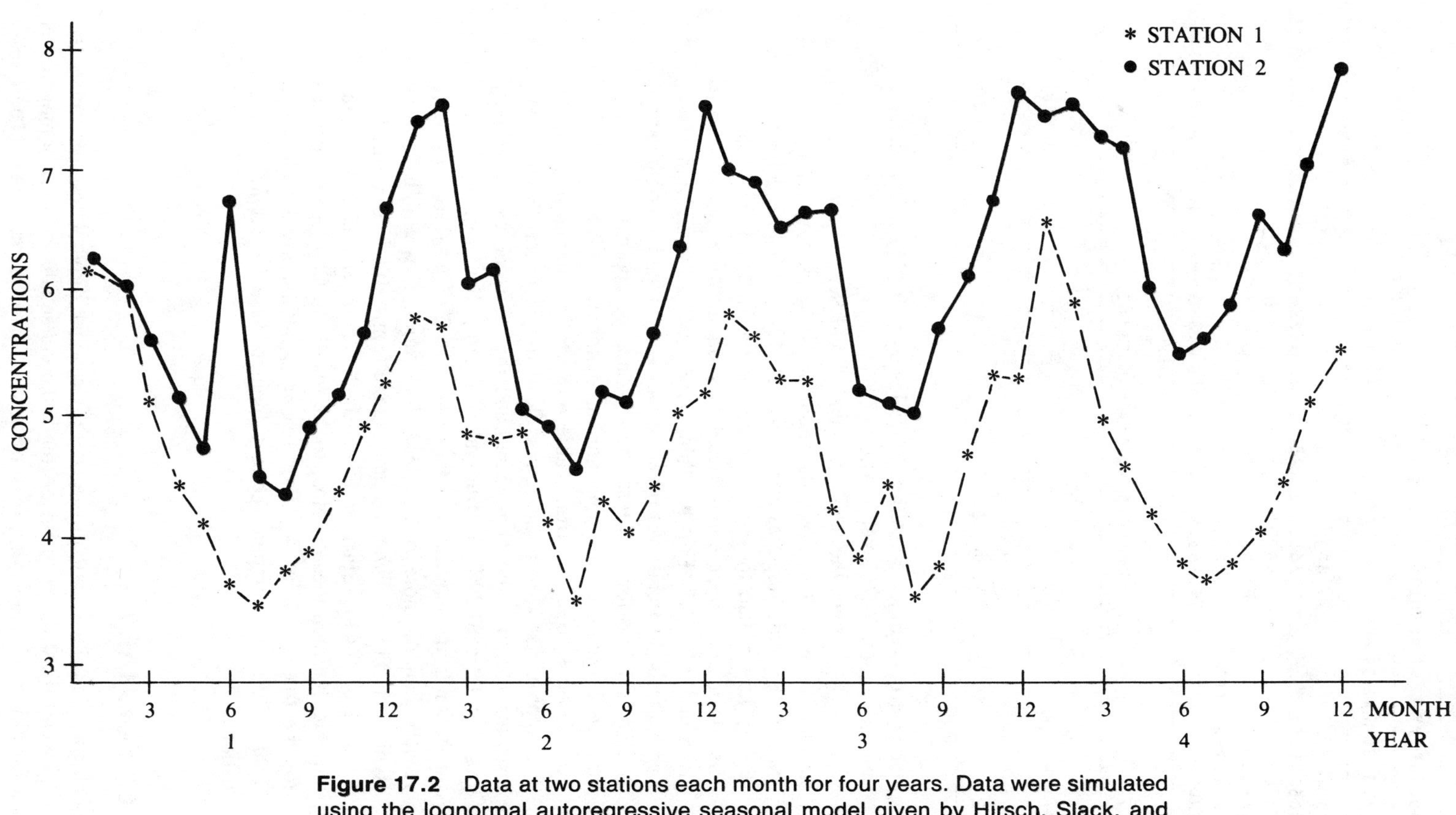

Figure 17.2 Data at two stations each month for four years. Data were simulated using the lognormal autoregressive seasonal model given by Hirsch, Slack, and Smith (1982, Eq. 14f). Simulated data were obtained by D. W. Engel.

Table 17.6 Chi-Square Tests for Homogeneity of Trends over Time between Seasons and between Stations

HOMOGENEITY TEST RESULTS

CHI-SQUARE STATISTICS		df	PROB. OF A LARGER VALUE	
TOTAL	45.02007	24	0.006	
HOMOGENEITY	19.26657	23	0.686	
SEASON	8.48201	11	0.670	
STATION	8.15667	1	0.004 ←	Trends not equal at the 2 stations
STATION-SEASON	2.62789	11	0.995	
TREND	25.75349	1	0.000 ←	Not meaningful

INDIVIDUAL STATION TREND

STATION	CHI-SQUARE	df	PROB. OF A LARGER VALUE
1	2.46154	1	0.117
2	31.44863	1	0.000

evidence of a trend at station 1 (P level = 0.117) and a definite trend at station 2 (P level = 0.000).

Table 17.7 gives the seasonal Kendall and Sen aligned rank tests at both stations. These results agree with the true situation. The seasonal Kendall slope estimates are 0.042 and 0.440, which are slightly larger than the actual values of 0.0 and 0.4, respectively. The lower and upper confidence limits on the true slope are also given in Table 17.7. Finally, Table 17.8 gives the individual Mann-Kendall tests for trend over time for each season-station combination. Since n is only 4 for each test, the P values are approximate because they were obtained from the normal distribution (Table A1). The exact P values obtained from Table A18 are also shown in the table. The approximate levels are quite close to the exact. None of the tests for station 1 are significant, and the 12 slope estimates vary from −0.08 to 0.208 (the true value is zero). Seven of the 12 tests for station 2 are significant at the $\alpha = 0.10$ 2-tailed level. If n were greater than 4, more of the tests for station 2 would have been significant. The 12 slope estimates range from −0.070 to 0.623 with a mean of 0.414. Since n is so small, these estimates are quite variable, but their mean is close to the true 0.40. Confidence intervals for the true slope for 4 station-season combinations are shown in Table 17.9. The computer code computes these for all KM combinations.

17.6 SUMMARY

This chapter described and illustrated the seasonal Kendall test for trend, the seasonal Kendall estimator of linear trend, the chi-square tests for homogeneous trends for different stations and seasons, and tests for global trends. These tests do not require the data to be normally distributed, they are not greatly affected

Table 17.7 Seasonal Kendall and Sen Aligned Ranks Tests for Trend over Time

STATION	SEASONAL KENDALL	n	PROB. OF EXCEEDING THE ABSOLUTE VALUE OF THE KENDALL STATISTIC (TWO-TAILED TEST)
1	1.47087	48	0.141
2	5.51784	48	0.000

STATION	SEN T	n	PROB. OF EXCEEDING THE ABSOLUTE VALUE OF THE SEN T STATISTIC (TWO-TAILED TEST)
1	1.02473	48	0.306
2	4.57814	48	0.000

SEASONAL-KENDALL SLOPE CONFIDENCE INTERVALS

STATION	ALPHA	LOWER LIMIT	SLOPE	UPPER LIMIT
1	0.010	-0.060	0.042	0.111
	0.050	-0.020	0.042	0.085
	0.100	-0.004	0.042	0.081
	0.200	0.007	0.042	0.070
2	0.010	0.345	0.440	0.525
	0.050	0.365	0.440	0.499
	0.100	0.377	0.440	0.486
	0.200	0.380	0.440	0.478

by outliers and gross errors, and missing data or ND values are allowed. However, the tests still require the data to be independent. If they are not, the tests tend to indicate that trends are present more than the allowed $100\alpha\%$ of the time.

EXERCISES

17.1 Use the following data to test for no trend versus a rising trend, using the seasonal Kendall test. Use $\alpha = 0.01$.

	Season					
Year	*1*	*2*	*3*	*4*	*5*	*6*
1	5.71	4.63	3.97	3.37	3.88	4.95
2	6.29	4.79	5.64	4.42	5.18	6.29
3	7.33	6.91	5.96	6.48	5.30	7.77

17.2 Plot the data in Exercise 17.1 in their time order, and estimate the slope of the rising trend, using the seasonal Kendall slope estimator.

Table 17.8 Mann-Kendall Tests for Trend over Time for Each Season at Each Station

STATION	SEASON	MANN-KENDALL S STATISTIC	Z^a STATISTIC	n	PROB. OF EXCEEDING THE ABSOLUTE VALUE OF THE Z STATISTIC (TWO-TAILED TEST) IF $n > 10$		SEN SLOPE
1	1	2	0.33968	4	0.734	(0.750)[b]	0.050
	2	-2	-0.33968	4	0.734	(0.750)	-0.080
	3	0	0.00000	4	1.000	(1.000)	-0.005
	4	2	0.33968	4	0.734	(0.750)	0.208
	5	0	0.00000	4	1.000	(1.000)	-0.002
	6	0	0.00000	4	1.000	(1.000)	-0.007
	7	4	1.01905	4	0.308	(0.334)	0.059
	8	-2	-0.33968	4	0.734	(0.750)	-0.057
	9	0	0.00000	4	1.000	(1.000)	0.016
	10	4	1.01905	4	0.308	(0.334)	0.052
	11	4	1.01905	4	0.308	(0.334)	0.090
	12	4	1.01905	4	0.308	(0.334)	0.107
2	1	4	1.01905	4	0.308	(0.334)	0.378
	2	3	0.72232	4	0.470	( )[c]	0.447
	3	6	1.69842	4	0.089	(0.084)	0.508
	4	6	1.69842	4	0.089	(0.084)	0.623
	5	4	1.01905	4	0.308	(0.334)	0.470
	6	0	0.00000	4	1.000	(1.000)	-0.070
	7	6	1.69842	4	0.089	(0.084)	0.445
	8	4	1.01905	4	0.308	(0.334)	0.442
	9	6	1.69842	4	0.089	(0.084)	0.578
	10	6	1.69842	4	0.089	(0.084)	0.435
	11	6	1.69842	4	0.089	(0.084)	0.413
	12	6	1.69842	4	0.089	(0.084)	0.300

[a] ± 1 correction factor used to compute the Z statistic.
[b] Exact two-tailed significance levels for the S statistic using Table A18.
[c] Cannot be determined from Table A18 since $S = 3$ resulted because of two tied data in the season.

17.3 Use the results in Exercises 17.1 and 17.2 to compute an 80% confidence interval about the true slope.

17.4 Test for equal trend directions in different seasons, using the data in Exercise 17.1. Use $\alpha = 0.01$. If the trends in the 6 seasons are homogeneous, use chi-square to test for a statistically significant trend at the $\alpha = 0.05$ level.

17.5 Suppose the data in Exercise 17.1 were collected at station 1 and the following data were collected at station 2.

	Season					
Year	*1*	2	3	4	5	6
1	9	8.5	8	7.5	8.3	10
2	12	11.5	11.2	11	12.5	15
3	17	16.5	16	15.5	16.3	17

Table 17.9 Sen Slope Estimates and Confidence Intervals for Each Station-Season Combination

STATION	SEASON	ALPHA	SEN SLOPE CONFIDENCE INTERVALS LOWER LIMIT	SLOPE	UPPER LIMIT
1	1	0.010	*n* too small[a]	0.050	*n* too small[a]
		0.050	*n* too small	0.050	0.087
		0.100	*n* too small	0.050	0.440
		0.200	-0.471	0.050	0.718
	2	0.010	*n* too small	-0.080	*n* too small
		0.050	*n* too small	-0.080	0.032
		0.100	*n* too small	-0.080	0.162
		0.200	-0.308	-0.080	0.258
2	1	0.010	*n* too small	0.378	*n* too small
		0.050	*n* too small	0.378	-0.171
		0.100	*n* too small	0.378	0.511
		0.200	-0.353	0.378	1.052
	2	0.010	*n* too small	0.447	*n* too small
		0.050	*n* too small	0.447	0.251
		0.100	*n* too small	0.447	0.984
		0.200	-0.488	0.447	1.265

[a] The lower and upper limits cannot be computed if *n* is too small.

Test for homogeneity of trend direction between seasons and between stations, using the chi-square tests in Table 17.4 with $\alpha = 0.01$. Test for a significant common trend at the 2 stations, if appropriate.

ANSWERS

17.1 $\mathrm{Var}(S_i) = 3(2)(11)/18 = 3.667$ for each season. $S' = \Sigma_{i=1}^{6} S_i = 18$, $\mathrm{Var}(S') = 6(3.667) = 22$. From Eq. 17.5, $Z = 17/\sqrt{22} = 3.62$. Since $\alpha = 0.01$ (one-tailed test), $Z_{0.99} = 2.326$. Since $3.62 > 2.326$, we accept the hypothesis of a rising trend.

17.2 The median of the 18 slope estimates is 1.09 units per year.

17.3 $Z_{1-\alpha/2} = Z_{0.90} = 1.282$, $\mathrm{Var}(S') = 22$ from Exercise 17.1. Therefore, $C_\alpha = 1.282\sqrt{22} = 6.0131$, $M_1 = 6$, $M_2 + 1 = 13$. Lower limit $=$ 0.81; upper limit $= 1.4$.

17.4 From Exercise 17.1 we have $Z_1 = 1.567 = Z_2 = Z_3 = Z_4 = Z_6$. Therefore $\bar{Z} = 1.567$; then $\chi^2_{\text{total}} = 14.7$, $\chi^2_{\text{trend}} = 14.7$, $\chi^2_{\text{homog}} = 0$. Since $\chi^2_{\text{homog}} < 15.09$ (from Table A19), we cannot reject the null hypothesis of homogeneous trend direction in all seasons. Hence, test for trend, using $\chi^2_{\text{trend}} = 14.7$. Since $14.7 > 3.84$ (from Table A19), we conclude that a significantly large trend is present.

17.5 $S_{im} = 3$, $\mathrm{Var}(S_{im}) = 3.667$, and $Z_{im} = 1.567$, for $i = 1, 2, \ldots, 6$ seasons and $m = 1, 2$ years. $\bar{Z}_{1.} = \bar{Z}_{2.} = \ldots = \bar{Z}_{6.} = \bar{Z}_{.1} = \bar{Z}_{.2} = \bar{Z}_{..} = 1.567$. $\chi^2_{\text{total}} = \chi^2_{\text{trend}} = 29.5$ and $\chi^2_{\text{station}} = \chi^2_{\text{season}} = 0$, $\chi^2_{\text{homog}} = \chi^2_{\text{station-season}} = 0$.

Since $\chi^2_{\text{season}} < 15.09$, we cannot reject H_0 of equal trend direction for seasons.
Since $\chi^2_{\text{station}} < 6.63$, we cannot reject H_0 of equal trend direction for stations.
Since $\chi^2_{\text{season-station}} < 15.09$, we cannot reject H_0 of no station-season interaction.
Since the foregoing tests are all nonsignificant and $\chi^2_{\text{trend}} > 6.63$, we conclude that a significant trend is present for both stations over all seasons.

18 Comparing Populations

An objective of many pollution monitoring and research studies is to make comparisons between pollution levels at different times or places or collected by different measurement techniques. This chapter provides simple nonparametric tests for making such comparisons. These tests do not require that data follow the normal distribution or any other specific distribution. Moreover, many of these tests can accommodate a few missing data or concentrations at the trace or ND (not detected) levels.

We begin with procedures for comparing two populations. The procedures are of two types: those for paired data, and those for independent data sets. Examples of paired data are (i) measurements of two pollutants on each of n field samples, (ii) measurements of a pollutant on air filters collected at two adjacent locations for n time periods, and (iii) measurements of a pollutant on both leaves and roots of the same n plants. The paired test we consider is the sign test. Friedman's test, an extension of the sign test to more than two populations, is also given.

Independent data sets are those for which there is no natural way to pair the data. For example, if n soil samples are collected at each of two hazardous waste sites, there may be no rational way to pair a pollution measurement from one site with a pollution measurement from the other site. For this type of data we illustrate Wilcoxon's rank sum test (also known as the Mann-Whitney test) for the comparison of two populations and the Kruskal-Wallis test for the comparison of more than two populations. The tests discussed in this chapter can be computed by using a statistical software computer package such as Biomedical Computer Programs P Series, (1983) and Statistical Package for the Social Sciences (1985). Additional information on the tests in this chapter and on related testing, parameter estimation, and confidence interval procedures are given in Lehmann (1975), Conover (1980) and/or Hollander and Wolfe (1973).

18.1 TESTS USING PAIRED DATA

Suppose n paired measurements have been made. Denote these pairs by (x_{11}, x_{21}), (x_{12}, x_{22}), . . . , (x_{1n}, x_{2n}), where x_{1i} is the ith observation from population 1 and x_{2i} is the paired ith observation from population 2. When data are paired, we could compare the two populations by looking at the sign or the magnitudes of the set of n differences $D_i = x_{2i} - x_{1i}$, $i = 1, 2, \ldots, n$. The sign test uses the signs and the Wilcoxon signed rank test uses the magnitudes. These two tests are alternatives to the commonly used paired t test described in many statistical methods books, e.g., Snedecor and Cochran (1980, p. 85). The latter test should be used if the differences are a random sample from a normal distribution.

18.1.1 Sign Test

The sign test is simple to compute and can be used no matter what the underlying distribution may be. It can also accommodate a few ND (not detected) concentrations. It is more versatile than the Wilcoxon signed rank test since the latter test requires that the underlying distribution be symmetric (though not necessarily normal) and that no NDs be present. However, the Wilcoxon test will usually have more power than the sign test to detect differences between the two populations. The sign test may be the better choice if ease of computation is an important consideration.

The sign test statistic, B, is the number of pairs (x_{1i}, x_{2i}) for which $x_{1i} < x_{2i}$, that is, the number of positive differences D_i. The magnitudes of the D_i are not considered; only their signs are. If any D_i is zero so that a + or − sign cannot be assigned, this data pair is dropped from the data set and n is reduced by 1. The statistic B is used to test the null hypothesis:

H_0: The median of the population of all possible differences is zero, that is, x_{1i} is as likely to be larger than x_{2i} as x_{2i} is likely to be larger than x_{1i} **18.1**

Clearly, if the number of + and − signs are about equal, there is little reason to reject H_0.

Two-Sided Test

If the number of paired data, n, is 75 or less, we may use Table A14 to test H_0 versus the alternative hypothesis

H_A: The median difference does not equal zero, that is, x_{1i} is more likely to exceed x_{2i} than x_{2i} is likely to exceed x_{1i}, or vice versa **18.2**

Then reject H_0 and accept H_A at the α significance level if

$$B \leq l - 1 \quad \text{or} \quad B \geq u$$

where l and u are integers taken from Table A14 for the appropriate n and chosen α.

For example, suppose there are $n = 34$ differences, and we choose to test at the $\alpha = 0.05$ level. Then we see from Table A14 that we reject H_0 and accept H_A if $B \leq 10$ or if $B \geq 24$.

EXAMPLE 18.1

Grivet (1980) reports average and maximum oxidant pollution concentrations at several air monitoring stations in California. The daily maximum (of hourly average) oxidant concentrations (parts per hundred million) at 2 stations for the first 20 days in July 1972 are given in Table 18.1. The data at the 2 stations are paired and perhaps correlated because they were taken on the same day. This type of correlation is permitted, but correlation between the pairs, that is, between observations taken on different days, should not be present for the test to be completely valid. If this latter type of positive correlation is present, the test would indicate more than the allowed $100\alpha\%$ of the time a significant difference between the 2 stations when none actually exists. This problem is discussed by Gastwirth and Rubin (1971) and by Albers (1978*a*).

We test the null hypothesis that the median difference in maximum concentrations between the two stations is zero, that is, there is no tendency for the oxidant concentrations at one station to be larger than at the other station. Since concentrations are tied on 3 days, n equals 17 rather than 20. The number of + signs is $B = 9$, the number of days that the maximum concentration at station 41541 exceeds that at station 28783. Suppose we use $\alpha = 0.05$. Then from Table A14 for $n = 17$ we find $l = 6$ and $u = 15$. Since B is not less than or equal to $l - 1 = 5$ nor greater than or equal to 15, we cannot reject H_0.

Table A14 gives values of l and u for $n \leq 75$. When $n > 20$, we may use the following approximate test procedure:

1. Compute

$$Z_B = \frac{B - n/2}{\sqrt{n/4}} = \frac{2B - n}{\sqrt{n}} \tag{18.3}$$

2. Reject H_0 and accept H_A (Eq. 18.2) if $Z_B \leq -Z_{1-\alpha/2}$ or if $Z_B \geq Z_{1-\alpha/2}$, where $Z_{1-\alpha/2}$ is obtained from Table A1.

EXAMPLE 18.2

Using the data in Example 18.1, we test H_0 versus H_A, using Z_B. We have $Z_B = (18 - 17)/\sqrt{17} = 0.243$. For $\alpha = 0.05$, Table A1 gives $Z_{0.975} = 1.96$. Since Z_B is not less than or equal to -1.96

Table 18.1 Maximum Oxidant Concentrations[a] at Two Stations in July 1972

Day	*Station 28783*	*Station 41541*	*Sign of Difference*	*Day*	*Station 28783*	*Station 41541*	*Sign of Difference*
1	8	10	+	11	11	13	+
2	5	7	+	12	12	14	+
3	6	7	+	13	13	20	+
4	7	7	[b]	14	14	28	+
5	4	6	+	15	12	6	−
6	4	6	+	16	12	7	−
7	3	3	[b]	17	13	7	−
8	5	4	−	18	14	6	−
9	5	5	[b]	19	12	4	−
10	6	4	−	20	15	5	−

[a]Data are parts per hundred million.
[b]Tied concentrations.

nor greater than or equal to 1.96, we cannot reject H_0. This conclusion is the same as that obtained by using Table A14 in Example 18.1.

One-Sided Test

Thus far we have considered only a two-sided alternative hypothesis (Eq. 18.2). One-sided tests may also be used. There are two such tests:

1. Test H_0 versus the alternative hypothesis, H_A, that the x_2 measurements tend to exceed the x_1 measurements more often than the reverse. In this case reject H_0 and accept H_A if $B \geq u$, where u is obtained from Table A14. Alternatively, if $n > 20$, reject H_0 and accept H_A if $Z_B \geq Z_{1-\alpha}$, where Z_B is computed by Eq. 18.3 and $Z_{1-\alpha}$ is from Table A1.
2. Test H_0 versus the alternative hypothesis that the x_1 measurements tend to exceed the x_2 measurements more often than the reverse. If $n \leq 75$, use Table A14 and reject H_0 and accept H_A if $B \leq l - 1$. Alternatively, if $n > 20$, reject H_0 and accept H_A if $Z_B \leq -Z_{1-\alpha}$, where Z_B is computed by Eq. 18.3.

When one-sided tests are conducted with Table A14, the α levels indicated in the table are divided by 2. Hence, Table A14 may only be used to make one-sided tests at the 0.025 and 0.005 significance levels.

Trace Concentrations

The sign test can be conducted even though some data are missing or are ND concentrations. See Table 18.2 for a summary of the types of data that can occur, whether or not the sign can be determined, and the effect on n. The effect of decreasing n is to lower the power of the test to indicate differences between the two populations.

18.1.2 Wilcoxon Signed Rank Test

The *Wilcoxon signed rank test* can be used instead of the sign test if the underlying distribution is symmetric, though it need not be a normal distribution. This Wilcoxon test (not to be confused with the Wilcoxon rank sum test discussed in Section 18.2.1) is more complicated to compute than the sign test

Table 18.2 Determination of the Sign Test

Type of Data	*Can Sign Be Computed?*	*Decreases n?*
One or both members of a pair are absent	No	Yes
$x_{2i} = x_{1i}$	No	Yes
One member of a pair is ND	Yes[a]	No
Both members of a pair are ND	No	Yes

[a]If the numerical value is greater than the detection limit of the ND value.

because it requires computing and ranking the D_i. In most situations it should have greater power to find differences in two populations than does the sign test. The null and alternative hypotheses are the same as for the sign test. The test is described by Hollander and Wolfe (1973).

18.1.3 Friedman's Test

Friedman's test is an extension of the sign test from two paired populations to k related populations. The underlying distribution need not be normal or even symmetric. Also, a moderate number of ND values can be accommodated without seriously affecting the test conclusions. However, no missing values are allowed. The null hypothesis is

H_0: There is no tendency for one population to have larger or smaller values than any other of the k populations

The usual alternative hypothesis is

H_A: At least one population tends to have larger values than one or more of the other populations

Examples of "populations" appropriate for Friedman's test are (i) measurements of $k = 3$ or more pollutants on each of n field samples, (ii) measurements of a single pollutant on air filters collected at $k = 3$ or more air monitoring stations for n time periods, or (iii) measurements obtained by $k = 3$ or more analytical laboratories on a set of n identical spiked samples. The data are laid out as follows:

	Block				
	1	2	3	....	n
Population 1	x_{11}	x_{12}	x_{13}	...	x_{1n}
Population 2	x_{21}	x_{22}	x_{23}	...	x_{2n}
⋮					
Population k	x_{k1}	x_{k2}	x_{k3}	...	x_{kn}

The steps in the testing procedure are as follows:

1. For each block, assign the rank 1 to the smallest measurement, the rank 2 to the next largest measurement, . . . , and the rank k to the largest measurement. If two or more measurements in the block are tied, then assign to each the midrank for that tied group (illustrated Example 18.3).
2. Compute R_j, the sum of the ranks for the jth population.
3. If no tied values occur within any block, compute the Friedman test statistic as follows:

$$F_r = \left[\frac{12}{nk(k+1)} \sum_{j=1}^{k} R_j^2\right] - 3n(k+1) \qquad \textbf{18.4}$$

4. If tied values are present within one or more blocks, compute the Friedman statistic as follows:

$$F_r = \frac{12 \sum_{j=1}^{k} \left[R_j - \frac{n(k+1)}{2} \right]^2}{nk(k+1) - \frac{1}{k-1} \sum_{i=1}^{n} \left\{ \left(\sum_{j=1}^{g_i} t_{i,j}^3 \right) - k \right\}} \qquad \textbf{18.5}$$

where g_i is the number of tied groups in block i and $t_{i,j}$ is the number of tied data in the jth tied group in block i. Each untied value within block i is considered to be a "group" of ties of size 1. The quantity in braces { } in the denominator of Eq. 18.5 is zero for any block that contains no ties. This method of handling ties is illustrated in Example 18.3. Equation 18.5 reduces to Eq. 18.4 when there are no ties in any block.

5. For an α level test, reject H_0 and accept H_A if $F_r \geq \chi^2_{1-\alpha,k-1}$, where $\chi^2_{1-\alpha,k-1}$ is the $1 - \alpha$ quantile of the chi-square distribution with $k - 1$ df, as obtained from Table A19, where k is the number of populations. The chi-square distribution is appropriate only if n is reasonably large. Hollander and Wolfe provide exact critical values (their Table A.15) for testing F_r for the following combinations of k and n: $k = 3$, $n = 2, 3, \ldots, 13$; $k = 4$, $n = 2, 3, \ldots, 8$; $k = 5$, $n = 3, 4, 5$. Odeh et al. (1977) extend these tables to $k = 5$, $n = 6, 7, 8$; $k = 6$, $n = 2, 3, 4, 5, 6$. These tables will give only an approximate test if ties are present. The use of F_r computed by Eq. 18.5 and evaluated using the chi-square tables may be preferred in this situation. The foregoing tests are completely valid only if the observations in different blocks are not correlated.

In step 1, if there is one ND value within a block, assign it the rank 1. If there are two or more ND values within a block, treat them as tied values and assign them the midrank. For example, if three NDs are present within a block, each is assigned the rank of 2, the average of 1, 2, and 3. This method of handling NDs assumes all measurements in the block are greater than the detection limit of all the ND values in the block.

EXAMPLE 18.3

The data in Table 18.3 are daily maximum oxidant air concentrations (parts per hundred million) at $k = 5$ monitoring stations in California for the first $n = 6$ days in July 1973 (from Grivet, 1980). We shall use Friedman's procedure to test at the $\alpha = 0.025$ significance level the null hypothesis, H_0, that there is no tendency for any station to

Table 18.3 Daily Maximum Air Concentrations[a] in California During July 1973

Station Number	Day (Block) 1	2	3	4	5	6	Sum of Ranks (R_j)
28783	7 (2)[b]	5 (4)	7 (3.5)	12 (2)	4 (3)	4 (4)	18.5
41541	5 (1)	3 (1.5)	3 (1)	8 (1)	3 (1.5)	2 (1)	7
43382	11 (4)	4 (3)	7 (3.5)	17 (4)	5 (4.5)	4 (4)	23
60335	13 (5)	6 (5)	12 (5)	21 (5)	5 (4.5)	4 (4)	28.5
60336	8 (3)	3 (1.5)	4 (2)	13 (3)	3 (1.5)	3 (2)	13

[a]Data are parts per hundred millions.
[b]Rank of the measurement.

Table 18.4 Computing the Correction for Ties in Eq. 18.5 for Friedman's Test

Blocks	g_i	$t_{i,j}$	$\sum_{j=1}^{g_i} t_{i,j}^3 - k$
2	4	$t_{2,1} = 2,\ t_{2,2} = t_{2,3} = t_{2,4} = 1$	11 − 5 = 6
3	4	$t_{3,1} = 2,\ t_{3,2} = t_{3,3} = t_{3,4} = 1$	11 − 5 = 6
5	3	$t_{5,1} = t_{5,2} = 2,\ t_{5,3} = 1$	17 − 5 = 12
6	3	$t_{6,1} = 3,\ t_{6,2} = t_{6,3} = 1$	29 − 5 = 24
			Sum = 48

have oxidant levels greater or smaller than any other station. Also shown in Table 18.3 are the ranks of the measurements obtained as in step 1 and the sum of the ranks for each station (step 2).

Since there are ties in blocks 2, 3, 5, and 6, we must use Eq. 18.5 to compute F_r. First compute the quantity in braces { } in the denominator of Eq. 18.5 for all blocks that contain ties. This computation is done in Table 18.4.

Note that the values of all the $t_{i,j}$'s in each block must sum to k, the number of populations (stations). Since $k = 5$, $n = 6$, and Sum = 48, Eq. 18.5 is

$$F_r = \frac{12[(18.5 - 18)^2 + (7 - 18)^2 + \cdots + (13 - 18)^2]}{6(5)(6) - 48/4}$$

$$= 20.1$$

For $\alpha = 0.025$ we find from Table A19 that $\chi^2_{0.975,4} = 11.14$. Since $F_r > 11.14$, we reject H_0 and accept the H_A that at least 1 station tends to have daily maximum oxidant concentrations at a different level than the other stations. From Table 18.3 it appears that stations 41541 and 60336 have consistently lower concentrations than the other stations.

18.2 INDEPENDENT DATA SETS

We discuss two nonparametric tests for independent data sets: the Wilcoxon rank sum test (not to be confused with the Wilcoxon signed rank test discussed in Section 18.1.2) and the Kruskal-Wallis rank test, which generalizes the Wilcoxon rank sum test to more than two populations.

18.2.1 Wilcoxon Rank Sum Test

The Wilcoxon rank sum test may be used to test for a shift in location between two independent populations, that is, the measurements from one population tend to be consistently larger (or smaller) than those from the other population. This test is an easily computed alternative to the usual independent-sample t test discussed in most statistics methods books (see, e.g., Snedecor and Cochran, 1980, p. 83). (Do not confuse the independent-sample t test with the paired t test for paired data. The latter is discussed by Snedecor and Cochran, 1980, p. 85.)

The rank sum test has two main advantages over the independent-sample t test: (i) The two data sets need not be drawn from normal distributions, and (ii) the rank sum test can handle a moderate number of ND values by treating them as ties (illustrated in Example 18.4). However, both tests assume that the distributions of the two populations are identical in shape (variance), but the distributions need not be symmetric. Modifications to the t test to account for unequal variances can be made as described in Snedecor and Cochran (1980, p. 96). Evidently, no such modification exists for the rank sum test. Reckhow and Chapra (1983) illustrate the use of the rank sum test on chlorophyll data in two lakes.

Suppose there are n_1 and n_2 data in data sets 1 and 2, respectively (n_1 need not equal n_2). We test

H_0: The populations from which the two data sets have been drawn have the same mean **18.6**

versus the alternative hypothesis

H_A: The populations have different means **18.7**

The Wilcoxon rank sum test procedure is as follows:

1. Consider all $m = n_1 + n_2$ data as one data set. Rank the m data from 1 to m, that is, assign the rank 1 to the smallest datum, the rank 2 to the next largest datum, . . . , and the rank m to the largest datum. If several data have the same value, assign them the midrank, that is, the average of the ranks that would otherwise be assigned to those data.
2. Sum the ranks assigned to the n_1 measurements from population 1 Denote this sum by W_{rs}.
3. If $n_1 \leq 10$ and $n_2 \leq 10$, the test of H_0 may be made by referring W_{rs} to the appropriate critical value in Table A.5 in Hollander and Wolfe (1973) (see their pages 67–74 for the test method).
4. If $n_1 > 10$ and $n_2 > 10$ and no ties are present, compute the large sample statistic

$$Z_{rs} = \frac{W_{rs} - n_1(m + 1)/2}{[n_1 n_2(m + 1)/12]^{1/2}} \qquad \mathbf{18.8}$$

5. If $n_1 > 10$ and $n_2 > 10$ and ties are present, do not compute Eq. 18.8. Instead, compute

$$Z_{rs} = \frac{W_{rs} - n_1(m + 1)/2}{\left\{\dfrac{n_1 n_2}{12}\left[m + 1 - \dfrac{\sum_{j=1}^{g} t_j(t_j^2 - 1)}{m(m - 1)}\right]\right\}^{1/2}} \qquad \mathbf{18.9}$$

where g is the number of tied groups and t_j is the number of tied data in the jth group. Equation 18.9 reduces to Eq. 18.8 when there are no ties.
6. For an α level two-tailed test, reject H_0 (Eq. 18.6) and accept H_A (Eq. 18.7) if $Z_{rs} \leq -Z_{1-\alpha/2}$ or if $Z_{rs} \geq Z_{1-\alpha/2}$.
7. For a one-tailed α level test of H_0 versus the H_A that the measurements from population 1 tend to exceed those from population 2, reject H_0 and accept H_A if $Z_{rs} \geq Z_{1-\alpha}$.

8. For a one-tailed α level test of H_0 versus the H_A that the measurements from population 2 tend to exceed those from population 1, reject H_0 and accept H_A if $Z_{rs} \leq -Z_{1-\alpha}$.

EXAMPLE 18.4

In Table 18.5 are ^{241}Am concentrations (pCi/g) in soil crust material collected within 2 plots, one near ("onsite") and one far ("offsite") from a nuclear reprocessing facility (Price, Gilbert, and Gano, 1981). Twenty measurements were obtained in each plot. We use the Wilcoxon rank sum test to test the null hypothesis that average concentrations at the 2 plots are equal versus the alternative hypothesis that the onsite plot (population 1) has larger concentrations than in the offsite plot (population 2). That is, we perform the test in step 7. We shall use $\alpha = 0.05$. The ranks of the combined data are shown in Table 18.5 and W_{rs} is computed to be 500.

There are $g = 6$ groups of ties. Four groups have length 2, that is, $t = 2$, and 2 groups have $t = 3$. Equation 18.9 gives

$$Z_{rs} = \frac{500 - 20(41)/2}{\{[20(20)/12]\,[41 - [(4)(2)(3) + (2)(3)(8)]/40(39)]\}^{1/2}} = 2.44$$

Performing the test in step 7, since $Z_{rs} > 1.645$, we reject H_0 and accept H_A that the onsite population has larger ^{241}Am concentrations than the offsite plot.

We note that the correction for ties, that is, using Eq. 18.9 instead of Eq. 18.8, will usually have a negligible effect on the value of Z_{rs}. The correction becomes more important if the t_j are large. Also, if NDs are present but occur in only one of the populations, it is still possible to rank all the data and perform the test. For instance in Example 18.4 if the negative concentrations had been reported by the analytical laboratory as ND values, they would still have been assigned the ranks 1, 2, and 3 if NDs were treated as being less in value than the smallest numerical value (0.0056). In addition, if the three ND values had been considered to be tied, all three would have been assigned the

Table 18.5 ^{241}Am Concentrations in (pCi/g) Soil Crust Material

Population 1 (onsite)				*Population 2 (offsite)*			
0.0059	(5)[a]	0.036	(28)	−0.011[b]	(1)	0.019	(16)
0.0074	(7)	0.040	(29)	−0.0088[b]	(2)	0.020	(18.5)
0.015	(9.5)	0.042	(30)	−0.0055[b]	(3)	0.020	(18.5)
0.018	(13.5)	0.045	(31)	0.0056	(4)	0.022	(20)
0.019	(16)	0.046	(32)	0.0063	(6)	0.025	(22)
0.019	(16)	0.053	(34)	0.013	(8)	0.030	(23)
0.024	(21)	0.062	(36)	0.015	(9.5)	0.031	(25)
0.031	(25)	0.066	(37)	0.016	(11.5)	0.050	(33)
0.031	(25)	0.069	(38)	0.016	(11.5)	0.057	(35)
0.034	(27)	0.081	(40)	0.018	(13.5)	0.073	(39)

$W_{rs} = 5 + 7 + 9.5 + \cdots + 38 + 40 = 500.$

[a]Rank of the datum.

[b]Negative measurements reported by the analytical laboratory.

average rank of 2, which would not have changed the value of W_{rs}. If NDs occur in both populations, they can be treated as tied values all less than the smallest numerical value in the combined data set. Hence, they would each receive the average rank value for that group of NDs, and the Wilcoxon test could still be conducted. (See Exercise 18.4.)

18.2.2 Kruskal-Wallis Test

The Kruskal-Wallis test is an extension of the Wilcoxon rank sum test from two to k independent data sets. These data sets need not be drawn from underlying distributions that are normal or even symmetric, but the k distributions are assumed to be identical in shape. A moderate number of tied and ND values can be accommodated. The null hypothesis is

H_0: The populations from which the k data sets have been drawn have the same mean **18.10**

The alternative hypothesis is

H_A: At least one population has a mean larger or smaller than at least one other population **18.11**

The data take the form

Population				
1	2	3	...	k
x_{11}	x_{21}	x_{31}	...	x_{k1}
x_{12}	x_{22}	x_{32}	...	x_{k2}
⋮	⋮	⋮	⋮	⋮
x_{1n_1}	x_{2n_2}	x_{3n_3}		x_{kn_k}

The total number of data is $m = n_1 + n_2 + \cdots + n_k$, where the n_i need not be equal. The steps in the testing procedure are as follows:

1. Rank the m data from smallest to largest, that is, assign the rank 1 to the smallest datum, the rank 2 to the next largest, and so on. If ties occur, assign the midrank (illustrated in Example 18.5). If NDs occur, treat these as a group of tied values that are less than the smallest numerical value in the data set (assuming the detection limit of the ND values is less than the smallest numerical value).
2. Compute the sum of the ranks for each data set. Denote this sum for the jth data set by R_j.
3. If there are no tied or ND values, compute the Kruskal-Wallis statistic as follows:

$$K_w = \left[\frac{12}{m(m+1)} \sum_{j=1}^{k} \frac{R_j^2}{n_j}\right] - 3(m+1) \qquad \textbf{18.12}$$

4. If there are ties or NDs treated as ties, compute a modified Kruskal-Wallis statistic by dividing K_w (Eq. 18.12) by a correction for ties, that is, compute

$$K'_w = \frac{K_w}{1 - \dfrac{1}{m(m^2 - 1)} \sum_{j=1}^{g} t_j(t_j^2 - 1)} \qquad \textbf{18.13}$$

where g is the number of tied groups and t_j is the number of tied data in the jth group. Equation 18.13 reduces to Eq. 18.12 when there are no ties.

5. For an α level test, reject H_0 and accept H_A if $K'_w \geq \chi^2_{1-\alpha,k-1}$, where $\chi^2_{1-\alpha,k-1}$ is the $1 - \alpha$ quantile of the chi-square distribution with $k - 1$ df, as obtained from Table A19, where k is the number of data sets. Iman, Quade, and Alexander (1975) provide exact significance levels for the following cases:

$k = 3 \quad n_i \leq 6$

$n_1 = n_2 = n_3 = 7$

$n_1 = n_2 = n_3 = 8$

$k = 4 \quad n_i \leq 4$

$k = 5 \quad n_i \leq 3$

Less extensive exact tables are given in Conover (1980) and Hollander and Wolfe (1973) for $k = 3$ data sets.

EXAMPLE 18.5

An aliquot-size variability study is conducted in which multiple soil aliquots of sizes 1 g, 10 g, 25 g, 50 g, and 100 g are analyzed for ^{241}Am. A portion of the data for aliquot sizes 1 g, 25 g, and 100 g is used in this example. (Two ND values are added for illustration.) The full data set is discussed by Gilbert and Doctor (1985). We test the null hypothesis that the concentrations from all 3 aliquot sizes have the same mean. The alternative hypothesis is that the concentrations for at least 1 aliquot size tend to be larger or smaller than those for at least 1 other aliquot size. We test at the $\alpha = 0.05$ level. The data, ranks, and rank sums are given in Table 18.6.

Table 18.6 Aliquot-Size Variability Study

^{241}Am Concentrations (nCi/g)		
1 g	*25 g*	*100 g*
1.45 (7)[a]	1.52 (8.5)	1.74 (13)
1.27 (6)	2.46 (22)	2.00 (17.5)
1.17 (4)	1.23 (5)	1.79 (14)
1.01 (3)	2.20 (20)	1.81 (15)
2.30 (21)	2.68 (23)	1.91 (16)
1.54 (10)	1.52 (8.5)	2.11 (19)
1.71 (11.5)	ND (1.5)	2.00 (17.5)
1.71 (11.5)		
ND (1.5)		
$R_1 = 75.5$	$R_2 = 88.5$	$R_3 = 112$
$n_1 = 9$	$n_2 = 7$	$n_3 = 7$

[a] Rank of the datum.
ND = not detected.

There are $g = 4$ groups of ties and $t = 2$ for each group. The modified Kruskal-Wallis statistic (Eq. 18.13) is

$$K_w' = \frac{[12/23(24)](75.5^2/9 + 88.5^2/7 + 112^2/7) - 3(24)}{1 - 4(2)(3)/23(528)} = 5.06$$

From Table A19 we find $\chi^2_{0.95,2} = 5.99$. Since $K_w' < 5.99$, we cannot reject H_0 at the $\alpha = 0.05$ level.

Note that the correction for ties made a negligible difference in the test statistic. However, a bigger correction is obtained if t is large for one or more groups of ties. This could happen if there are many NDs, where t is the number of NDs.

18.3 SUMMARY

This chapter discussed simple nonparametric tests to determine whether observed differences in two or more populations are statistically significant, that is, of a greater magnitude than would be expected to occur by chance. We emphasize the correction for ties that these nonparametric tests provide, since a moderate number of trace or ND measurements can be accommodated by assuming they are a group of tied values. Hollander and Wolfe (1973) and Conover (1980) provide other uses for these tests and discuss related estimation and confidence interval procedures.

EXERCISES

18.1 Use the first 10 days of oxidant data in Example 18.1 to conduct a one-tailed sign test at the $\alpha = 0.025$ level. Use the alternative hypothesis H_A: Maximum oxidant concentrations at station 41541 tend to exceed those at station 28781 more than the reverse.

18.2 Suppose the following paired measurements have been obtained (ND = not detected; M = missing data):

	Pair										
	1	*2*	*3*	*4*	*5*	*6*	*7*	*8*	*9*	*10*	*11*
x_1	ND	7	ND	M	3	M	3	7	12	10	15
x_2	ND	6	6	6	1	M	2	1	11	8	3

Conduct a one-tailed sign test of H_0 versus the H_A that x_1 measurements tend to exceed x_2 measurements more often than the reverse. Use $\alpha = 0.025$.

18.3 Compute Friedman's test, using the data in Example 18.3 and $\alpha = 0.025$. Ignore the correction for ties.

18.4 Suppose all ^{241}Am concentrations less than 0.02 pCi/g in the 2 populations in Example 18.4 were reported by the analytical laboratory as ND. Use the Wilcoxon rank sum test to test H_0: means of both populations are equal versus H_A: the offsite population has a smaller mean than the onsite population. Use $\alpha = 0.025$. What effect do the large number of NDs have on Z_{rs}? Is it more difficult to reject H_0 if NDs are present?

18.5 Suppose that all ^{241}Am measurements less than 1.5 nCi/g in Example 18.5 were reported by the laboratory as ND. Use the Kruskal-Wallis test on the resulting data set. Use $\alpha = 0.05$. (Retain the 2 NDs in Example 18.5.)

ANSWERS

18.1 Denote station 28781 data as x_1 data, and station 41541 data as x_2 data. $B = 5$. From Table A14, $u = 7$. Since $B < 7$, we cannot reject H_0 at the $\alpha = 0.025$ level.

18.2 Delete pairs 1, 4, and 6. $n = 8$, $B = 1$. From Table A14, $l - 1 = 0$. Since $B = 1$, we cannot reject H_0.

18.3 Equation 18.4 gives $F_r = 18.77$. Reject H_0 and accept H_A, since $18.77 > 11.14$, the same result as when the correction for ties was made.

18.4 $W_{rs} = 487$. $g = 3$ with $t = 17$, 2, and 3. $Z_{rs} = 77/35.5163 = 2.168$. Since $Z_{rs} > 1.645$, reject H_0 and accept H_A. The NDs reduced Z_{rs} from 2.436 in Example 18.4 to 2.168. Yes!

18.5 $R_1 = 74$, $R_2 = 90$, $R_3 = 112$, $m = 23$, $K'_w = 5.339/0.97085 = 5.50$. Since $K'_w < 5.99$, we cannot reject H_0 at the $\alpha = 0.05$ level.

Appendix A

Statistical Tables

Table A1 Cumulative Normal Distribution (Values of p Corresponding to Z_p for the Normal Curve)

Z_p	.00	.01	.02	.03	.04	.05	.06	.07	.08	.09
.0	.5000	.5040	.5080	.5120	.5160	.5199	.5239	.5279	.5319	.5359
.1	.5398	.5438	.5478	.5517	.5557	.5596	.5636	.5674	.4714	.5753
.2	.5793	.5832	.5871	.5910	.5948	.5967	.6026	.6064	.6103	.6141
.3	.6179	.6217	.6255	.6293	.6331	.6368	.6406	.6443	.6480	.6517
.4	.6554	.6591	.6628	.6664	.6700	.6736	.6772	.6808	.6844	.6879
.5	.6915	.6950	.6985	.7019	.7054	.7088	.7123	.7157	.7190	.7224
.6	.7257	.7291	.7324	.7357	.7389	.7422	.7454	.7486	.7517	.7549
.7	.7580	.7611	.7642	.7673	.7704	.7734	.7764	.7794	.7823	.7852
.8	.7881	.7910	.7939	.7967	.7995	.8023	.8051	.8078	.8106	.8133
.9	.8159	.8186	.8212	.8238	.8264	.8289	.8315	.8340	.8365	.8389
1.0	.8413	.8438	.8461	.8485	.8508	.8531	.8554	.8577	.8599	.8621
1.1	.8643	.8665	.8686	.8708	.8729	.8749	.8770	.8790	.8810	.8830
1.2	.8849	.8869	.8888	.8907	.8925	.8944	.8962	.8980	.8997	.9015
1.3	.9032	.9049	.9066	.9082	.9099	.9115	.9131	.9147	.9162	.9177
1.4	.9192	.9207	.9222	.9236	.9251	.9265	.9279	.9292	.9306	.9319
1.5	.9332	.9345	.9357	.9370	.9382	.9394	.9406	.9418	.9429	.9441
1.6	.9452	.9463	.9474	.9484	.9495	.9505	.9515	.9525	.9535	.9545
1.7	.9554	.9564	.9573	.9582	.9591	.9599	.9608	.9616	.9625	.9633
1.8	.9641	.9649	.9656	.9664	.9671	.9678	.9686	.9693	.9699	.9706
1.9	.9713	.9719	.9726	.9732	.9738	.9744	.9750	.9756	.9761	.9767
2.0	.9772	.9778	.9783	.9788	.9793	.9798	.9803	.9808	.9812	.9817
2.1	.9821	.9826	.9830	.9834	.9838	.9842	.9846	.9850	.9854	.9857
2.2	.9861	.9864	.9868	.9871	.9875	.9878	.9881	.9884	.9887	.9890
2.3	.9893	.9896	.9898	.9901	.9904	.9906	.9909	.9911	.9913	.9916
2.4	.9918	.9920	.9922	.9925	.9927	.9929	.9931	.9932	.9934	.9936
2.5	.9938	.9940	.9941	.9943	.9945	.9946	.9948	.9949	.9951	.9952
2.6	.9953	.9955	.9956	.9957	.9959	.9960	.9961	.9962	.9963	.9964
2.7	.9965	.9966	.9967	.9968	.9969	.9970	.9971	.9972	.9973	.9974
2.8	.9974	.9975	.9976	.9977	.9977	.9978	.9979	.9979	.9980	.9981
2.9	.9981	.9982	.9982	.9983	.9984	.9984	.9985	.9985	.9986	.9986
3.0	.9987	.9987	.9987	.9988	.9986	.9989	.9989	.9989	.9990	.9990
3.1	.9990	.9991	.9991	.9991	.9992	.9992	.9992	.9992	.9993	.9993
3.2	.9993	.9993	.9994	.9994	.9994	.9994	.9994	.9995	.9995	.9995
3.3	.9995	.9995	.9995	.9996	.9996	.9996	.9996	.9996	.9996	.9997
3.4	.9997	.9997	.9997	.9997	.9997	.9997	.9997	.9997	.9997	.9998

Source: After Pearson and Hartley, 1966.
This table is first used in Section 4.4.2.

Table A2 Quantiles of the t Distribution (Values of t Such That $100p\%$ of the Distribution Is Less Than t_p)

Degrees of Freedom	$t_{0.60}$	$t_{0.70}$	$t_{0.80}$	$t_{0.90}$	$t_{0.95}$	$t_{0.975}$	$t_{0.990}$	$t_{0.995}$
1	.325	.727	1.376	3.078	6.314	12.706	31.821	63.657
2	.289	.617	1.061	1.886	2.920	4.303	6.965	9.925
3	.277	.584	.978	1.638	2.353	3.182	4.541	5.841
4	.271	.569	.941	1.533	2.132	2.776	3.747	4.604
5	.267	.559	.920	1.476	2.015	2.571	3.365	4.032
6	.265	.553	.906	1.440	1.943	2.447	3.143	3.707
7	.263	.549	.896	1.415	1.895	2.365	2.998	3.499
8	.262	.546	.889	1.397	1.860	2.306	2.896	3.355
9	.261	.543	.883	1.383	1.833	2.262	2.821	3.250
10	.260	.542	.879	1.372	1.812	2.228	2.764	3.169
11	.260	.540	.876	1.363	1.796	2.201	2.718	3.106
12	.259	.539	.873	1.356	1.782	2.179	2.681	3.055
13	.259	.538	.870	1.350	1.771	2.160	2.650	3.012
14	.258	.537	.868	1.345	1.761	2.145	2.624	2.977
15	.258	.536	.866	1.341	1.753	2.131	2.602	2.947
16	.258	.535	.865	1.337	1.746	2.120	2.583	2.921
17	.257	.534	.863	1.333	1.740	2.110	2.567	2.898
18	.257	.534	.862	1.330	1.734	2.101	2.552	2.878
19	.257	.533	.861	1.328	1.729	2.093	2.539	2.861
20	.257	.533	.860	1.325	1.725	2.086	2.528	2.845
21	.257	.532	.859	1.323	1.721	2.080	2.518	2.831
22	.256	.532	.858	1.321	1.717	2.074	2.508	2.819
23	.256	.532	.858	1.319	1.714	2.069	2.500	2.807
24	.256	.531	.857	1.318	1.711	2.064	2.492	2.797
25	.256	.531	.856	1.316	1.708	2.060	2.485	2.787
26	.256	.531	.856	1.315	1.706	2.056	2.479	2.779
27	.256	.531	.855	1.314	1.703	2.052	2.473	2.771
28	.256	.530	.855	1.313	1.701	2.048	2.467	2.763
29	.256	.530	.854	1.311	1.699	2.045	2.462	2.756
30	.256	.530	.854	1.310	1.697	2.042	2.457	2.750
40	.255	.529	.851	1.303	1.684	2.021	2.423	2.704
60	.254	.527	.848	1.296	1.671	2.000	2.390	2.660
120	.254	.526	.845	1.289	1.658	1.980	2.358	2.617
∞	.253	.524	.842	1.282	1.645	1.960	2.326	2.576

Source: From Fisher and Yates, 1974. Used by permission.
This table is first used in Section 4.4.2.

Table A3 Factors $K_{1-\alpha,p}$ for Estimating an Upper $100(1-\alpha)\%$ Confidence Limit on the *p*th Quantile of a Normal Distribution

	1 - α = 0.90				
	p				
n	0.900	0.950	0.975	0.990	0.999
2	10.253	13.090	15.586	18.500	24.582
3	4.258	5.311	6.244	7.340	9.651
4	3.188	3.957	4.637	5.438	7.129
5	2.744	3.401	3.983	4.668	6.113
6	2.494	3.093	3.621	4.243	5.556
7	2.333	2.893	3.389	3.972	5.201
8	2.219	2.754	3.227	3.783	4.955
9	2.133	2.650	3.106	3.641	4.771
10	2.066	2.568	3.011	3.532	4.628
11	2.012	2.503	2.936	3.444	4.515
12	1.966	2.448	2.872	3.371	4.420
13	1.928	2.403	2.820	3.310	4.341
14	1.895	2.363	2.774	3.257	4.274
15	1.866	2.329	2.735	3.212	4.215
16	1.842	2.299	2.700	3.172	4.164
17	1.819	2.272	2.670	3.137	4.118
18	1.800	2.249	2.643	3.106	4.078
19	1.781	2.228	2.618	3.078	4.041
20	1.765	2.208	2.597	3.052	4.009
21	1.750	2.190	2.575	3.028	3.979
22	1.736	2.174	2.557	3.007	3.952
23	1.724	2.159	2.540	2.987	3.927
24	1.712	2.145	2.525	2.969	3.904
25	1.702	2.132	2.510	2.952	3.882
30	1.657	2.080	2.450	2.884	3.794
35	1.623	2.041	2.406	2.833	3.730
40	1.598	2.010	2.371	2.793	3.679
45	1.577	1.986	2.344	3.762	3.638
50	1.560	1.965	2.320	2.735	3.604
60	1.532	1.933	2.284	2.694	3.552
70	1.511	1.909	2.257	2.663	3.513
80	1.495	1.890	2.235	2.638	3.482
90	1.481	1.874	2.217	2.618	3.456
100	1.470	1.861	2.203	2.601	3.435
120	1.452	1.841	2.179	2.574	3.402
145	1.436	1.821	2.158	2.550	3.371
300	1.386	1.765	2.094	2.477	3.280
500	1.362	1.736	2.062	2.442	3.235
∞	1.282	1.645	1.960	2.326	3.090

	1 - α = 0.95				
	p				
n	0.900	0.950	0.975	0.990	0.999
2	20.581	26.260	31.257	37.094	49.276
3	6.155	7.656	8.986	10.553	13.857
4	4.162	5.144	6.015	7.042	9.214
5	3.413	4.210	4.916	5.749	7.509
6	3.008	3.711	4.332	5.065	6.614
7	2.756	3.401	3.971	4.643	6.064
8	2.582	3.188	3.724	4.355	5.689
9	2.454	3.032	3.543	4.144	5.414
10	2.355	2.911	3.403	3.981	5.204
11	2.275	2.815	3.291	3.852	5.036
12	2.210	2.736	3.201	3.747	4.900
13	2.155	2.670	3.125	3.659	4.787
14	2.108	2.614	3.060	3.585	4.690
15	2.068	2.566	3.005	3.520	4.607
16	2.032	2.523	2.956	3.463	4.534
17	2.002	2.486	2.913	3.414	4.471
18	1.974	2.453	2.875	3.370	4.415
19	1.949	2.423	2.840	3.331	4.364
20	1.926	2.396	2.809	3.295	4.319
21	1.905	2.371	2.781	3.262	4.276
22	1.887	2.350	2.756	3.233	4.238
23	1.869	2.329	2.732	3.206	4.204
24	1.853	2.309	2.711	3.181	4.171
25	1.838	2.292	2.691	3.158	4.143
30	1.778	2.220	2.608	3.064	4.022
35	1.732	2.166	2.548	2.994	3.934
40	1.697	2.126	2.501	2.941	3.866
45	1.669	2.092	2.463	2.897	3.811
50	1.646	2.065	2.432	2.863	3.766
60	1.609	2.022	2.384	2.807	3.695
70	1.581	1.990	2.348	2.766	3.643
80	1.560	1.965	2.319	2.733	3.601
90	1.542	1.944	2.295	2.706	3.567
100	1.527	1.927	2.276	2.684	3.539
120	1.503	1.899	2.245	2.649	3.495
145	1.481	1.874	2.217	2.617	3.455
300	1.417	1.800	2.133	2.522	3.335
500	1.385	1.763	2.092	2.475	3.277
∞	1.282	1.645	1.960	2.326	3.090

Source: From Owen, 1962. Used by permission.
This table is used in Section 11.3.

Table A4 Nonparametric 95% and 99% Confidence Intervals on a Proportion

u	n = 1	n = 2	n = 3	n = 4	n = 5	n = 6	u
0	0 0 .95 .99	0 0 .78 .90	0 0 .63 .78	0 0 .53 .68	0 0 .50 .60	0 0 .41 .54	0
1	.01 .05 1 1	.01 .03 .97 .99	.00 .02 .86 .94	.00 .01 .75 .86	.00 .01 .66 .78	.00 .01 .59 .71	1
2		.10 .22 1 1	.06 .14 .98 1	.04 .10 .90 .96	.03 .08 .81 .89	.03 .06 .73 .83	2
3			.22 .37 1 1	.14 .25 .99 1	.11 .19 .92 .97	.08 .15 .85 .92	3

u	n = 7	n = 8	n = 9	n = 10	n = 11	n = 12	u
0	0 0 .38 .50	0 0 .36 .45	0 0 .32 .43	0 0 .29 .38	0 0 .26 .36	0 0 .24 .25	0
1	.00 .01 .55 .64	.00 .01 .50 .59	.00 .01 .44 .57	.00 .01 .44 .51	.00 .00 .40 .50	.00 .00 .37 .45	1
2	.02 .05 .66 .76	.02 .05 .64 .71	.02 .04 .56 .66	.02 .04 .56 .62	.01 .03 .50 .59	.01 .03 .46 .55	2
3	.07 .13 .77 .86	.06 .11 .71 .80	.05 .10 .68 .75	.05 .09 .62 .70	.04 .08 .60 .66	.04 .07 .54 .65	3
4	.14 .23 .87 .93	.12 .19 .81 .88	.11 .17 .75 .83	.09 .15 .70 .78	.08 .14 .67 .74	.08 .12 .63 .70	4
5	.24 .34 .95 .98	.20 .29 .89 .94	.17 .25 .83 .89	.15 .22 .78 .85	.13 .20 .74 .81	.12 .18 .71 .77	5
6	.36 .45 .99 1	.29 .36 .95 .98	.25 .32 .90 .95	.22 .29 .85 .91	.19 .26 .80 .87	.17 .24 .76 .83	6

u	n = 13	n = 14	n = 15	n = 16	n = 17	n = 18	u
0	0 0 .23 .32	0 0 .23 .30	0 0 .22 .28	0 0 .20 .26	0 0 .19 .26	0 0 .18 .25	0
1	.00 .00 .34 .43	.00 .00 .32 .42	.00 .00 .30 .39	.00 .00 .30 .36	.00 .00 .28 .35	.00 .00 .27 .34	1
2	.01 .03 .43 .52	.01 .03 .42 .50	.01 .02 .39 .46	.01 .02 .37 .45	.01 .02 .35 .43	.01 .02 .33 .41	2
3	.04 .07 .52 .59	.03 .06 .50 .58	.03 .06 .47 .54	.03 .05 .44 .52	.03 .05 .42 .50	.03 .05 .41 .47	3
4	.07 .11 .59 .68	.06 .10 .58 .64	.06 .10 .53 .61	.06 .09 .50 .58	.05 .08 .49 .57	.05 .08 .47 .53	4
5	.11 .17 .66 .73	.10 .15 .63 .70	.09 .14 .61 .67	.09 .13 .56 .64	.08 .12 .54 .62	.08 .12 .53 .59	5
6	.16 .22 .74 .79	.15 .21 .68 .75	.13 .19 .67 .72	.13 .18 .63 .70	.12 .17 .59 .66	.11 .16 .59 .66	6
7	.21 .26 .78 .84	.19 .24 .76 .81	.18 .22 .71 .77	.17 .20 .70 .74	.16 .19 .65 .73	.15 .18 .63 .69	7
8	.27 .34 .83 .89	.25 .32 .79 .85	.23 .29 .78 .82	.21 .27 .73 .79	.20 .25 .72 .76	.18 .24 .67 .75	8
9	.32 .41 .89 .93	.30 .37 .85 .90	.28 .33 .81 .87	.26 .30 .80 .83	.24 .28 .75 .80	.23 .27 .73 .77	9

u	n = 19	n = 20	n = 21	n = 22	n = 23	n = 24	u
0	0 0 .17 .24	0 0 .16 .22	0 0 .15 .21	0 0 .15 .20	0 0 .14 .19	0 0 .13 .19	0
1	.00 .00 .25 .32	.00 .00 .24 .31	.00 .00 .23 .29	.00 .00 .22 .28	.00 .00 .21 .27	.00 .00 .20 .26	1
2	.01 .02 .32 .39	.01 .02 .32 .37	.01 .02 .30 .37	.01 .02 .29 .35	.01 .02 .27 .33	.01 .02 .26 .32	2
3	.02 .04 .39 .46	.02 .04 .37 .44	.02 .04 .35 .42	.02 .04 .34 .40	.02 .04 .32 .39	.02 .03 .31 .39	3
4	.05 .08 .45 .52	.04 .07 .42 .50	.04 .07 .40 .47	.04 .06 .39 .45	.04 .06 .39 .45	.04 .06 .37 .43	4
5	.07 .11 .50 .56	.07 .10 .47 .56	.07 .10 .46 .53	.06 .09 .45 .50	.06 .09 .43 .50	.06 .09 .41 .48	5
6	.10 .15 .55 .61	.10 .14 .53 .60	.09 .13 .51 .58	.09 .13 .50 .55	.08 .12 .48 .55	.08 .11 .46 .52	6
7	.14 .17 .61 .68	.13 .16 .58 .64	.12 .15 .55 .63	.12 .15 .55 .60	.11 .14 .52 .58	.11 .13 .50 .57	7
8	.17 .22 .66 .71	.16 .21 .63 .69	.15 .20 .60 .66	.15 .19 .58 .65	.14 .18 .57 .62	.13 .17 .54 .61	8
9	.21 .25 .69 .76	.20 .24 .68 .73	.19 .23 .65 .71	.18 .22 .62 .68	.17 .21 .61 .67	.16 .20 .59 .64	9
10	.24 .31 .75 .79	.22 .29 .71 .78	.21 .28 .70 .74	.20 .26 .66 .72	.19 .25 .64 .70	.19 .23 .63 .68	10
11	.29 .34 .78 .83	.27 .32 .76 .80	.26 .30 .72 .79	.24 .29 .71 .76	.23 .27 .68 .73	.22 .26 .66 .72	11
12	.32 .39 .83 .86	.31 .37 .79 .84	.29 .35 .77 .81	.28 .34 .74 .80	.27 .32 .73 .77	.26 .31 .69 .74	12

u	n = 25	n = 26	n = 27	n = 28	n = 29	n = 30	u
0	0 0 .13 .18	0 0 .12 .17	0 0 .12 .17	0 0 .12 .16	0 0 .11 .16	0 0 .11 .16	0
1	.00 .00 .19 .26	.00 .00 .19 .25	.00 .00 .18 .24	.00 .00 .17 .23	.00 .00 .17 .22	.00 .00 .16 .22	1
2	.01 .01 .25 .31	.01 .01 .24 .30	.01 .01 .23 .30	.01 .01 .23 .29	.01 .01 .22 .28	.01 .01 .21 .27	2
3	.02 .03 .30 .37	.02 .03 .30 .36	.02 .03 .29 .34	.02 .03 .28 .33	.02 .03 .27 .32	.01 .03 .26 .31	3
4	.03 .06 .36 .41	.03 .05 .34 .40	.03 .05 .33 .38	.03 .05 .32 .38	.03 .05 .31 .37	.03 .05 .30 .36	4
5	.05 .08 .40 .46	.05 .08 .38 .44	.05 .08 .37 .44	.05 .07 .36 .42	.05 .07 .36 .41	.04 .07 .35 .39	5
6	.08 .11 .44 .50	.07 .11 .42 .49	.07 .10 .41 .48	.07 .10 .41 .46	.07 .09 .39 .44	.06 .09 .38 .43	6
7	.10 .13 .48 .54	.10 .12 .47 .53	.09 .12 .46 .52	.09 .12 .44 .50	.09 .11 .43 .48	.08 .11 .41 .47	7
8	.13 .16 .52 .59	.12 .15 .51 .56	.12 .15 .50 .56	.11 .14 .48 .54	.11 .14 .46 .52	.10 .13 .45 .51	8
9	.16 .19 .56 .63	.15 .19 .54 .60	.14 .18 .54 .59	.14 .17 .52 .58	.13 .17 .50 .56	.13 .16 .48 .54	9
10	.18 .22 .60 .66	.17 .21 .58 .64	.17 .20 .57 .62	.16 .19 .56 .62	.16 .18 .54 .59	.15 .18 .52 .57	10
11	.21 .25 .64 .69	.19 .24 .62 .68	.18 .23 .60 .66	.18 .23 .59 .64	.17 .22 .57 .63	.16 .21 .55 .61	11
12	.25 .30 .68 .74	.23 .28 .66 .70	.22 .27 .63 .70	.21 .26 .62 .67	.21 .25 .61 .65	.20 .24 .59 .64	12
13	.26 .32 .70 .75	.25 .30 .70 .75	.24 .29 .67 .72	.23 .28 .65 .71	.22 .27 .64 .68	.22 .26 .62 .67	13
14	.31 .36 .75 .79	.30 .34 .72 .77	.28 .33 .71 .76	.27 .32 .68 .73	.26 .31 .66 .72	.25 .30 .65 .69	14
15	.34 .40 .78 .82	.32 .38 .76 .81	.30 .37 .73 .78	.29 .35 .72 .77	.28 .34 .69 .74	.27 .32 .68 .73	15

Source: After Blyth and Still, 1983.

Inner entries give the 95% interval, and outer entries the 99% interval. For example, for $n = 13$, $u = 3$, the 95% interval is (0.07, 0.52) and the 99% interval is (0.04, 0.59). n = number of observations. u = number of those that exceed some specified value x_c.

This table is used in Section 11.11.

Table A5 Quantiles of the Rank von Neumann Statistic R_v

T	$R_{0.005}$	$R_{0.010}$	$R_{0.025}$	$R_{0.050}$	$R_{0.100}$
10	0.62	0.72	0.89	1.04	1.23
11	0.67	0.77	0.93	1.08	1.26
12	0.71	0.81	0.96	1.11	1.29
13	0.74	0.84	1.00	1.14	1.32
14	0.78	0.87	1.03	1.17	1.34
15	0.81	0.90	1.05	1.19	1.36
16	0.84	0.93	1.08	1.21	1.38
17	0.87	0.96	1.10	1.24	1.40
18	0.89	0.98	1.13	1.26	1.41
19	0.92	1.01	1.15	1.27	1.43
20	0.94	1.03	1.17	1.29	1.44
21	0.96	1.05	1.18	1.31	1.45
22	0.98	1.07	1.20	1.32	1.46
23	1.00	1.09	1.22	1.33	1.48
24	1.02	1.10	1.23	1.35	1.49
25	1.04	1.12	1.25	1.36	1.50
26	1.05	1.13	1.26	1.37	1.51
27	1.07	1.15	1.27	1.38	1.51
28	1.08	1.16	1.28	1.39	1.52
29	1.10	1.18	1.30	1.40	1.53
30	1.11	1.19	1.31	1.41	1.54
32	1.13	1.21	1.33	1.43	1.55
34	1.16	1.23	1.35	1.45	1.57
36	1.18	1.25	1.36	1.46	1.58
38	1.20	1.27	1.38	1.48	1.59
40	1.22	1.29	1.39	1.49	1.60
42	1.24	1.30	1.41	1.50	1.61
44	1.25	1.32	1.42	1.51	1.62
46	1.27	1.33	1.43	1.52	1.63
48	1.28	1.35	1.45	1.53	1.63
50	1.29	1.36	1.46	1.54	1.64
55	1.33	1.39	1.48	1.56	1.66
60	1.35	1.41	1.50	1.58	1.67
65	1.38	1.43	1.52	1.60	1.68
70	1.40	1.45	1.54	1.61	1.70
75	1.42	1.47	1.55	1.62	1.71
80	1.44	1.49	1.57	1.64	1.71
85	1.45	1.50	1.58	1.65	1.72
90	1.47	1.52	1.59	1.66	1.73
95	1.48	1.53	1.60	1.66	1.74
100	1.49	1.54	1.61	1.67	1.74

Source: From Bartels, 1982. Used by permission.
This table is used in Section 11.13.

Table A6 Coefficients a_i for the Shapiro-Wilk W Test for Normality

i \ n	2	3	4	5	6	7	8	9	10
1	0.7071	0.7071	0.6872	0.6646	0.6431	0.6233	0.6052	0.5888	0.5739
2	-	0.0000	0.1677	0.2413	0.2806	0.3031	0.3164	0.3244	0.3291
3	-	-	-	0.0000	0.0875	0.1401	0.1743	0.1976	0.2141
4	-	-	-	-	-	0.0000	0.0561	0.0947	0.1224
5	-	-	-	-	-	-	-	0.0000	0.0399

i \ n	11	12	13	14	15	16	17	18	19	20
1	0.5601	0.5475	0.5359	0.5251	0.5150	0.5056	0.4968	0.4886	0.4808	0.4734
2	0.3315	0.3325	0.3325	0.3318	0.3306	0.3290	0.3273	0.3253	0.3232	0.3211
3	0.2260	0.2347	0.2412	0.2460	0.2495	0.2521	0.2540	0.2553	0.2561	0.2565
4	0.1429	0.1586	0.1707	0.1802	0.1878	0.1939	0.1988	0.2027	0.2059	0.2085
5	0.0695	0.0922	0.1099	0.1240	0.1353	0.1447	0.1524	0.1587	0.1641	0.1686
6	0.0000	0.0303	0.0539	0.0727	0.0880	0.1005	0.1109	0.1197	0.1271	0.1334
7	-	-	0.0000	0.0240	0.0433	0.0593	0.0725	0.0837	0.0932	0.1013
8	-	-	-	-	0.0000	0.0196	0.0359	0.0496	0.0612	0.0711
9	-	-	-	-	-	-	0.0000	0.0163	0.0303	0.0422
10	-	-	-	-	-	-	-	-	0.0000	0.0140

i \ n	21	22	23	24	25	26	27	28	29	30
1	0.4643	0.4590	0.4542	0.4493	0.4450	0.4407	0.4366	0.4328	0.4291	0.4254
2	0.3185	0.3156	0.3126	0.3098	0.3069	0.3043	0.3018	0.2992	0.2968	0.2944
3	0.2578	0.2571	0.2563	0.2554	0.2543	0.2533	0.2522	0.2510	0.2499	0.2487
4	0.2119	0.2131	0.2139	0.2145	0.2148	0.2151	0.2152	0.2151	0.2150	0.2148
5	0.1736	0.1764	0.1787	0.1807	0.1822	0.1836	0.1848	0.1857	0.1864	0.1870
6	0.1399	0.1443	0.1480	0.1512	0.1539	0.1563	0.1584	0.1601	0.1616	0.1630
7	0.1092	0.1150	0.1201	0.1245	0.1283	0.1316	0.1346	0.1372	0.1395	0.1415
8	0.0804	0.0878	0.0941	0.0997	0.1046	0.1089	0.1128	0.1162	0.1192	0.1219
9	0.0530	0.0618	0.0696	0.0764	0.0823	0.0876	0.0923	0.0965	0.1002	0.1036
10	0.0263	0.0368	0.0459	0.0539	0.0610	0.0672	0.0728	0.0778	0.0822	0.0862
11	0.0000	0.0122	0.0228	0.0321	0.0403	0.0476	0.0540	0.0598	0.0650	0.0697
12	-	-	0.0000	0.0107	0.0200	0.0284	0.0358	0.0424	0.0483	0.0537
13	-	-	-	-	0.0000	0.0094	0.0178	0.0253	0.0320	0.0381
14	-	-	-	-	-	-	0.0000	0.0084	0.0159	0.0227
15	-	-	-	-	-	-	-	-	0.0000	0.0076

Source: From Shapiro and Wilk, 1965. Used by permission.
This table is used in Section 12.3.1.

Table A6 (continued)

i \ n	31	32	33	34	35	36	37	38	39	40
1	0.4220	0.4188	0.4156	0.4127	0.4096	0.4068	0.4040	0.4015	0.3989	0.3964
2	0.2921	0.2898	0.2876	0.2854	0.2834	0.2813	0.2794	0.2774	0.2755	0.2737
3	0.2475	0.2462	0.2451	0.2439	0.2427	0.2415	0.2403	0.2391	0.2380	0.2368
4	0.2145	0.2141	0.2137	0.2132	0.2127	0.2121	0.2116	0.2110	0.2104	0.2098
5	0.1874	0.1878	0.1880	0.1882	0.1883	0.1883	0.1883	0.1881	0.1880	0.1878
6	0.1641	0.1651	0.1660	0.1667	0.1673	0.1678	0.1683	0.1686	0.1689	0.1691
7	0.1433	0.1449	0.1463	0.1475	0.1487	0.1496	0.1505	0.1513	0.1520	0.1526
8	0.1243	0.1265	0.1284	0.1301	0.1317	0.1331	0.1344	0.1356	0.1366	0.1376
9	0.1066	0.1093	0.1118	0.1140	0.1160	0.1179	0.1196	0.1211	0.1225	0.1237
10	0.0899	0.0931	0.0961	0.0988	0.1013	0.1036	0.1056	0.1075	0.1092	0.1108
11	0.0739	0.0777	0.0812	0.0844	0.0873	0.0900	0.0924	0.0947	0.0967	0.0986
12	0.0585	0.0629	0.0669	0.0706	0.0739	0.0770	0.0798	0.0824	0.0848	0.0870
13	0.0435	0.0485	0.0530	0.0572	0.0610	0.0645	0.0677	0.0706	0.0733	0.0759
14	0.0289	0.0344	0.0395	0.0441	0.0484	0.0523	0.0559	0.0592	0.0622	0.0651
15	0.0144	0.0206	0.0262	0.0314	0.0361	0.0404	0.0444	0.0481	0.0515	0.0546
16	0.0000	0.0068	0.0131	0.0187	0.0239	0.0287	0.0331	0.0372	0.0409	0.0444
17	-	-	0.0000	0.0062	0.0119	0.0172	0.0220	0.0264	0.0305	0.0343
18	-	-	-	-	0.0000	0.0057	0.0110	0.0158	0.0203	0.0244
19	-	-	-	-	-	-	0.0000	0.0053	0.0101	0.0146
20	-	-	-	-	-	-	-	-	0.0000	0.0049

i \ n	41	42	43	44	45	46	47	48	49	50
1	0.3940	0.3917	0.3894	0.3872	0.3850	0.3830	0.3808	0.3789	0.3770	0.3751
2	0.2719	0.2701	0.2684	0.2667	0.2651	0.2635	0.2620	0.2604	0.2589	0.2574
3	0.2357	0.2345	0.2334	0.2323	0.2313	0.2302	0.2291	0.2281	0.2271	0.2260
4	0.2091	0.2085	0.2078	0.2072	0.2065	0.2058	0.2052	0.2045	0.2038	0.2032
5	0.1876	0.1874	0.1871	0.1868	0.1865	0.1862	0.1859	0.1855	0.1851	0.1847
6	0.1693	0.1694	0.1695	0.1695	0.1695	0.1695	0.1695	0.1693	0.1692	0.1691
7	0.1531	0.1535	0.1539	0.1542	0.1545	0.1548	0.1550	0.1551	0.1553	0.1554
8	0.1384	0.1392	0.1398	0.1405	0.1410	0.1415	0.1420	0.1423	0.1427	0.1430
9	0.1249	0.1259	0.1269	0.1278	0.1286	0.1293	0.1300	0.1306	0.1312	0.1317
10	0.1123	0.1136	0.1149	0.1160	0.1170	0.1180	0.1189	0.1197	0.1205	0.1212
11	0.1004	0.1020	0.1035	0.1049	0.1062	0.1073	0.1085	0.1095	0.1105	0.1113
12	0.0891	0.0909	0.0927	0.0943	0.0959	0.0972	0.0986	0.0998	0.1010	0.1020
13	0.0782	0.0804	0.0824	0.0842	0.0860	0.0876	0.0892	0.0906	0.0919	0.0932
14	0.0677	0.0701	0.0724	0.0745	0.0765	0.0783	0.0801	0.0817	0.0832	0.0846
15	0.0575	0.0602	0.0628	0.0651	0.0673	0.0694	0.0713	0.0731	0.0748	0.0764
16	0.0476	0.0506	0.0534	0.0560	0.0584	0.0607	0.0628	0.0648	0.0667	0.0685
17	0.0379	0.0411	0.0442	0.0471	0.0497	0.0522	0.0546	0.0568	0.0588	0.0608
18	0.0283	0.0318	0.0352	0.0383	0.0412	0.0439	0.0465	0.0489	0.0511	0.0532
19	0.0188	0.0227	0.0263	0.0296	0.0328	0.0357	0.0385	0.0411	0.0436	0.0459
20	0.0094	0.0136	0.0175	0.0211	0.0245	0.0277	0.0307	0.0335	0.0361	0.0386
21	0.0000	0.0045	0.0087	0.0126	0.0163	0.0197	0.0229	0.0259	0.0288	0.0314
22	-	-	0.0000	0.0042	0.0081	0.0118	0.0153	0.0185	0.0215	0.0244
23	-	-	-	-	0.0000	0.0039	0.0076	0.0111	0.0143	0.0174
24	-	-	-	-	-	-	0.0000	0.0037	0.0071	0.0104
25	-	-	-	-	-	-	-	-	0.0000	0.0035

Table A7 Quantiles of the Shapiro-Wilk *W* Test for Normality (Values of *W* Such That 100*p*% of the Distribution of *W* Is Less Than W_p)

n	$W_{0.01}$	$W_{0.02}$	$W_{0.05}$	$W_{0.10}$	$W_{0.50}$
3	0.753	0.756	0.767	0.789	0.959
4	0.687	0.707	0.748	0.792	0.935
5	0.686	0.715	0.762	0.806	0.927
6	0.713	0.743	0.788	0.826	0.927
7	0.730	0.760	0.803	0.838	0.928
8	0.749	0.778	0.818	0.851	0.932
9	0.764	0.791	0.829	0.859	0.935
10	0.781	0.806	0.842	0.869	0.938
11	0.792	0.817	0.850	0.876	0.940
12	0.805	0.828	0.859	0.883	0.943
13	0.814	0.837	0.866	0.889	0.945
14	0.825	0.846	0.874	0.895	0.947
15	0.835	0.855	0.881	0.901	0.950
16	0.844	0.863	0.887	0.906	0.952
17	0.851	0.869	0.892	0.910	0.954
18	0.858	0.874	0.897	0.914	0.956
19	0.863	0.879	0.901	0.917	0.957
20	0.868	0.884	0.905	0.920	0.959
21	0.873	0.888	0.908	0.923	0.960
22	0.878	0.892	0.911	0.926	0.961
23	0.881	0.895	0.914	0.928	0.962
24	0.884	0.898	0.916	0.930	0.963
25	0.888	0.901	0.918	0.931	0.964
26	0.891	0.904	0.920	0.933	0.965
27	0.894	0.906	0.923	0.935	0.965
28	0.896	0.908	0.924	0.936	0.966
29	0.898	0.910	0.926	0.937	0.966
30	0.900	0.912	0.927	0.939	0.967
31	0.902	0.914	0.929	0.940	0.967
32	0.904	0.915	0.930	0.941	0.968
33	0.906	0.917	0.931	0.942	0.968
34	0.908	0.919	0.933	0.943	0.969
35	0.910	0.920	0.934	0.944	0.969
36	0.912	0.922	0.935	0.945	0.970
37	0.914	0.924	0.936	0.946	0.970
38	0.916	0.925	0.938	0.947	0.971
39	0.917	0.927	0.939	0.948	0.971
40	0.919	0.928	0.940	0.949	0.972
41	0.920	0.929	0.941	0.950	0.972
42	0.922	0.930	0.942	0.951	0.972
43	0.923	0.932	0.943	0.951	0.973
44	0.924	0.933	0.944	0.952	0.973
45	0.926	0.934	0.945	0.953	0.973
46	0.927	0.935	0.945	0.953	0.974
47	0.928	0.936	0.946	0.954	0.974
48	0.929	0.937	0.947	0.954	0.974
49	0.929	0.937	0.947	0.955	0.974
50	0.930	0.938	0.947	0.955	0.974

Source: After Shapiro and Wilk, 1965.

The null hypothesis of a normal distribution is rejected at the α significance level if the calculated W is less than W_α.

This table is used in Section 12.3.1.

Table A8 Quantiles of D'Agostino's Test for Normality (Values of Y Such That $100p\%$ of the Distribution of Y is Less Than Y_p)

n	$Y_{0.005}$	$Y_{0.01}$	$Y_{0.025}$	$Y_{0.05}$	$Y_{0.10}$	$Y_{0.90}$	$Y_{0.95}$	$Y_{0.975}$	$Y_{0.99}$	$Y_{0.995}$
50	-3.949	-3.442	-2.757	-2.220	-1.661	0.759	0.923	1.038	1.140	1.192
60	-3.846	-3.360	-2.699	-2.179	-1.634	0.807	0.986	1.115	1.236	1.301
70	-3.762	-3.293	-2.652	-2.146	-1.612	0.844	1.036	1.176	1.312	1.388
80	-3.693	-3.237	-2.613	-2.118	-1.594	0.874	1.076	1.226	1.374	1.459
90	-3.635	-3.100	-2.580	-2.095	-1.579	0.899	1.109	1.268	1.426	1.518
100	-3.584	-3.150	-2.552	-2.075	-1.566	0.920	1.137	1.303	1.470	1.569
150	-3.409	-3.009	-2.452	-2.004	-1.520	0.990	1.233	1.423	1.623	1.746
200	-3.302	-2.922	-2.391	-1.960	-1.491	1.032	1.290	1.496	1.715	1.853
250	-3.227	-2.861	-2.348	-1.926	-1.471	1.060	1.328	1.545	1.779	1.927
300	-3.172	-2.816	-2.316	-1.906	-1.456	1.080	1.357	1.528	1.826	1.983
350	-3.129	-2.781	-2.291	-1.888	-1.444	1.096	1.379	1.610	1.863	2.026
400	-3.094	-2.753	-2.270	-1.873	-1.434	1.108	1.396	1.633	1.893	2.061
450	-3.064	-2.729	-2.253	-1.861	-1.426	1.119	1.411	1.652	1.918	2.090
500	-3.040	-2.709	-2.239	-1.850	-1.419	1.127	1.423	1.668	1.938	2.114
550	-3.019	-2.691	-2.226	-1.841	-1.413	1.135	1.434	1.682	1.957	2.136
600	-3.000	-2.676	-2.215	-1.833	-1.408	1.141	1.443	1.694	1.972	2.154
650	-2.984	-2.663	-2.206	-1.826	-1.403	1.147	1.451	1.704	1.986	2.171
700	-2.969	-2.651	-2.197	-1.820	-1.399	1.152	1.458	1.714	1.999	2.185
750	-2.956	-2.640	-2.189	-1.814	-1.395	1.157	1.465	1.722	2.010	2.199
800	-2.944	-2.630	-2.182	-1.809	-1.392	1.161	1.471	1.730	2.020	2.211
850	-2.933	-2.621	-2.176	-1.804	-1.389	1.165	1.476	1.737	2.029	2.221
900	-2.923	-2.613	-2.170	-1.800	-1.386	1.168	1.481	1.743	2.037	2.231
950	-2.914	-2.605	-2.164	-1.796	-1.383	1.171	1.485	1.749	2.045	2.241
1000	-2.906	-2.599	-2.159	-1.792	-1.381	1.174	1.489	1.754	2.052	2.249

Source: From D'Agostino, 1971. Used by permission.

The null hypothesis of a normal distribution is rejected at the α significance level if the D'Agostino test statistic Y is less than $Y_{\alpha/2}$ or greater than $Y_{1-\alpha/2}$.

This table is used in Section 12.3.2.

Table A9 Multiplying Factor $\Psi_n(t)$ for Estimating the Lognormal Mean and Variance

	Number of Samples (n)																
t	2	5	8	10	13	15	20	25	30	50	70	90	100	150	200	500	∞
0.05	1.025	1.041	1.045	1.046	1.047	1.048	1.048	1.049	1.049	1.050	1.050	1.051	1.051	1.051	1.051	1.051	1.051
0.10	1.050	1.082	1.091	1.093	1.096	1.097	1.099	1.100	1.101	1.103	1.103	1.104	1.104	1.104	1.105	1.105	1.105
0.15	1.076	1.125	1.138	1.143	1.147	1.149	1.152	1.154	1.155	1.158	1.159	1.160	1.160	1.160	1.161	1.161	1.162
0.20	1.102	1.169	1.187	1.194	1.200	1.203	1.207	1.210	1.212	1.216	1.217	1.218	1.218	1.219	1.220	1.221	1.221
0.25	1.128	1.214	1.238	1.247	1.255	1.259	1.265	1.268	1.271	1.276	1.278	1.280	1.280	1.281	1.282	1.283	1.284
0.30	1.154	1.260	1.291	1.302	1.312	1.317	1.325	1.330	1.333	1.340	1.342	1.344	1.345	1.346	1.347	1.349	1.350
0.35	1.180	1.307	1.345	1.359	1.372	1.378	1.387	1.393	1.398	1.406	1.410	1.412	1.412	1.415	1.416	1.418	1.419
0.40	1.207	1.356	1.401	1.418	1.433	1.441	1.453	1.460	1.465	1.476	1.480	1.483	1.484	1.486	1.488	1.490	1.492
0.45	1.234	1.406	1.459	1.479	1.498	1.506	1.521	1.530	1.536	1.548	1.554	1.557	1.558	1.562	1.563	1.566	1.568
0.50	1.261	1.457	1.519	1.542	1.564	1.574	1.592	1.602	1.610	1.625	1.631	1.635	1.637	1.641	1.643	1.646	1.649
0.55	1.288	1.509	1.581	1.608	1.633	1.645	1.666	1.678	1.687	1.705	1.713	1.717	1.719	1.724	1.726	1.730	1.733
0.60	1.315	1.563	1.645	1.675	1.705	1.719	1.743	1.757	1.768	1.789	1.798	1.803	1.805	1.811	1.814	1.819	1.822
0.65	1.343	1.618	1.711	1.746	1.780	1.796	1.823	1.840	1.852	1.876	1.887	1.893	1.896	1.902	1.905	1.912	1.916
0.70	1.371	1.675	1.779	1.818	1.857	1.876	1.907	1.926	1.940	1.968	1.981	1.988	1.990	1.998	2.002	2.009	2.014
0.75	1.399	1.733	1.849	1.894	1.938	1.958	1.994	2.016	2.032	2.064	2.079	2.087	2.090	2.099	2.103	2.111	2.117
0.80	1.427	1.792	1.922	1.971	2.021	2.045	2.085	2.110	2.128	2.165	2.182	2.191	2.194	2.205	2.210	2.219	2.226
0.85	1.456	1.853	1.996	2.052	2.108	2.134	2.179	2.208	2.228	2.270	2.289	2.300	2.304	2.316	2.322	2.332	2.340
0.90	1.485	1.915	2.074	2.135	2.197	2.227	2.278	2.310	2.333	2.381	2.402	2.414	2.419	2.432	2.439	2.451	2.460
0.95	1.514	1.979	2.153	2.221	2.291	2.323	2.380	2.417	2.442	2.496	2.521	2.534	2.540	2.554	2.562	2.708	2.586
1.00	1.543	2.044	2.235	2.310	2.387	2.424	2.487	2.528	2.556	2.617	2.644	2.660	2.666	2.683	2.692	2.576	2.718
1.05	1.573	2.111	2.320	2.403	2.487	2.528	2.598	2.644	2.676	2.744	2.774	2.792	2.798	2.818	2.828	2.845	2.858
1.10	1.602	2.180	2.407	2.498	2.591	2.636	2.714	2.765	2.800	2.876	2.911	2.930	2.938	2.959	2.970	2.990	3.004
1.15	1.632	2.250	2.497	2.596	2.698	2.748	2.834	2.891	2.930	3.014	3.053	3.076	3.083	3.108	3.120	3.143	3.158
1.20	1.662	2.321	2.589	2.698	2.810	2.864	2.960	3.022	3.066	3.159	3.203	3.228	3.237	3.263	3.277	3.303	3.320
1.25	1.693	2.395	2.685	2.803	2.926	2.985	3.090	3.159	3.207	3.311	3.359	3.387	3.397	3.427	3.442	3.471	3.490
1.30	1.724	2.470	2.783	2.911	3.045	3.111	3.226	3.301	3.354	3.470	3.523	3.554	3.565	3.599	3.616	3.648	3.669
1.35	1.754	2.547	2.884	3.023	3.169	3.241	3.367	3.450	3.508	3.636	3.695	3.729	3.741	3.779	3.798	3.833	3.857
1.40	1.786	2.626	2.988	3.139	3.298	3.376	3.514	3.604	3.669	3.809	3.875	3.912	3.926	3.968	3.989	4.028	4.055
1.45	1.817	2.706	3.096	3.259	3.431	3.515	3.666	3.766	3.836	3.991	4.063	4.105	4.120	4.166	4.189	4.233	4.263
1.50	1.849	2.788	3.206	3.382	3.569	3.661	3.825	3.933	4.011	4.181	4.260	4.306	4.323	4.374	4.400	4.448	4.482
1.55	1.880	2.873	3.320	3.510	3.711	3.811	3.990	4.108	4.193	4.379	4.467	4.518	4.536	4.592	4.621	4.675	4.712
1.60	1.913	2.959	3.437	3.642	3.859	3.967	4.161	4.291	4.383	4.587	4.683	4.739	4.759	4.821	4.853	4.912	4.953
1.65	1.945	3.047	3.558	3.777	4.012	4.129	4.339	4.480	4.581	4.804	4.910	4.971	4.993	5.062	5.097	5.162	5.207
1.70	1.977	3.137	3.682	3.918	4.171	4.297	4.525	4.678	4.787	5.031	5.147	5.215	5.239	5.314	5.352	5.424	5.474
1.75	2.010	3.229	3.810	4.062	4.334	4.471	4.717	4.883	5.003	5.269	5.395	5.469	5.496	5.578	5.621	5.700	5.755
1.80	2.043	3.323	3.942	4.212	4.504	4.651	4.917	5.097	5.228	5.517	5.655	5.736	5.766	5.856	5.903	5.990	6.050
1.85	2.077	3.420	4.077	4.366	4.680	4.838	5.125	5.320	5.461	5.776	5.928	6.016	6.048	6.147	6.198	6.294	6.360
1.90	2.110	3.518	4.216	4.525	4.861	5.031	5.341	5.552	5.705	6.048	6.212	6.309	6.344	6.453	6.509	6.613	6.686
1.95	2.144	3.619	4.359	4.688	5.049	5.232	5.566	5.794	5.959	6.331	6.511	6.616	6.655	6.773	6.834	6.949	7.029
2.00	2.178	3.721	4.506	4.857	5.243	5.439	5.799	6.045	6.224	6.628	6.823	6.938	6.980	7.109	7.176	7.302	7.389

Source: After Koch and Link, 1980 and Aitchison and Brown, 1968.

This table is used in Section 13.1.1.

Table A10 Values of $H_{1-\alpha} = H_{0.90}$ for Computing a One-Sided Upper 90% Confidence Limit on a Lognormal Mean

s_y	n: 3	5	7	10	12	15	21	31	51	101
0.10	1.686	1.438	1.381	1.349	1.338	1.328	1.317	1.308	1.301	1.295
0.20	1.885	1.522	1.442	1.396	1.380	1.365	1.348	1.335	1.324	1.314
0.30	2.156	1.627	1.517	1.453	1.432	1.411	1.388	1.370	1.354	1.339
0.40	2.521	1.755	1.607	1.523	1.494	1.467	1.437	1.412	1.390	1.371
0.50	2.990	1.907	1.712	1.604	1.567	1.532	1.494	1.462	1.434	1.409
0.60	3.542	2.084	1.834	1.696	1.650	1.606	1.558	1.519	1.485	1.454
0.70	4.136	2.284	1.970	1.800	1.743	1.690	1.631	1.583	1.541	1.504
0.80	4.742	2.503	2.119	1.914	1.845	1.781	1.710	1.654	1.604	1.560
0.90	5.349	2.736	2.280	2.036	1.955	1.880	1.797	1.731	1.672	1.621
1.00	5.955	2.980	2.450	2.167	2.073	1.985	1.889	1.812	1.745	1.686
1.25	7.466	3.617	2.904	2.518	2.391	2.271	2.141	2.036	1.946	1.866
1.50	8.973	4.276	3.383	2.896	2.733	2.581	2.415	2.282	2.166	2.066
1.75	10.48	4.944	3.877	3.289	3.092	2.907	2.705	2.543	2.402	2.279
2.00	11.98	5.619	4.380	3.693	3.461	3.244	3.005	2.814	2.648	2.503
2.50	14.99	6.979	5.401	4.518	4.220	3.938	3.629	3.380	3.163	2.974
3.00	18.00	8.346	6.434	5.359	4.994	4.650	4.270	3.964	3.697	3.463
3.50	21.00	9.717	7.473	6.208	5.778	5.370	4.921	4.559	4.242	3.965
4.00	24.00	11.09	8.516	7.062	6.566	6.097	5.580	5.161	4.796	4.474
4.50	27.01	12.47	9.562	7.919	7.360	6.829	6.243	5.769	5.354	4.989
5.00	30.01	13.84	10.61	8.779	8.155	7.563	6.909	6.379	5.916	5.508
6.00	36.02	16.60	12.71	10.50	9.751	9.037	8.248	7.607	7.048	6.555
7.00	42.02	19.35	14.81	12.23	11.35	10.52	9.592	8.842	8.186	7.607
8.00	48.03	22.11	16.91	13.96	12.96	12.00	10.94	10.08	9.329	8.665
9.00	54.03	24.87	19.02	15.70	14.56	13.48	12.29	11.32	10.48	9.725
10.00	60.04	27.63	21.12	17.43	16.17	14.97	13.64	12.56	11.62	10.79

Source: After Land, 1975.
This table is used in Section 13.2.

Table A11 Values of $H_{\alpha} = H_{0.10}$ for Computing a One-Sided Lower 10% Confidence Limit on a Lognormal Mean

s_y	n: 3	5	7	10	12	15	21	31	51	101
0.10	-1.431	-1.320	-1.296	-1.285	-1.281	-1.279	-1.277	-1.277	-1.278	-1.279
0.20	-1.350	-1.281	-1.268	-1.266	-1.266	-1.266	-1.268	-1.272	-1.275	-1.280
0.30	-1.289	-1.252	-1.250	-1.254	-1.257	-1.260	-1.266	-1.272	-1.280	-1.287
0.40	-1.245	-1.233	-1.239	-1.249	-1.254	-1.261	-1.270	-1.279	-1.289	-1.301
0.50	-1.213	-1.221	-1.234	-1.250	-1.257	-1.266	-1.279	-1.291	-1.304	-1.319
0.60	-1.190	-1.215	-1.235	-1.256	-1.266	-1.277	-1.292	-1.307	-1.324	-1.342
0.70	-1.176	-1.215	-1.241	-1.266	-1.278	-1.292	-1.310	-1.329	-1.349	-1.370
0.80	-1.168	-1.219	-1.251	-1.280	-1.294	-1.311	-1.332	-1.354	-1.377	-1.403
0.90	-1.165	-1.227	-1.264	-1.298	-1.314	-1.333	-1.358	-1.383	-1.409	-1.439
1.00	-1.166	-1.239	-1.281	-1.320	-1.337	-1.358	-1.387	-1.414	-1.445	-1.478
1.25	-1.184	-1.280	-1.334	-1.384	-1.407	-1.434	-1.470	-1.507	-1.547	-1.589
1.50	-1.217	-1.334	-1.400	-1.462	-1.491	-1.523	-1.568	-1.613	-1.063	-1.716
1.75	-1.260	-1.398	-1.477	-1.551	-1.585	-1.624	-1.677	-1.732	-1.790	-1.855
2.00	-1.310	-1.470	-1.562	-1.647	-1.688	-1.733	-1.795	-1.859	-1.928	-2.003
2.50	-1.426	-1.634	-1.751	-1.862	-1.913	-1.971	-2.051	-2.133	-2.223	-2.321
3.00	-1.560	-1.817	-1.960	-2.095	-2.157	-2.229	-2.326	-2.427	-2.536	-2.657
3.50	-1.710	-2.014	-2.183	-2.341	-2.415	-2.499	-2.615	-2.733	-2.864	-3.007
4.00	-1.871	-2.221	-2.415	-2.596	-2.681	-2.778	-2.913	-3.050	-3.200	-3.366
4.50	-2.041	-2.435	-2.653	-2.858	-2.955	-3.064	-3.217	-3.372	-3.542	-3.731
5.00	-2.217	-2.654	-2.897	-3.126	-3.233	-3.356	-3.525	-3.698	-3.889	-4.100
6.00	-2.581	-3.104	-3.396	-3.671	-3.800	-3.949	-4.153	-4.363	-4.594	-4.849
7.00	-2.955	-3.564	-3.904	-4.226	-4.377	-4.549	-4.790	-5.037	-5.307	-5.607
8.00	-3.336	-4.030	-4.418	-4.787	-4.960	-5.159	-5.433	-5.715	-6.026	-6.370
9.00	-3.721	-4.500	-4.937	-5.352	-5.547	-5.771	-6.080	-6.399	-6.748	-7.136
10.00	-4.109	-4.973	-5.459	-5.920	-6.137	-6.386	-6.730	-7.085	-7.474	-7.906

Source: After Land, 1975.
This table is used in Section 13.2.

Table A12 Values of $H_{1-\alpha} = H_{0.95}$ for Computing a One-Sided Upper 95% Confidence Limit on a Lognormal Mean

s_y	n 3	5	7	10	12	15	21	31	51	101
0.10	2.750	2.035	1.886	1.802	1.775	1.749	1.722	1.701	1.684	1.670
0.20	3.295	2.198	1.992	1.881	1.843	1.809	1.771	1.742	1.718	1.697
0.30	4.109	2.402	2.125	1.977	1.927	1.882	1.833	1.793	1.761	1.733
0.40	5.220	2.651	2.282	2.089	2.026	1.968	1.905	1.856	1.813	1.777
0.50	6.495	2.947	2.465	2.220	2.141	2.068	1.989	1.928	1.876	1.830
0.60	7.807	3.287	2.673	2.368	2.271	2.181	2.085	2.010	1.946	1.891
0.70	9.120	3.662	2.904	2.532	2.414	2.306	2.191	2.102	2.025	1.960
0.80	10.43	4.062	3.155	2.710	2.570	2.443	2.307	2.202	2.112	2.035
0.90	11.74	4.478	3.420	2.902	2.738	2.589	2.432	2.310	2.206	2.117
1.00	13.05	4.905	3.698	3.103	2.915	2.744	2.564	2.423	2.306	2.205
1.25	16.33	6.001	4.426	3.639	3.389	3.163	2.923	2.737	2.580	2.447
1.50	19.60	7.120	5.184	4.207	3.896	3.612	3.311	3.077	2.881	2.713
1.75	22.87	8.250	5.960	4.795	4.422	4.081	3.719	3.437	3.200	2.997
2.00	26.14	9.387	6.747	5.396	4.962	4.564	4.141	3.812	3.533	3.295
2.50	32.69	11.67	8.339	6.621	6.067	5.557	5.013	4.588	4.228	3.920
3.00	39.23	13.97	9.945	7.864	7.191	6.570	5.907	5.388	4.947	4.569
3.50	45.77	16.27	11.56	9.118	8.326	7.596	6.815	6.201	5.681	5.233
4.00	52.31	18.58	13.18	10.38	9.469	8.630	7.731	7.024	6.424	5.908
4.50	58.85	20.88	14.80	11.64	10.62	9.669	8.652	7.854	7.174	6.590
5.00	65.39	23.19	16.43	12.91	11.77	10.71	9.579	8.688	7.929	7.277
6.00	78.47	27.81	19.68	15.45	14.08	12.81	11.44	10.36	9.449	8.661
7.00	91.55	32.43	22.94	18.00	16.39	14.90	13.31	12.05	10.98	10.05
8.00	104.6	37.06	26.20	20.55	18.71	17.01	15.18	13.74	12.51	11.45
9.00	117.7	41.68	29.46	23.10	21.03	19.11	17.05	15.43	14.05	12.85
10.00	130.8	46.31	32.73	25.66	23.35	21.22	18.93	17.13	15.59	14.26

Source: After Land, 1975.
This table is used in Section 13.2.

Table A13 Values of $H_{\alpha} = H_{0.05}$ for Computing a One-Sided Lower 5% Confidence Limit on a Lognormal Mean

s_y	n 3	5	7	10	12	15	21	31	51	101
0.10	-2.130	-1.806	-1.731	-1.690	-1.677	-1.666	-1.655	-1.648	-1.644	-1.642
0.20	-1.949	-1.729	-1.678	-1.653	-1.646	-1.640	-1.636	-1.636	-1.637	-1.641
0.30	-1.816	-1.669	-1.639	-1.627	-1.625	-1.625	-1.627	-1.632	-1.638	-1.648
0.40	-1.717	-1.625	-1.611	-1.611	-1.613	-1.617	-1.625	-1.635	-1.647	-1.662
0.50	-1.644	-1.594	-1.594	-1.603	-1.609	-1.618	-1.631	-1.646	-1.663	-1.683
0.60	-1.589	-1.573	-1.584	-1.602	-1.612	-1.625	-1.643	-1.662	-1.685	-1.711
0.70	-1.549	-1.560	-1.582	-1.608	-1.622	-1.638	-1.661	-1.686	-1.713	-1.744
0.80	-1.521	-1.555	-1.586	-1.620	-1.636	-1.656	-1.685	-1.714	-1.747	-1.783
0.90	-1.502	-1.556	-1.595	-1.637	-1.656	-1.680	-1.713	-1.747	-1.785	-1.826
1.00	-1.490	-1.562	-1.610	-1.658	-1.681	-1.707	-1.745	-1.784	-1.827	-1.874
1.25	-1.486	-1.596	-1.662	-1.727	-1.758	-1.793	-1.842	-1.893	-1.949	-2.012
1.50	-1.508	-1.650	-1.733	-1.814	-1.853	-1.896	-1.958	-2.020	-2.091	-2.169
1.75	-1.547	-1.719	-1.819	-1.916	-1.962	-2.015	-2.088	-2.164	-2.247	-2.341
2.00	-1.598	-1.799	-1.917	-2.029	-2.083	-2.144	-2.230	-2.318	-2.416	-2.526
2.50	-1.727	-1.986	-2.138	-2.283	-2.351	-2.430	-2.540	-2.654	-2.780	-2.921
3.00	-1.880	-2.199	-2.384	-2.560	-2.644	-2.740	-2.874	-3.014	-3.169	-3.342
3.50	-2.051	-2.429	-2.647	-2.855	-2.953	-3.067	-3.226	-3.391	-3.574	-3.780
4.00	-2.237	-2.672	-2.922	-3.161	-3.275	-3.406	-3.589	-3.779	-3.990	-4.228
4.50	-2.434	-2.924	-3.206	-3.476	-3.605	-3.753	-3.960	-4.176	-4.416	-4.685
5.00	-2.638	-3.183	-3.497	-3.798	-3.941	-4.107	-4.338	-4.579	-4.847	-5.148
6.00	-3.062	-3.715	-4.092	-4.455	-4.627	-4.827	-5.106	-5.397	-5.721	-6.086
7.00	-3.499	-4.260	-4.699	-5.123	-5.325	-5.559	-5.886	-6.227	-6.608	-7.036
8.00	-3.945	-4.812	-5.315	-5.800	-6.031	-6.300	-6.674	-7.066	-7.502	-7.992
9.00	-4.397	-5.371	-5.936	-6.482	-6.742	-7.045	-7.468	-7.909	-8.401	-8.953
10.00	-4.852	-5.933	-6.560	-7.168	-7.458	-7.794	-8.264	-8.755	-9.302	-9.918

Source: After Land, 1975.
This table is used in Section 13.2.

Table A14 Confidence Limits for the Median of Any Continuous Distribution

n	α = 0.05 ℓ	α = 0.05 u	α = 0.01 ℓ	α = 0.01 u	n	α = 0.05 ℓ	α = 0.05 u	α = 0.01 ℓ	α = 0.01 u
5	-	-	-	-	41	14	28	12	30
6	1	6	-	-	42	15	28	13	30
7	1	7	-	-	43	15	29	13	31
8	1	8	1	8	44	16	29	14	31
9	2	8	1	9	45	16	30	14	32
10	2	9	1	10	46	16	31	14	33
11	2	10	1	11	47	17	31	15	33
12	3	10	2	11	48	17	32	15	34
13	3	11	2	12	49	18	32	16	34
14	3	12	2	13	50	18	33	16	35
15	4	12	3	13	51	19	33	16	36
16	4	13	3	14	52	19	34	17	36
17	5	13	3	15	53	19	35	17	37
18	5	14	4	15	54	20	35	18	37
19	5	15	4	16	55	20	36	18	38
20	6	15	4	17	56	21	36	18	39
21	6	16	5	17	57	21	37	19	39
22	6	17	5	18	58	22	37	19	40
23	7	17	5	19	59	22	38	20	40
24	7	18	6	19	60	22	39	20	41
25	8	18	6	20	61	23	39	21	41
26	8	19	7	20	62	23	40	21	42
27	8	20	7	21	63	24	40	21	43
28	9	20	7	22	64	24	41	22	43
29	9	21	8	22	65	25	41	22	44
30	10	21	8	23	66	25	42	23	44
31	10	22	8	24	67	26	42	23	45
32	10	23	9	24	68	26	43	23	46
33	11	23	9	25	69	26	44	24	46
34	11	24	10	25	70	27	44	24	47
35	12	24	10	26	71	27	45	25	47
36	12	25	10	27	72	28	45	25	48
37	13	25	11	27	73	28	46	26	48
38	13	26	11	28	74	29	46	26	49
39	13	27	12	28	75	29	47	26	50
40	14	27	12	29					

Source: After Geigy, 1982.
Given are the values l and u such that for the order statistics $x_{[l]}$ and $x_{[u]}$, Prob[$x_{[l]}$ < true median < $x_{[u]}$] ≥ $1 - \alpha$.
This table is first used in Section 13.4.

Table A15 Values of λ for Estimating the Mean and Variance of a Normal Distribution Using a Singly Censored Data Set

γ \ h	.01	.02	.03	.04	.05	.06	.07	.08	0.9	.10	.15	.20	
.00	.010100	.020400	.030902	.041583	.052507	.063627	.074953	.086488	.09824	.11020	.17342	.24268	.
.05	.010551	.021294	.032225	.043350	.054670	.066189	.077909	.089834	.10197	.11431	.17935	.25033	.
.10	.010950	.022082	.033398	.044902	.056596	.068483	.080568	.092852	.10534	.11804	.18479	.25741	.
.15	.011310	.022798	.034466	.046318	.058356	.070586	.083009	.095629	.10845	.12148	.18985	.26405	.
.20	.011642	.023459	.035453	.047629	.059990	.072539	.085280	.098216	.11135	.12469	.19460	.27031	.
.25	.011952	.024076	.036377	.048858	.061522	.074372	.087413	.10065	.11408	.12772	.19910	.27626	.
.30	.012243	.024658	.037249	.050018	.062969	.076106	.089433	.10295	.11667	.13059	.20338	.28193	.
.35	.012520	.025211	.038077	.051120	.064345	.077756	.091355	.10515	.11914	.13333	.20747	.28737	.
.40	.012784	.025738	.038866	.052173	.065660	.079332	.093193	.10725	.12150	.13595	.21139	.29260	.
.45	.013036	.026243	.039624	.053182	.066921	.080845	.094958	.10926	.12377	.13847	.21517	.29765	.
.50	.013279	.026728	.040352	.054153	.068135	.082301	.096657	.11121	.12595	.14090	.21882	.30253	.
.55	.013513	.027196	.041054	.055089	.069306	.083708	.098298	.11308	.12806	.14325	.22235	.30725	.
.60	.013739	.027649	.021733	.055995	.070439	.085068	.099887	.11490	.13011	.14552	.22578	.31184	.
.65	.013958	.028087	.042391	.056874	.071538	.086388	.10143	.11666	.13209	.14773	.22910	.31630	.
.70	.014171	.028513	.043030	.057726	.072605	.087670	.10292	.11837	.13402	.14987	.23234	.32065	.
.75	.104378	.028927	.043652	.058556	.073643	.088917	.10438	.12004	.13590	.15196	.23550	.32489	.
.80	.014579	.029330	.044258	.059364	.074655	.090133	.10580	.12167	.13773	.15400	.23858	.32903	.
.85	.014775	.029723	.044848	.060153	.075642	.091319	.10719	.12325	.13952	.15599	.24158	.33307	.
.90	.014967	.030107	.045425	.060923	.076606	.092477	.10854	.12480	.14126	.15793	.24452	.33703	.
.95	.015154	.030483	.045989	.061676	.077549	.093611	.10987	.12632	.14297	.15983	.24740	.34091	.
1.00	.015338	.030850	.046540	.062413	.078471	.094720	.11116	.12780	.14465	.16170	.25022	.34471	1.

γ \ h	.25	.30	.35	.40	.45	.50	.55	.60	.65	.70	.80	.90	
.00	.31862	.4021	.4941	.5961	.7096	.8368	.9808	1.145	1.336	1.561	2.176	3.283	
.05	.32793	.4130	.5066	.6101	.7252	.8540	.9994	1.166	1.358	1.585	2.203	3.314	
.10	.33662	.4233	.5184	.6234	.7400	.8703	1.017	1.185	1.379	1.608	2.229	3.345	
.15	.34480	.4330	.5296	.6361	.7542	.8860	1.035	1.204	1.400	1.630	2.255	3.376	
.20	.35255	.4422	.5403	.6483	.7678	.9012	1.051	1.222	1.419	1.651	2.280	3.405	
.25	.35993	.4510	.5506	.6600	.7810	.9158	1.067	1.240	1.439	1.672	2.305	3.435	
.30	.36700	.4595	.5604	.6713	.7937	.9300	1.083	1.257	1.457	1.693	2.329	3.464	
.35	.37379	.4676	.5699	.6821	.8060	.9437	1.098	1.274	1.476	1.713	2.353	3.492	
.40	.38033	.4755	.5791	.6927	.8169	.9570	1.113	1.290	1.494	1.732	2.376	3.520	
.45	.38665	.4831	.5880	.7029	.8295	.9700	1.127	1.306	1.511	1.751	2.399	3.547	
.50	.39276	.4904	.5967	.7129	.8408	.9826	1.141	1.321	1.528	1.770	2.421	3.575	
.55	.39870	.4976	.6051	.7225	.8517	.9950	1.155	1.337	1.545	1.788	2.443	3.601	
.60	.40447	.5045	.6133	.7320	.8625	1.007	1.169	1.351	1.561	1.806	2.465	3.628	
.65	.41008	.5114	.6213	.7412	.8729	1.019	1.182	1.366	1.557	1.824	2.486	3.654	
.70	.41555	.5180	.6291	.7502	.8832	1.030	1.195	1.380	1.593	1.841	2.507	3.679	
.75	.42090	.5245	.6367	.7590	.8932	1.042	1.207	1.394	1.608	1.858	2.528	3.705	
.80	.42612	.5308	.6441	.7676	.9031	1.053	1.220	1.408	1.624	1.875	2.548	3.730	
.85	.43122	.5370	.6515	.7761	.9127	1.064	1.232	1.422	1.639	1.892	2.568	3.754	
.90	.43622	.5430	.6586	.7844	.9222	1.074	1.244	1.435	1.653	1.908	2.588	3.779	
.95	.44112	.5490	.6656	.7925	.9314	1.085	1.255	1.448	1.668	1.924	2.607	3.803	
1.00	.44592	.5548	.6724	.8005	.9406	1.095	1.267	1.461	1.682	1.940	2.626	3.827	1

Source: From Cohen, 1961. Used by permission.

γ and h are defined in Section 14.3.2 where this table is used.

Table A16 Approximate Critical Values λ_{i+1} for Rosner's Generalized ESD Many-Outlier Procedure

		α		
n	i+1	0.05	0.01	0.005
25	1	2.82	3.14	3.25
	2	2.80	3.11	3.23
	3	2.78	3.09	3.20
	4	2.76	3.06	3.17
	5	2.73	3.03	3.14
	10	2.59	2.85	2.95
26	1	2.84	3.16	3.28
	2	2.82	3.14	3.25
	3	2.80	3.11	3.23
	4	2.78	3.09	3.20
	5	2.76	3.06	3.17
	10	2.62	2.89	2.99
27	1	2.86	3.18	3.30
	2	2.84	3.16	3.28
	3	2.82	3.14	3.25
	4	2.80	3.11	3.23
	5	2.78	3.09	3.20
	10	2.65	2.93	3.03
28	1	2.88	3.20	3.32
	2	2.86	3.18	3.30
	3	2.84	3.16	3.28
	4	2.82	3.14	3.25
	5	2.80	3.11	3.23
	10	2.68	2.97	3.07
29	1	2.89	3.22	3.34
	2	2.88	3.20	3.32
	3	2.86	3.18	3.30
	4	2.84	3.16	3.28
	5	2.82	3.14	3.25
	10	2.71	3.00	3.11
30	1	2.91	3.24	3.36
	2	2.89	3.22	3.34
	3	2.88	3.20	3.32
	4	2.86	3.18	3.30
	5	2.84	3.16	3.28
	10	2.73	3.03	3.14
31	1	2.92	3.25	3.38
	2	2.91	3.24	3.36
	3	2.89	3.22	3.34
	4	2.88	3.20	3.32
	5	2.86	3.18	3.30
	10	2.76	3.06	3.17
32	1	2.94	3.27	3.40
	2	2.92	3.25	3.38
	3	2.91	3.24	3.36
	4	2.89	3.22	3.34
	5	2.88	3.20	3.32
	10	2.78	3.09	3.20
33	1	2.95	3.29	3.41
	2	2.94	3.27	3.40
	3	2.92	3.25	3.38
	4	2.91	3.24	3.36
	5	2.89	3.22	3.34
	10	2.80	3.11	3.23
34	1	2.97	3.30	3.43
	2	2.95	3.29	3.41
	3	2.94	3.27	3.40
	4	2.92	3.25	3.38
	5	2.91	3.24	3.36
	10	2.82	3.14	3.25
35	1	2.98	3.32	3.44
	2	2.97	3.30	3.43
	3	2.95	3.29	3.41
	4	2.94	3.27	3.40
	5	2.92	3.25	3.38
	10	2.84	3.16	3.28
36	1	2.99	3.33	3.46
	2	2.98	3.32	3.44
	3	2.97	3.30	3.43
	4	2.95	3.29	3.41
	5	2.94	3.27	3.40
	10	2.86	3.18	3.30

Table A16 (continued)

n	i+1	α 0.05	0.01	0.005	n	i+1	α 0.05	0.01	0.005
37	1	3.00	3.34	3.47	44	1	3.08	3.43	3.56
	2	2.99	3.33	3.46		2	3.07	3.41	3.55
	3	2.98	3.32	3.44		3	3.06	3.40	3.54
	4	2.97	3.30	3.43		4	3.05	3.39	3.52
	5	2.95	3.29	3.41		5	3.04	3.38	3.51
	10	2.88	3.20	3.32		10	2.98	3.32	3.44
38	1	3.01	3.36	3.49	45	1	3.09	3.44	3.57
	2	3.00	3.34	3.47		2	3.08	3.43	3.56
	3	2.99	3.33	3.46		3	3.07	3.41	3.55
	4	2.98	3.32	3.44		4	3.06	3.40	3.54
	5	2.97	3.30	3.43		5	3.05	3.39	3.52
	10	2.89	3.22	3.34		10	2.99	3.33	3.46
39	1	3.03	3.37	3.50	46	1	3.09	3.45	3.58
	2	3.01	3.36	3.49		2	3.09	3.44	3.57
	3	3.00	3.34	3.47		3	3.08	3.43	3.56
	4	2.99	3.33	3.46		4	3.07	3.41	3.55
	5	2.98	3.32	3.44		5	3.06	3.40	3.54
	10	2.91	3.24	3.36		10	3.00	3.34	3.47
40	1	3.04	3.38	3.51	47	1	3.10	3.46	3.59
	2	3.03	3.37	3.50		2	3.09	3.45	3.58
	3	3.01	3.36	3.49		3	3.09	3.44	3.57
	4	3.00	3.34	3.47		4	3.08	3.43	3.56
	5	2.99	3.33	3.46		5	3.07	3.41	3.55
	10	2.92	3.25	3.38		10	3.01	3.36	3.49
41	1	3.05	3.39	3.52	48	1	3.11	3.46	3.60
	2	3.04	3.38	3.51		2	3.10	3.46	3.59
	3	3.03	3.37	3.50		3	3.09	3.45	3.58
	4	3.01	3.36	3.49		4	3.09	3.44	3.57
	5	3.00	3.34	3.47		5	3.08	3.43	3.56
	10	2.94	3.27	3.40		10	3.03	3.37	3.50
42	1	3.06	3.40	3.54	49	1	3.12	3.47	3.61
	2	3.05	3.39	3.52		2	3.11	3.46	3.60
	3	3.04	3.38	3.51		3	3.10	3.46	3.59
	4	3.03	3.37	3.50		4	3.09	3.45	3.58
	5	3.01	3.36	3.49		5	3.09	3.44	3.57
	10	2.95	3.29	3.41		10	3.04	3.38	3.51
43	1	3.07	3.41	3.55	50	1	3.13	3.48	3.62
	2	3.06	3.40	3.54		2	3.12	3.47	3.61
	3	3.05	3.39	3.52		3	3.11	3.46	3.60
	4	3.04	3.38	3.51		4	3.10	3.46	3.59
	5	3.03	3.37	3.50		5	3.09	3.45	3.58
	10	2.97	3.30	3.43		10	3.05	3.39	3.52

Table A16 (continued)

n	i+1	α		
		0.05	0.01	0.005
60	1	3.20	3.56	3.70
	2	3.19	3.55	3.69
	3	3.19	3.55	3.69
	4	3.18	3.54	3.68
	5	3.17	3.53	3.67
	10	3.14	3.49	3.63
70	1	3.26	3.62	3.76
	2	3.25	3.62	3.76
	3	3.25	3.61	3.75
	4	3.24	3.60	3.75
	5	3.24	3.60	3.74
	10	3.21	3.57	3.71
80	1	3.31	3.67	3.82
	2	3.30	3.67	3.81
	3	3.30	3.66	3.81
	4	3.29	3.66	3.80
	5	3.29	3.65	3.80
	10	3.26	3.63	3.77
90	1	3.35	3.72	3.86
	2	3.34	3.71	3.86
	3	3.34	3.71	3.85
	4	3.34	3.70	3.85
	5	3.33	3.70	3.84
	10	3.31	3.68	3.82
100	1	3.38	3.75	3.90
	2	3.38	3.75	3.90
	3	3.38	3.75	3.89
	4	3.37	3.74	3.89
	5	3.37	3.74	3.89
	10	3.35	3.72	3.87
150	1	3.52	3.89	4.04
	2	3.51	3.89	4.04
	3	3.51	3.89	4.03
	4	3.51	3.88	4.03
	5	3.51	3.88	4.03
	10	3.50	3.87	4.02
200	1	3.61	3.98	4.13
	2	3.60	3.98	4.13
	3	3.60	3.97	4.12
	4	3.60	3.97	4.12
	5	3.60	3.97	4.12
	10	3.59	3.96	4.11
250	1	3.67	4.04	4.19
	5	3.67	4.04	4.19
	10	3.66	4.03	4.18
300	1	3.72	4.09	4.24
	5	3.72	4.09	4.24
	10	3.71	4.09	4.23
350	1	3.77	4.14	4.28
	5	3.76	4.13	4.28
	10	3.76	4.13	4.28
400	1	3.80	4.17	4.32
	5	3.80	4.17	4.32
	10	3.80	4.16	4.31
450	1	3.84	4.20	4.35
	5	3.83	4.20	4.35
	10	3.83	4.20	4.34
500	1	3.86	4.23	4.38
	5	3.86	4.23	4.37
	10	3.86	4.22	4.37
750	1-10	3.95	4.30	4.44
1000	1-10	4.02	4.37	4.52
2000	1-10	4.20	4.54	4.68
3000	1-10	4.29	4.63	4.77
4000	1-10	4.36	4.70	4.83
5000	1-10	4.41	4.75	4.88

Source: Entries for $n \leq 500$ are from Table 3 in Rosner, 1983 and are used by permission.
For $n > 500$, the approximate percentage points were computed as $Z_p(n - i - 1)/[(n - i - 2 + Z_p^2)(n - i)]^{1/2}$, where $p = 1 - [(\alpha/2)/(n - i)]$ and Z_p is the pth quantile of the $N(0, 1)$ distribution (from Rosner, 1983).
This table is used in Section 15.3.2.

Table A17 Factors for Computing Control Chart Lines

n_i	d_2	d_3	c_4
2	1.128	0.853	0.7979
3	1.693	0.888	0.8862
4	2.059	0.880	0.9213
5	2.326	0.864	0.9400
6	2.534	0.848	0.9515
7	2.704	0.833	0.9594
8	2.847	0.820	0.9650
9	2.970	0.808	0.9693
10	3.078	0.797	0.9727
11	3.173	0.787	0.9754
12	3.258	0.778	0.9776
13	3.336	0.770	0.9794
14	3.407	0.763	0.9810
15	3.472	0.756	0.9823
16	3.532	0.750	0.9835
17	3.588	0.744	0.9845
18	3.640	0.739	0.9854
19	3.689	0.734	0.9862
20	3.735	0.729	0.9869
21	3.778	0.724	0.9876
22	3.819	0.720	0.9882
23	3.858	0.716	0.9887
24	3.895	0.712	0.9892
25	3.931	0.708	0.9896

Source: From Burr, 1976. Used by permission.
n_i = number of data in the subgroup.
c_4 approaches 1 as n_i becomes large.
This table is used in Section 15.6.3.

Table A18 Probabilities for the Mann-Kendall Nonparametric Test for Trend

S	Values of n				S	Values of n		
	4	5	8	9		6	7	10
0	0.625	0.592	0.548	0.540	1	0.500	0.500	0.500
2	0.375	0.408	0.452	0.460	3	0.360	0.386	0.431
4	0.167	0.242	0.360	0.381	5	0.235	0.281	0.364
6	0.042	0.117	0.274	0.306	7	0.136	0.191	0.300
8		0.042	0.199	0.238	9	0.068	0.119	0.242
10		$0.0^{2}83$	0.138	0.179	11	0.028	0.068	0.190
12			0.089	0.130	13	$0.0^{2}83$	0.035	0.146
14			0.054	0.090	15	$0.0^{2}14$	0.015	0.108
16			0.031	0.060	17		$0.0^{2}54$	0.078
18			0.016	0.038	19		$0.0^{2}14$	0.054
20			$0.0^{2}71$	0.022	21		$0.0^{3}20$	0.036
22			$0.0^{2}28$	0.012	23			0.023
24			$0.0^{3}87$	$0.0^{2}63$	25			0.014
26			$0.0^{3}19$	$0.0^{2}29$	27			$0.0^{2}83$
28			$0.0^{4}25$	$0.0^{2}12$	29			$0.0^{2}46$
30				$0.0^{3}43$	31			$0.0^{2}23$
32				$0.0^{3}12$	33			$0.0^{2}11$
34				$0.0^{4}25$	35			$0.0^{3}47$
36				$0.0^{5}28$	37			$0.0^{3}18$
					39			$0.0^{4}58$
					41			$0.0^{4}15$
					43			$0.0^{5}28$
					45			$0.0^{6}28$

Source: From Kendall, 1975. Used by permission.

Repeated zeros are indicated by powers; for example, $0.0^{3}47$ stands for 0.00047.

Each table entry is the probability that the Mann-Kendall statistic S equals or exceeds the specified value of S when no trend is present.

This table is used in Section 16.4.1.

Table A19 Quantiles of the Chi-Square Distribution with *v* Degrees of Freedom

Degrees of Freedom	Probability of obtaining a value of χ^2 smaller than the tabled value													
v	0.005	0.001	0.025	0.050	0.100	0.250	0.50	0.750	0.900	0.950	0.975	0.990	0.995	0.999
1	....	....	....	....	0.02	0.10	0.45	1.32	2.71	3.84	5.02	6.63	7.88	10.83
2	0.01	0.02	0.05	0.10	0.21	0.58	1.39	2.77	4.61	5.99	7.38	9.21	10.60	13.82
3	0.07	0.11	0.22	0.35	0.58	1.21	2.37	4.11	6.25	7.81	9.35	11.34	12.84	16.27
4	0.21	0.30	0.48	0.71	1.06	1.92	3.36	5.39	7.78	9.49	11.14	13.28	14.86	18.47
5	0.41	0.55	0.83	1.15	1.61	2.67	4.35	6.63	9.24	11.07	12.83	15.09	16.75	20.52
6	0.68	0.87	1.24	1.64	2.20	3.45	5.35	7.84	10.64	12.59	14.45	16.81	18.55	22.46
7	0.99	1.24	1.69	2.17	2.83	4.25	6.35	9.04	12.02	14.07	16.01	18.48	20.28	24.32
8	1.34	1.65	2.18	2.73	3.49	5.07	7.34	10.22	13.36	15.51	17.53	20.09	21.96	26.12
9	1.73	2.09	2.70	3.33	4.17	5.90	8.34	11.39	14.68	16.92	19.02	21.67	23.59	27.88
10	2.16	2.56	3.25	3.94	4.87	6.74	9.34	12.55	15.99	18.31	20.48	23.21	25.19	29.59
11	2.60	3.05	3.82	4.57	5.58	7.58	10.34	13.70	17.28	19.68	21.92	24.72	26.76	31.26
12	3.07	3.57	4.40	5.23	6.30	8.44	11.34	14.85	18.55	21.03	23.34	26.22	28.30	32.91
13	3.57	4.11	5.01	5.89	7.04	9.30	12.34	15.98	19.81	22.36	24.74	27.69	29.82	34.53
14	4.07	4.66	5.63	6.57	7.79	10.17	13.34	17.12	21.06	23.68	26.12	29.14	31.32	36.12
15	4.60	5.23	6.27	7.26	8.55	11.04	14.34	18.25	22.31	25.00	27.49	30.58	32.80	37.70
16	5.14	5.81	6.91	7.96	9.31	11.91	15.34	19.37	23.54	26.30	28.85	32.00	34.27	39.25
17	5.70	6.41	7.56	8.67	10.09	12.79	16.34	20.49	24.77	27.59	30.19	33.41	35.72	40.79
18	6.26	7.01	8.23	9.39	10.86	13.68	17.34	21.60	25.99	28.87	31.53	34.81	37.16	42.31
19	6.84	7.63	8.91	10.12	11.65	14.56	18.34	22.72	27.20	30.14	32.85	36.19	38.58	43.82
20	7.43	8.26	9.59	10.85	12.44	15.45	19.34	23.83	28.41	31.41	34.17	37.57	40.00	45.32
21	8.03	8.90	10.28	11.59	13.24	16.34	20.34	24.93	29.62	32.67	35.48	38.93	41.40	46.80
22	8.64	9.54	10.98	12.34	14.04	17.24	21.34	26.04	30.81	33.92	36.78	40.29	42.80	48.27
23	9.26	10.20	11.69	13.09	14.85	18.14	22.34	27.14	32.01	35.17	38.08	41.64	44.18	49.73
24	9.89	10.86	12.40	13.85	15.66	19.04	23.34	28.24	33.20	36.42	39.36	42.98	45.56	51.18
25	10.52	11.52	13.12	14.61	16.47	19.94	24.34	29.34	34.38	37.65	40.65	44.31	46.93	52.62
26	11.16	12.20	13.84	15.38	17.29	20.84	25.34	30.43	35.56	38.89	41.92	45.64	48.29	54.05
27	11.81	12.88	14.57	16.15	18.11	21.75	26.34	31.53	36.74	40.11	43.19	46.96	49.64	55.48
28	12.46	13.56	15.31	16.93	18.94	22.66	27.34	32.62	37.92	41.34	44.46	48.28	50.99	56.89
29	13.12	14.26	16.05	17.71	19.77	23.57	28.34	33.71	39.09	42.56	45.72	49.59	52.34	58.30
30	13.79	14.95	16.79	18.49	20.60	24.48	29.34	34.80	40.26	43.77	46.98	50.89	53.67	59.70
40	20.71	22.16	24.43	26.51	29.05	33.66	39.34	45.62	51.80	55.76	59.34	63.69	66.77	73.40
50	27.99	29.71	32.36	34.76	37.69	42.94	49.33	56.33	63.17	67.50	71.42	76.15	79.49	86.66
60	35.53	37.48	40.48	43.19	46.46	52.29	59.33	66.98	74.40	79.08	83.30	88.38	91.95	99.61
70	43.28	45.44	48.76	51.74	55.33	61.70	69.33	77.58	85.53	90.53	95.02	100.42	104.22	112.32
80	51.17	53.54	57.15	60.39	64.28	71.14	79.33	88.13	96.58	101.88	106.63	112.33	116.32	124.84
90	59.20	61.75	65.65	69.13	73.29	80.62	89.33	98.64	107.56	113.14	118.14	124.12	128.30	137.21
100	67.33	70.06	74.22	77.93	82.36	90.13	99.33	109.14	118.50	124.34	129.56	135.81	140.17	149.45
X	-2.576	-2.326	-1.96	-1.645	-1.282	-0.674	0.0	0.674	1.282	1.645	1.96	2.326	2.576	3.090

Source: After Pearson and Hartley, 1966.
For $v > 100$, take $\chi^2 = v[1 - 2/9v + X\sqrt{2/9v}]^3$ or $\chi^2 = 1/2\ [X + \sqrt{2v - 1}]^2$ if less accuracy is needed, where X is given in the last row of the table.
This table is first used in Section 16.4.4.

Appendix B

TREND

TESTING FOR MONOTONIC TRENDS USING MANN-KENDALL, SEASONAL KENDALL, AND RELATED NONPARAMETRIC TECHNIQUES

IMPLEMENTATION

Written in FORTRAN 77.

The I/O units that may need to be changed are:

```
IN    = 5   ! Input from the terminal
IOUT  = 6   ! Output to the terminal
IFIN  = 1   ! Input data from a file
IFOUT = 3   ! Output results to a file
```

There are a few requirements if a driver program is substituted for the provided driver (TREND).

1. Parameters in the parameter statement must be large enough to fit the data. The parameters are

```
a. NYRS  = 15      ! Number of possible years
b. NSEAS = 12      ! Number of possible seasons
c. NSIT  = 10      ! Number of possible stations
                   ! (sites)
d. NTOT  = 180     ! Number of possible data points
                   ! NTOT = NYRS * NSEAS
e. NTTOT = 16110   ! Number of possible differences for
                   ! the Mann-Kendall slopes
                   ! NTTOT = NTOT * (NTOT - 1)/2
f. NYS   = 2500    ! Number of possible differences for
                   ! the seasonal Kendall slopes
```

2. The common blocks and parameter statement must be present as listed in the driver program. These same common blocks and parameters are used in the subroutines. Thus, if either a common block or a parameter is changed in the driver program, it must be changed in the subsequent routines.

Note: Appendix B was written by D. W. Engel, technical specialist, Pacific Northwest Laboratory.

DESCRIPTION OF ROUTINES

Driver routine

1. Establishes parameters and common blocks
2. Inputs the following data parameters
   - `NYEAR` Number of years of data
   - `NSEASON` Number of seasons of data
   - `NSITE` Number of sites or stations of data
   - `ALPHA` Acceptance level of significance of the homogeneity statistics
3. Inputs data into the following arrays
   - `YEAR` Years
   - `SEASON` Seasons
   - `DATA` Observed values
4. Calls the subroutines to calculate the different trend statistics after inputting one data set (station) at a time. These statistics are stored for output and for calculating statistics between stations (sites).
5. Output results in the form of a table.

Major Subroutines

1. `THOMO` Calculates and stores the homogeneity statistics
2. `SENT` Calculates the Sen T statistic
3. `KTEST` Calculates the seasonal Kendall statistic or the Mann-Kendall statistic. Also calculates the Sen and seasonal Kendall slopes. Confidence intervals are also calculated by calling the subroutine `CONINT`

COMMENTS

All of the statistics (homogeneity, Sen T, Kendall) are calculated with each run of the program. The Mann-Kendall statistic is calculated if the number of seasons that is input (`NSEASON`) is zero; otherwise the seasonal Kendall statistic is calculated.

All of the homogeneity statistics are calculated in the calls to the subroutine `THOMO`. On output, if the homogeneity statistics are significant (i.e., $P <$ `ALPHA`), then some of the related statistics are not printed.

Data does not have to be sorted on input. The driver program sorts the data by seasons. The Sen T statistic cannot be calculated if there are missing years, seasons, or stations (sites). The program calculates the homogeneity and Kendall statistics, but only outputs a message for the Sen T statistic if there are missing data (years, seasons, stations). Replicate data values are averaged when calculating the Sen T statistic. Data is input from one data file for each station (site).

When running the Mann-Kendall test (`NSEASON` = 0), only time and data values are needed. But to keep the program `TREND` general enough to run all of the tests, a year, season, and data value must be input. This may be accomplished by reading the time variable twice, using a T (tab) format.

The current version of `TREND` calculates a confidence interval about the estimated slope by using four different alpha (significance) levels. Different alpha levels may be added, or those presently in **TREND** may be deleted by changing two statements in the subroutine **CONINT**. Simply add or delete the alpha level (**ALP**) and its corresponding Z value (**ZA**) in the data statement.

If the number of alpha levels (**NCI**) is changed, the line setting the number must also be changed. At present, the line reads **NCI** = 4.

Input file names and data format are input to the driver program (**TREND**). The results for each run are written to the disk file **TREND.OUT**.

If the investigator wants to treat ND, trace, or LT values as missing observations, then no data are entered into the program for those cases. If they will be treated as one half the detection limit or some other values, those values should be entered as part of the data set.

```
      PROGRAM TREND
C
C       THIS IS THE DRIVING PROGRAM FOR THE SUBROUTINES THAT
C     CALCULATE THE FOUR STATISTICS TESTING FOR TREND.  THE
C     SUBROUTINES ARE,
C
C       THOMO .... CALCULATES THE CHI-SQUARED STATISTICS FOR THE
C                  HOMOGENEITY TESTS.
C       SEN ...... CALCULATES THE ALIGNED RANK (SEN) STATISTIC.
C       KTEST .... CALCULATES THE SEASONAL KENDALL STATISTIC, OR
C                  CALCULATES THE MANN KENDALL STATISTIC IF THE NUMBER
C                  OF SEASONS = 0.
C
C     INPUT PARAMETERS:
C         NYEAR .... NUMBER OF YEARS.
C         NSEASON .. NUMBER OF SEASONS.
C         NSITE .... NUMBER OF SITES (STATIONS).
C         ALPHA .... ACCEPTANCE LEVEL FOR THE STATISTICS.
C
C     INPUT DATA ARRAYS:
C         YEAR ..... ARRAY OF YEARS.
C         SEASON ... ARRAY OF SEASONS.
C         DATA ..... ARRAY OF DATA.
C
C     NSEASON = 0  -->  CALCULATE THE MANN KENDALL STATISTIC.  NO
C                       SEASON AFFECT.
C
      PARAMETER NYRS  = 30, NSEAS = 12, NTOT = 180, NSIT = 10,
     1     NTTOT = 16110, NYS = 2500, NA = 10, NE = 2
C
      COMMON /DATA/  NYEAR,           YEAR(NTOT),
     1         NSEASON,  SEASON(NTOT),
     2         NDATA(NSIT),  DATA(NTOT)
      COMMON /SORT/  SORTY(NTOT), SORTS(NTOT), SORTD(NTOT), NCR(NTOT)
      COMMON /SLOPE/ ZSLOPE(NSEAS,NSIT), SEASSL(NSEAS), SITESL(NSIT),
     1          YSSLOPE(NYS,NSEAS), NYSSLOPE(NSEAS)
      COMMON /ZST/   ZSTAT(NSEAS,NSIT), NSTAT(NSEAS,NSIT),
     1          SSTAT(NSEAS,NSIT)
      COMMON /HOMO/  TOTAL,          NTOTAL,         PTOTAL,
     1          HOMOGEN,        NHOMOGEN,       PHOMOGEN,
     2          SEASONAL,       NSEASONAL,      PSEASONAL,
     3          SITE,           NSITES,         PSITES,
     4          SITESEAS,       NSITESEAS,      PSITESEAS,
     5          TRENDS,         NTRENDS,        PTRENDS,
     6          ZSEASON(NSEAS),  PSEASON(NSEAS),
     7          ZSITE(NSIT),      PSITE(NSIT)
      COMMON /CI1/ ALP(NA), ZA(NA)
      COMMON /CI2/ CIMKS(NA,NE,NSEAS,NSIT), CISIS(NA,NE,NSIT),
     1         VSEA(NSEAS), NUME,         CISES(NA,NE,NSEAS)
      COMMON /WR1/ SENT(NSIT), PSENT(NSIT), ZKEN(NSIT), PKENZ(NSIT)
      COMMON /WR2/ NC(NTOT), NCRS(NTOT,NSIT), NCRC(NSIT)
```

```
C
      REAL RANKS(NTOT)
      BYTE INFILE(80), OUTFILE(80), FMT(80), IFPRT
      LOGICAL LAST
C
      DATA LAST / .FALSE. /
C
C     SETTING I/O UNITS
      IN    = 5        ! INPUT FROM TERMINAL
      IOUT  = 6        ! OUTPUT TO  TERMINAL
      IFIN  = 1        ! INPUT FORM FILE
      IFOUT = 3        ! OUTPUT TO FILE
C
C     OUTPUT TO FILE  TREND.OUT
      OPEN(UNIT=IFOUT,NAME='TREND.OUT',TYPE='NEW')
C
C     INPUT INFORMATION
C
      WRITE(IOUT,110)
110   FORMAT(' ENTER MAX NUMBER OF YEARS, SEASONS, AND STATIONS.',
     1 '  NUMBER SEASON = 0  -->  CALCULATE MANN KENDALL STATISTIC')
      READ(IN,*) NYEAR, NSEASON, NSITE
C
      WRITE(IOUT,*) ' ENTER ALPHA (ACCEPTANCE) LEVEL'
      READ(IN,*) ALPHA
C
      WRITE(IOUT,*) ' DO YOU WANT DATA PRINTED ON OUTPUT?  Y or N'
      READ(IN,120) IFPRT
120   FORMAT(A1)
C
C     HEADER FOR OUTPUT FILE
      IF (NSEASON.EQ.0) THEN
          WRITE(IFOUT,130) NYEAR, NSITE, ALPHA
130       FORMAT(' NUMBER OF ITMES     =',I4/
     1           ' NUMBER OF STATIONS =',I4/
     2           ' ALPHA LEVEL         =',F6.3)
      ELSE
          WRITE(IFOUT,140) NYEAR, NSEASON, NSITE, ALPHA
140       FORMAT(' NUMBER OF YEARS     =',I4/
     1           ' NUMBER OF SEASONS  =',I4/
     2           ' NUMBER OF STATIONS =',I4/
     3           ' ALPHA LEVEL         =',F6.3)
      ENDIF
C
C     MAIN LOOP FOR DIFFERENT SITES (STATIONS)
C
      DO 300 J=1,NSITE
C
C         INPUT DATA FILE SPECS
          WRITE(IOUT,*) ' ENTER NAME OF INPUT DATA FILE'
          READ(IN,150) KCI, (INFILE(I),I=1,KCI)
150       FORMAT(Q,80A1)
          INFILE(KCI+1)=0.0
          OPEN(UNIT=IFIN,NAME=INFILE,TYPE='OLD',READONLY)
C
          WRITE(IOUT,*) ' ENTER INPUT FILE FORMAT;  YEAR, SEASON,',
     1           ' DATA.  Ex.  (3F10.0)'
          READ(IN,150) KC, (FMT(I),I=1,KC)
          FMT(KC+1)=0.0
C
C         INITIALIZE DATA TO 0.0
          DO 160 I=1,NTOT
           YEAR(I)   = 0.0
           SEASON(I) = 0.0
           DATA(I)   = 0.0
160       CONTINUE
C
C         READ DATA, FIND MINIMUM YEAR, AND SET UP FOR MANN TEST
          ND = 0
          YM = 2000.0
```

```
          DO 170 I=1,2000
           READ(IFIN,FMT,END=180) YR, SSN, DATA(I)
           ND = ND + 1
           IF (YR.LT.YM .AND. YR.NE.0.0) YM = YR
           YEAR(I) = YR
C
           IF(NSEASON.EQ.0) THEN
               NC(I) = 0
               RANKS(I)  = I
               SEASON(I) = I
           ELSE
               SEASON(I) = SSN
           ENDIF
170       CONTINUE
180       IF (NSEASON.EQ.0) NC(1) = ND
C
C         HEADER FOR OUTPUT FILE
          IF (J.GT.1) WRITE(IFOUT,190)
190       FORMAT(1H1)
          WRITE(IFOUT,200)
200       FORMAT(//' STATION  NUMBER DATA POINTS    INPUT FILE NAME'/)
C
          WRITE(IFOUT,210) J, ND, (INFILE(I),I=1,KCI)
210       FORMAT(X,I4,T12,I6,T33,80A1)
          NDATA(J) = ND
          CLOSE(UNIT=IFIN)
C
C         OUTPUT DATA IF DESIRED
          IF (IFPRT.EQ.'Y' .OR. IFPRT.EQ.'y') THEN
           IF (NSEASON.EQ.0) THEN
               WRITE(IFOUT,220) J
220            FORMAT(//'          TIME    STATION ',I2)
               DO 240 I=1,ND
                IYEAR = YEAR(I)
                ISEAS = SEASON(I)
                WRITE(IFOUT,230) IYEAR, DATA(I)
230             FORMAT(5X,I8,2X,F10.2)
240               CONTINUE
           ELSE
               WRITE(IFOUT,250) J
250            FORMAT(//'  YEAR   SEASON   STATION ',I2)
               DO 270 I=1,ND
                IYEAR = YEAR(I)
                ISEAS = SEASON(I)
                WRITE(IFOUT,260) IYEAR, ISEAS, DATA(I)
260             FORMAT(I5,I8,2X,F10.2)
270               CONTINUE
           ENDIF
          ENDIF
C
C         SORT DATA BY SEASONS AND SCALE YEAR FROM 1 TO NYEAR
          IF (NSEASON.NE.0) CALL RANK(SEASON,RANKS,ND,NC)
          INCR = 0
          NCRT = 0
          DO 280 I=1,ND
           NCI  = NC(I)
           NCRT = NCRT + NCI
           IF (NCI.NE.0) THEN
               INCR = INCR + 1
               NCR(INCR) = NCI
               NCRS(INCR,J)= NCI
C              IF (NCI.LT.2) STOP ' NOT ENOUGH DATA IN SEASON'
           ENDIF
           YEAR(I) = YEAR(I) - YM + 1
280       CONTINUE
C
          DO 290 I=1,ND
           RS  = RANKS(I)
           NR  = NC(RS)
           RI  = RS - (NR - 1.0)/2.0
```

```
          NC(RS)    = NR - 2
          IF (NSEASON.EQ.0) RI = I
          SORTY(RI) = YEAR(I)
          SORTS(RI) = SEASON(I)
          SORTD(RI) = DATA(I)
290      CONTINUE
         IF (NSEASON.EQ.0) NCR(1) = ND
C
C        TESTING FOR HOMOGENEITY OF TREND.  IF THIS IS THE
C        LAST CALL TO THOMO, THEN CALCULATE ALL STATISTICS.
         IF (J.EQ.NSITE) LAST = .TRUE.
         CALL THOMO(J,LAST)
C
C        CALCULATING THE SEN T STATISTIC IF THERE IS MORE
C        THAN 1 SEASONS
         IF (NSEASON.GT.0) THEN
          CALL SEN(T,ND)
          CALL PNORM(T,PT)
          SENT(J)  = T
          IF (T.GT.0.0) THEN
                PSENT(J) = 2*(1.0 - PT)
          ELSE
              PSENT(J) = 2*PT
          ENDIF
         ENDIF
C
C        CALCULATING THE SEASONAL KENDALL (NUMBER SEASONS > 0) OR
C        THE MANN KENDALL (NUMBER SEASONS = 0) STATISTIC.
         CALL KTEST(Z,J,LAST)
         CALL PNORM(Z,PZ)
         ZKEN(J)  = Z
         IF (Z.GT.0.0) THEN
             PKENZ(J) = 2*(1.0 - PZ)
         ELSE
          PKENZ(J) = 2*PZ
         ENDIF
         NCRC(J) = NCRT
300   CONTINUE
C
C     OUTPUT THE RESULTS OF THE TREND TESTS
C
      CALL WRITE1(IFOUT,NSITE,ALPHA)
C
      CLOSE(UNIT=IFOUT)
      STOP
      END


C     ***************************************************************
      SUBROUTINE THOMO(IJ,LAST)
C     ***************************************************************
C
C       THIS SUBROUTINE CALCULATES THE CHI-SQUARE STATISTICS Z**2
C     FROM THE KENDALL S STATISTIC.
C
C     Z = S / SQRT(VAR(S))
C
C         S      = SUM[Sj]
C         VAR(S) = SUM[VAR(Sj)]
C                  Sj       = SUM(SUM(SIGN(Xj - Xk)))
C                      k=1 to N-1, j= k+1 to N, and N = NUMBER
C                                             OF YEARS
C               VAR(Sj) = [N(N-1)(2N+5) - TIE]/18 + CORRECT
C                              TIE = SUM(t(t-1)(2t+5))
C                                   t = NUMBER INVOLVED IN TIE.
C               CORRECT IS A CORRECTION FACTOR FOR REPLICATE DATA.
C
C     IJ ..... SITE (STATION) INDEX.
C     LAST ... LOGICAL VARIABLE.  IF LAST = .TRUE. -->  THIS IS LAST
```

```
C                    SITE, SO CALCULATE THE FINAL STATISTICS.
C
      PARAMETER NYRS  = 30, NSEAS = 12, NTOT = 180, NSIT = 10,
     1      NTTOT = 16110
C
      COMMON /DATA/  NYEAR,          YEAR(NTOT),
     1              NSEASON,         SEASON(NTOT),
     2              NDATA(NSIT), DATA(NTOT)
      COMMON /SORT/  SORTY(NTOT), SORTS(NTOT), SORTD(NTOT), NCR(NTOT)
      COMMON /HOMO/  TOTAL,          NTOTAL,         PTOTAL,
     1              HOMOGEN,     NHOMOGEN,     PHOMOGEN,
     2              SEASONAL,    NSEASONAL,    PSEASONAL,
     3              SITE,        NSITES,       PSITES,
     4              SITESEAS,    NSITESEAS,    PSITESEAS,
     5              TRENDS,      NTRENDS,      PTRENDS,
     6              ZSEASON(NSEAS),  PSEASON(NSEAS),
     7              ZSITE(NSIT),     PSITE(NSIT)
C
      REAL    SLOPES(NTTOT)
      INTEGER NSTAT(NSEAS,NSIT)
      LOGICAL LAST
C
      DATA TOTAL, TOT2, ZSEASON / 0.0, 0.0, NSEAS*0.0 /
      NSITE = IJ
C
C     LOOP FOR EACH SEASON.  DEFINE  N = NUMBER SEASONS, BUT IF N = 0
C     (ie MANN KENDALL TESTS) THEN SET N = 1 FOR THE CALCULATIONS.
C
         N  = NSEASON
      IF (N.EQ.0) N = 1
      IS = 1
      ZSITES = 0.0
      DO 120 I=1,N
         NC = NCR(I)
C
C        CALCULATE THE KENDALL STATISTIC Sj
         CALL KEND(SORTD(IS),SORTY(IS),SJ,NC,SLOPES,SMED,NS)
C
C        CALCULATE THE VARIANCE OF Sj
         CALL TIES(SORTD(IS),SORTY(IS),NC,VAR)
         IS  = IS + NC
C
C        CALCULATE THE Z-STATISTIC.
         ZSTATS = 0.0
         IF(VAR.EQ.0.0) GOTO 120
C
         SJV = SJ/SQRT(VAR)
         TOTAL  = TOTAL  + SJV*SJV
         ZSITES = ZSITES + SJV
         ZSEASON(I) = ZSEASON(I) + SJV
120   CONTINUE
      ZSITE(IJ) = ZSITES
C
C     IF NOT LAST CALL (ie. IJ .NE. NSITE) THEN RETURN
      IF (.NOT.LAST) RETURN
C
C     IF LAST CALL THEN CALCULATE THE FINAL CHI-SQUARE STATISTICS
C
C     CALCULATE Z-STAT OVER THE DIFFERENT SITES, SEASONS MEANS
      ZDD = 0.0
      DO 220 I=1,NSITE
         ZSITES   = ZSITE(I) / N
         ZSITE(I) = ZSITES
         ZDD = ZDD + ZSITES
220   CONTINUE
C
      ZDDS = 0.0
      DO 230 I=1,N
         ZSEASONS   = ZSEASON(I) / NSITE
         ZSEASON(I) = ZSEASONS
         ZDDS = ZDDS + ZSEASONS
```

```
230  CONTINUE
C
     ZDD  = ZDD  / NSITE
     ZDDS = ZDDS / N
C
C    CALCULATING CHI-SQUARE STATISTICS
C
     NTOTAL = N*NSITE
     CALL CHICDF(TOTAL,NTOTAL,PTOTAL)
C
     TRENDS  = NSITE*N*ZDD*ZDD
     NTRENDS = 1
     CALL CHICDF(TRENDS,NTRENDS,PTRENDS)
C
     HOMOGEN  = TOTAL - TRENDS
     NHOMOGEN = NTOTAL - 1
     CALL CHICDF(HOMOGEN;NHOMOGEN,PHOMOGEN)
C
     SEASONAL = 0.0
     DO 320 I=1,N
        SUM = ZSEASON(I) - ZDD
        SEASONAL = SEASONAL + SUM*SUM
320  CONTINUE
     SEASONAL  = NSITE*SEASONAL
     NSEASONAL = N - 1
     CALL CHICDF(SEASONAL,NSEASONAL,PSEASONAL)
C
     SITE = 0.0
     DO 330 I=1,NSITE
        SUM = ZSITE(I) - ZDD
        SITE = SITE + SUM*SUM
330  CONTINUE
     SITE   = N*SITE
     NSITES = NSITE - 1
     CALL CHICDF(SITE,NSITES,PSITES)
C
     SITESEAS  = HOMOGEN - SEASONAL - SITE
     NSITESEAS = NSITES*NSEASONAL
     IF (SITESEAS.LT.0.0) SITESEAS = 0.0
     CALL CHICDF(SITESEAS,NSITESEAS,PSITESEAS)
C
C    CALCULATE CHI-SQUARE FOR INDIVIDUAL SEASONS AND SITES
     DO 340 I=1,N
        ZS = ZSEASON(I)
        ZS = NSITE*ZS*ZS
        CALL CHICDF(ZS,1,PSEASON(I))
        ZSEASON(I) = ZS
340  CONTINUE
C
     DO 350 I=1,NSITE
        ZS = ZSITE(I)
        ZS = N*ZS*ZS
        CALL CHICDF(ZS,1,PSITE(I))
        ZSITE(I) = ZS
350  CONTINUE
C
360  RETURN
     END



C    *****************************************************
     SUBROUTINE SEN(T,ND)
C    *****************************************************
C
C      THIS ROUTINE CALCULATES AND RETURNS THE SEN T STATISTIC
C    FOR TESTING OF TRENDS.  THIS ROUTINE AVERAGES REPLICATE
C    DATA.
C      FOR LARGE ENOUGH M (SEASONS) T IS DISTRIBUTED N(0,1).
C
```

```
C     T = SQRT[12M*M/(N(N+1)SUM(SUM(Rij-R.j)))]*
C           SUM[(i-(N+1)/2)(Ri.-(NM+1)/2)]
C
C         R = RANKED DATA WITH SEASON EFFECT REMOVED.  (Xij - X.j)
C         M = NUMBER SEASON, N = NUMBER YEARS, i = 1 to N, j = 1 to M
C
C
      PARAMETER NYRS = 30, NSEAS = 12, NTOT = 180, NSIT = 10
      COMMON /DATA/  NYEAR,         YEAR(NTOT),
      1         NSEASON,  SEASON(NTOT),
      2         NDATA(NSIT),   DATA(NTOT)
      REAL RIDOT(NYRS), RDOTJ(NSEAS), RIJ(NYRS,NSEAS)
      REAL XDATA(NTOT), RDATA(NTOT)
      INTEGER NC(NTOT), ICOUNT(NYRS,NSEAS)
      DOUBLE PRECISION SUMI, SUMJ
C
C     MISSING DATA IF ICOUNT(I,J) = 0
      DO 110 J=1,NSEAS
      DO 110 I=1,NYRS
          RIJ(I,J)    = 0.0
          ICOUNT(I,J) = 0
110   CONTINUE
C
C     PUT DATA INTO   YEAR x SEASON     MATRIX  AND
C     AVERAGE REPLICATE DATA
      DO 120 I=1,ND
          IIND = YEAR(I)
          JIND = SEASON(I)
          RIJ(IIND,JIND)    = RIJ(IIND,JIND)    + DATA(I)
          ICOUNT(IIND,JIND) = ICOUNT(IIND,JIND) + 1
120   CONTINUE
C
      DO 130 J=1,NSEASON
      DO 130 I=1,NYEAR
          ICOUN = ICOUNT(I,J)
          IF (ICOUN.GT.1) THEN
           RIJ(I,J) = RIJ(I,J)/ICOUN
          ELSE
           IF (ICOUN.EQ.0) GOTO 260
          ENDIF
130   CONTINUE
C
C     REMOVE SEASONAL EFFECT BY SUBTRACTING THE SEASON
C     AVERAGES FROM THE DATA MATRIX RIJ.
      DO 140 I=1,NYEAR
          RIDOT(I) = 0.0
140   CONTINUE
C
      K = 1
      DO 170 J=1,NSEASON
          RDOTJS = 0.0
          DO 150 I=1,NYEAR
           X = RIJ(I,J)
           RDOTJS = RDOTJS + X
150       CONTINUE
          RDOTJS = RDOTJS/NYEAR
          DO 160 I=1,NYEAR
           RIJIJ = RIJ(I,J) - RDOTJS
           RIJ(I,J) = RIJIJ
           XDATA(K) = RIJIJ
           K = K + 1
160       CONTINUE
170   CONTINUE
C
C     RANK DIFFERENCES ALL TOGETHER
      CALL RANK(XDATA,RDATA,ND,NC)
C
C     PUT RANKS BACK INTO   YEAR x SEASON     MATRIX
      IND = 1
      DO 180 J=1,NSEASON
      DO 180 I=1,NYEAR
          RIJ(I,J) = RDATA(IND)
          IND = IND + 1
```

```
180   CONTINUE
C
C     CALCULATE YEAR AND SEASON AVERAGES FOR THE RANKED MATRIX RIJ.
      DO 190 I=1,NYEAR
          RIDOT(I) = 0.0
190   CONTINUE
C
      DO 210 J=1,NSEASON
          RDOTJS = 0.0
          DO 200 I=1,NYEAR
           X = RIJ(I,J)
           RDOTJS    = RDOTJS    + X
           RIDOT(I) = RIDOT(I) + X
200       CONTINUE
          RDOTJ(J) = RDOTJS/NYEAR
210   CONTINUE
C
      DO 220 I=1,NYEAR
          RIDOT(I) = RIDOT(I)/NSEASON
220   CONTINUE
C
C     CALCULATE SEN STATISTIC
      SUMI = 0.0
      SUMJ = 0.0
      YADD1 = (NYEAR + 1)/2.0
      YADD2 = (NYEAR*NSEASON + 1)/2.0
      DO 240 J=1,NSEASON
          RDOTJS = RDOTJ(J)
          DO 230 I=1,NYEAR
           SUMJ = SUMJ + (RIJ(I,J) - RDOTJS)**2
230       CONTINUE
240   CONTINUE
C
      DO 250 I=1,NYEAR
          SUMI = SUMI + (I - YADD1)*(RIDOT(I) - YADD2)
250   CONTINUE
C
      IF (SUMJ.EQ.0.0 .OR. SUMI.EQ.0.0) GOTO 260
      T = SQRT(12*NSEASON*NSEASON/(NYEAR*(NYEAR+1)*SUMJ))*SUMI
      GOTO 270
C
C     MISSING DATA IN DATA SET, CAN NOT COMPUTE SEN T STATISTIC
260   T = 0.0
C
270   RETURN
      END



C     ***************************************************************
      SUBROUTINE KTEST(Z,J,LAST)
C     ***************************************************************
C
C       THIS ROUTINE CALCULATES THE Z (STANDARD NORMAL) STATISTIC
C     FOR THE SEASONAL KENDALL TEST OF TREND.  IF NSEASON = 0
C     CALCULATES THE MANN KENDALL STATISTIC.  ALSO CALCULATES THE
C     KENDALL SLOPES AND CONFIDENCE INTERVALS ABOUT THE SLOPE.
C
C     Z .... RETURN KENDALL STATISTIC.
C     J .... NUMBER OF SITE (STATION) FOR THIS CALL.
C
C     Z = (S - 1)/SQRT(VAR(S)) IF  S > 0
C     Z = 0                    IF  S = 0
C     Z = (S + 1)/SQRT(VAR(S)) IF  S < 0
C
C         S      = SUM[Sj]
C         VAR(S) = SUM[VAR(Sj)]
C                  Sj      = SUM(SUM(SIGN(Xj - Xk)))
C                      k=1 to N-1, j= k+1 to N, and N = NUMBER
C                                        OF YEARS
C               VAR(Sj) = [N(N-1)(2N+5) - TIE]/18 + CORRECT
C                         TIE = SUM(t(t-1)(2t+5))
```

```
C                         t = NUMBER INVOLVED IN TIE.
C                  CORRECT IS THE CORRECTION FACTOR FOR REPLICATE DATA
C
C     SLOPE = MEDIAN(Xj - Xk)/(j-k);  j = 1 to N-1 by 1 and  k = j+1 to N by 1.
C
C
      PARAMETER NYRS  = 30, NSEAS  = 12, NTOT = 180, NSIT = 10,
     1      NTTOT = 16110, NYS = 2500, NA = 10, NE = 2
      COMMON /DATA/  NYEAR,        YEAR(NTOT),
     1          NSEASON,  SEASON(NTOT),
     2          NDATA(NSIT),   DATA(NTOT)
      COMMON /SORT/ SORTY(NTOT), SORTS(NTOT), SORTD(NTOT), NCR(NTOT)
      COMMON /ZST/  ZSTAT(NSEAS,NSIT), NSTAT(NSEAS,NSIT),
     1          SSTAT(NSEAS,NSIT)
      COMMON /SLOPE/ ZSLOPE(NSEAS,NSIT), SEASSL(NSEAS), SITESL(NSIT),
     1          YSSLOPE(NYS,NSEAS), NYSSLOPE(NSEAS)
      COMMON /CI1/ ALP(NA), ZA(NA)
      COMMON /CI2/ CIMKS(NA,NE,NSEAS,NSIT), CISIS(NA,NE,NSIT),
     1          VSEA(NSEAS), NUME,       CISES(NA,NE,NSEAS)
C
      REAL SLOPES(NTTOT)
      LOGICAL LAST
C
C     LOOP FOR EACH SEASON.  DEFINE  N = NUMBER SEASONS, BUT IF N = 0
C     (ie MANN KENDALL TESTS) THEN SET N = 1 FOR THE CALCULATIONS.
C
      N  = NSEASON
      IF (N.EQ.0) N = 1
      IS = 1
      IN = 1
      NSS  = 0
      SSUM = 0.0
      VSUM = 0.0
      DO 110 I=1,N
         NC = NCR(I)
C
C        CALCULATE THE KENDALL STATISTIC Sj
         CALL KEND(SORTD(IS),SORTY(IS),SJ,NC,SLOPES(IN),SMED,NS)
         SSUM = SSUM + SJ
         ZSLOPE(I,J) = SMED
C
C        SAVE SLOPE VALUES
         IYS = NYSSLOPE(I) + 1
         NSS = NSS + NS
         IN1 = IN
         IN2 = IN + NS - 1
         DO 100 K=IN1,IN2
          YSSLOPE(IYS,I) = SLOPES(K)
          IYS = IYS + 1
100      CONTINUE
         NYSSLOPE(I) = IYS - 1
C
C        CALCULATE THE VARIANCE OF Sj
         CALL TIES(SORTD(IS),SORTY(IS),NC,VAR)
         IS = IS + NC
         IN = IN + NS
         VSUM = VSUM + VAR
         VSEA(I) = VSEA(I) + VAR
C
C        CALCULATE CONFIDENCE INTERVALS ABOUT THE SLOPE
         CALL CONINT(NS,VAR,SLOPES(IN1),CIMKS(1,1,I,J),
     1                       NUME,CIMKS(1,2,I,J))
C
C        STORE ALL Z'S USING THE CONTINUITY CORRECTION
         IF (VAR.EQ.0.0) GOTO 110
         IF (SJ.GT.0.0) THEN
          ZSTAT(I,J) = (SJ - 1)/SQRT(VAR)
         ELSE
          IF (SJ.LT.0.0) ZSTAT(I,J) = (SJ + 1)/SQRT(VAR)
         ENDIF
         NSTAT(I,J)  = NC
         SSTAT(I,J) = SJ
```

```
110   CONTINUE
C
C     Z-STATISTIC  USING THE CONTINUITY CORRECTION
      Z = 0.0
      IF (VSUM.EQ.0.0) GOTO 120
      IF (SSUM.GT.0.0) THEN
         Z = (SSUM - 1)/SQRT(VSUM)
      ELSE
         IF (SSUM.LT.0.0) Z = (SSUM + 1)/SQRT(VSUM)
      ENDIF
C
C     FIND SEASONAL-KENDALL SLOPE
C
120   IN = IN - 1
      CALL SORT(SLOPES,IN,SMED)
      SITESL(J) = SMED
C
C     CALCULATE CONFIDENCE INTERVALS ABOUT THE SLOPE
      CALL CONINT(NSS,VSUM,SLOPES,CISIS(1,1,J),NUME,CISIS(1,2,J))
C
C     IF LAST CALL THEN CALCULATE FINAL KENDALL SLOPE
C
      IF (.NOT.LAST) GOTO 140
      DO 130 I=1,N
         IS = NYSSLOPE(I)
         CALL SORT(YSSLOPE(1,I),IS,SMED)
         SEASSL(I) = SMED
         CALL CONINT(IS,VSEA(I),YSSLOPE(1,I),CISES(1,1,I),
     *       NUME,CISES(1,2,I))
130   CONTINUE
C
140   RETURN
      END
C


C     ************************************************************
      SUBROUTINE KEND(X,Y,S,N,SLOPE,SMED,NS)
C     ************************************************************
C
C       THIS ROUTINE CALCULATES THE KENDALL STATISTIC S
C
C     X ....... ARRAY OF DATA FOR ONE SEASON AND N YEARS.
C     Y ....... ARRAY OF YEARS CORRESPONDING TO THE DATA (USED
C               FOR MULTIPLE OBSERVATIONS).
C     S ....... OUTPUT KENDALL STATISTIC.
C     N ....... NUMBER DATA POINTS IN THIS SEASON.
C     SLOPE ... ARRAY OF CALCULATED SEASONAL KENDALL SLOPE ESTIMATORS.
C             THE SLOPE ESTIMATOR IS THE MEDIAN OF ALL THE SLOPE
C             ESTIMATES THAT GO WITH S, INSTEAD OF JUST THE SIGN.
C     SMED .... MEDIAN SLOPE FOR EACH CALL TO THIS SUBROUTINE.
C     NS ...... NUMBER OF SLOPE ESTIMATES CALCULATED.
C
C     S = SUM[SIGN(X(J)-X(I))]      WHERE I=1,N-1  J=I+1,N
C          SIGN(X) = +1 IF X > 0
C          SIGN(X) =  0 IF X = 0
C          SIGN(X) = -1 IF X < 0
C
C       IF MULTIPLE OBSERVATIONS OCCUR IN TIME, THEN S = 0.
C
      REAL X(1), Y(1), SLOPE(1)
C
      S  = 0.0
      NS = 0
      DO 120 I=1,N-1
          XI = X(I)
          YI = Y(I)
          DO 110 J=I+1,N
           XD = X(J) - XI
           YD = Y(J) - YI
```

```
          IF (YD.NE.0.0) THEN
              IF (XD.GT.0.0) THEN
               S = S + 1.0
              ELSE
               IF (XD.LT.0.0) S = S - 1.0
              ENDIF
              NS = NS + 1
              SLOPE(NS) = XD/YD
          ENDIF
110     CONTINUE
120   CONTINUE
C
C     FIND MEDIAN SLOPE.  1ST SORT AND THEN PICK MEDIAN
      CALL SORT(SLOPE,NS,SMED)
C
      RETURN
      END



C     ******************************************************
      SUBROUTINE TIES(X,Y,N,VAR)
C     ******************************************************
C
C       SUBROUTINE TO CALCULATE THE CORRECTION FACTOR DUE TO TIES
C     AND CALCULATE THE VARIANCE OF THE KENDALL S STATISTIC.
C
C     VAR(S) = [N(N-1)(2N-5) - NT1 - NU1]/18 +
C          NT2*NU2/[9N(N-1)(N-2)] + NT3*NU3/[2N(N-1)]
C
C           N   = NUMBER DATA (INPUT TO SUBROUTINE)
C           NT1 = SUM[Ti*(Ti - 1)*(2*Ti + 5)]      i = 1 to g
C           NU1 = SUM[Uj*(Uj - 1)*(2*Uj + 5)]      j = 1 to h
C           NT2 = SUM[Ti*(Ti - 1)*(Ti - 2)]  i = 1 to g
C           NU2 = SUM[Uj*(Uj - 1)*(Uj - 2)]  j = 1 to h
C           NT3 = SUM[Ti*(Ti - 1)]           i = 1 to g
C           NU3 = SUM[Uj*(Uj - 1)]           j = 1 to h
C
C     INPUT TO SUBROUTINE:
C
C     X .... VECTOR CONTAINING DATA
C     Y .... VECTOR CONTAINING YEARS (TIME)
C     N .... NUMBER OF VALUES IN X AND Y
C     VAR .. OUTPUT VARIANCE OF THE KENDALL Z STATISTIC
C
      REAL X(1), Y(1)
      INTEGER*2 INDEXT(100), INDEXU(100)
C
C     COUNT TIES
      INDT = 0
      INDU = 0
      DO 140 I=1,N
         XI = X(I)
         YI = Y(I)
         INDT = 0.0
         INDU = 0.0
         IF (I.EQ.N) GOTO 130
C
C        CHECK TO SEE IF THIS TIE HAS ALREADY BEEN COUNTED
         IF (I.GT.1) THEN
             DO 110 K=1,I-1
               IF (X(K).EQ.XI) XI = -999.99
               IF (Y(K).EQ.YI) YI = -999.99
110          CONTINUE
         ENDIF
C
         DO 120 J=I,N
          IF (XI.EQ.X(J)) INDT = INDT + 1
          IF (YI.EQ.Y(J)) INDU = INDU + 1
120      CONTINUE
```

```
130       INDEXT(I) = INDT
          INDEXU(I) = INDU
140   CONTINUE
C
C     CALCULATE CORRECTION FACTORS
      NT1 = 0
      NT2 = 0
      NT3 = 0
      NU1 = 0
      NU2 = 0
      NU3 = 0
      DO 150 I=1,N
          NT  = INDEXT(I)
          NU  = INDEXU(I)
          NT1 = NT1 + NT*(NT - 1)*(2*NT + 5)
          NU1 = NU1 + NU*(NU - 1)*(2*NU + 5)
          NT2 = NT2 + NT*(NT - 1)*(NT - 2)
          NU2 = NU2 + NU*(NU - 1)*(NU - 2)
          NT3 = NT3 + NT*(NT - 1)
          NU3 = NU3 + NU*(NU - 1)
150   CONTINUE
C
C     CALCULATE VAR(S)
C
      VAR = N*(N - 1.0)*(2.0*N + 5.0)/18.0
      IF (N.LE.2) GOTO 160
      VAR = (N*(N - 1.0)*(2.0*N + 5.0) - NT1 - NU1)/18.0 +
     1    NT2*NU2/(9.0*N*(N - 1.0)*(N - 2.0)) +
     2    NT3*NU3/(2.0*N*(N - 1.0))
C
160   RETURN
      END
C



C     *****************************************************
      SUBROUTINE RANK(X,R,N,NC)
C     *****************************************************
C
C       THIS ROUTINE RANKS THE DATA IN VECTOR X
C
C     X .... ARRAY TO BE RANKED
C     R .... OUTPUT ARRAY OF RANKS
C     N .... NUMBER OF VALUES IN X
C     NC ... ARRAY OF THE NUMBER IN EACH RANKING
C
C
      DIMENSION X(1), R(1), NC(1)
C
      DO 110 I=1,N
          NC(I) = 0
          R(I)  = 0.0
110   CONTINUE
C
      DO 130 I=1,N
          XI = X(I)
          DO 120 J=I,N
            IF (ABS(XI-X(J)).LT.1.0E-6) THEN
                R(J)  = R(J)  + 0.5
                R(I)  = R(I)  + 0.5
            ELSE
                IF (XI.GT.X(J)) THEN
                 R(I) = R(I) + 1.0
                ELSE
                 R(J) = R(J) + 1.0
                ENDIF
            ENDIF
120       CONTINUE
130   CONTINUE
```

```
C
C     COUNTING NUMBER IN RANKS (NUMBER IF THERE WERE TIES)
C
      DO 140 I=1,N
          IR = R(I)
          NC(IR) = NC(IR) + 1
140   CONTINUE
C
      RETURN
      END
C
```

```
C     ********************************************************
      SUBROUTINE SORT(X,N,SMED)
C     ********************************************************
C
C       THIS SUBROUTINE SORTS THE ARRAY X AND ALSO RETURNS
C     MEDIAN.  USES BUBBLE SORTING.
C       THIS ROUTINE CALLS THE ROUTINE RANK TO RANK THE DATA.
C
C     X ..... ON INPUT  X IS THE ARRAY TO SORT,
C          ON OUTPUT X IS THE SORTED ARRAY.
C     N ..... NUMBER IN ARRAY X.
C     SMED .. RETURN MEDIAN VALUE OF X.
C
C     FOR THIS ROUTINE AS IS, N MUST BE LESS THAN 7410.
C
      PARAMETER NTTOT = 10000
      COMMON /T/ ITO
      DIMENSION X(1), WORK(NTTOT), RANKS(NTTOT), NC(NTTOT)
C
      SMED = 0.0
      IF (N.LT.1) GOTO 150
      SMED = X(1)
      IF (N.EQ.1) GOTO 150
C
      ITEMP = 1
      DO 140 I=2,N
          IF (X(ITEMP).LE.X(I)) GOTO 130
          TEMP = X(I)
          DO 110 J=ITEMP,1,-1
           JI = J
           IF (TEMP.GE.X(J)) GOTO 120
           X(J+1) = X(J)
110       CONTINUE
          JI = 0
C
120             X(JI+1) = TEMP
130             ITEMP   = I
140   CONTINUE
C
C     FIND MEDIAN
C
      NH = N/2
      NF = 2*NH
      IF (NF.NE.N) THEN
          SMED = X(NH+1)
      ELSE
          SMED = (X(NH+1) + X(NH))/2.0
      ENDIF
C
150   RETURN
      END
```

```
C     ******************************************************
      SUBROUTINE CONINT(N,VAR,SLOPE,CIL,NCI,CIU)
C     ******************************************************
C
C       THIS ROUTINE CALCULATES THE CONFIDENCE INTERVALS
C     ABOUT THE KENDALL SLOPE.
C
C       N ...... NUMBER DIFFERENCES IN CALCULATING SLOPE.
C       VAR .... VARIANCE OF THE KENDALL S STATISTIC.
C       SLOPE .. ARRAY CONTAINING SLOPES.
C       CIL .... ARRAY OF LOWER LIMITS.  ONE FOR EACH ALPHA LEVEL.
C       CIU .... ARRAY OF UPPER LIMITS.  ONE FOR EACH ALPHA LEVEL.
C
C       LOWER LIMIT = M1 th LARGEST SLOPE ESTIMATE.
C       UPPER LIMIT = M2 + 1 th LARGEST SLOPE ESTIMATE.
C
C          M1 = (N' - C)/2
C          M2 = (N' + C)/2
C
C                 N' = NUMBER OF ORDERED SLOPE ESTIMATES
C                 C  = Z * SQRT[VAR(S)]
C
C                        Z IS FROM A NORMAL TABLE FOR ALPHA/2
C
C       THIS ROUTINE IS SET UP TO HANDLE ALPHA = .01, .05, .10, .20.
C     TO ADD NEW ALPHAS OR CHANGE THE EXISTING ALPHAS, TWO LINES OF
C     CODE MUST BE CHANGED.  ADD TO OR CHANGE THE DATA STATEMENT BELOW.
C     ALSO, IF THE NUMBER OF ALPHAS IS CHANGED, CHANGE THE STATEMENT
C                        NCI = 4.
C
      PARAMETER NA = 10, NE = 2, NSEAS = 12, NSIT = 10
      COMMON /CI1/ ALP(NA), ZA(NA)
C
      REAL SLOPE(1), CIL(1), CIU(1)
      DATA ALP, ZA /  .01,   .05,   .10,   .20, 6*0.0,
     1            2.576, 1.960, 1.645, 1.282, 6*0.0 /
C
      NCI = 4
C
C     CALCULATE CONFIDENCE INTERVALS ABOUT THE SLOPE
      DO 100 I=1,NCI
          CA  = ZA(I)*SQRT(VAR)
          XM1 = .5*(N - CA)
          XM2 = .5*(N + CA) + 1
C
          M1  = XM1
          M2  = XM2
          XD1 = XM1 - M1
          XD2 = XM2 - M2
C
C         CHECK TO SEE IF ENOUGH DATA TO CALCULATE CI.
C         -99.99 MEANS THAT N IS TOO SMALL TO CALCULATE THE CI.
          IF (M1.GE.1) THEN
           CIL(I) = XD1*SLOPE(M1+1) + (1.0 - XD1)*SLOPE(M1)
          ELSE
           CIL(I) = -99.99
          ENDIF
C
          IF (M2.LE.N) THEN
           CIU(I) = XD2*SLOPE(M2+1) + (1.0 - XD2)*SLOPE(M2)
          ELSE
           CIU(I) = -99.99
          ENDIF
C
100   CONTINUE
C
      RETURN
      END
C
```

```
C     ********************************************************
      SUBROUTINE CHICDF(C,N,P)
C     ********************************************************
C
C       THIS ROUTINE COMPUTES THE CUMULATIVE CHI-SQUARE
C     PROBABILITY.
C
C     C ... COMPUTED CHISQUARE VALUE.
C     N ... DEGREES OF FREEDOM.
C     P ... RETURNED CUMULATIVE CHI-SQUARED PROBABILITY.
C
      P = 0.0
      IF (C.LE.0.00001) GOTO 50
      IF (N.LE.0) RETURN
      A = FLOAT(N)
      GAMMA = 1.0
10    GAMMA = GAMMA*C/A
      A = A - 2
      IF(A) 20,30,10
20    GAMMA = GAMMA*SQRT(2./C)/1.7724578
30    C2 = 1.0
      C3 = 1.0
      D  = FLOAT(N)
40    D  = D + 2.0
      C3 = C3*C/D
      C2 = C2 + C3
      IF (C3.GE.0.5E-7) GO TO 40
      P = EXP(-C/2.)*C2*GAMMA
50    P = 1.0 - P
C
      RETURN
      END
C
```

```
C     ********************************************
      SUBROUTINE PNORM(X,P)
C     ********************************************
C
C       THIS ROUTINE COMPUTES A VERY QUICK AND CRUDE
C     CUMULATIVE NORMAL PROBABILITY.
C
C     X ... COMPUTED NORMAL VALUE.
C     P ... RETURNED NORMAL CUMULATIVE PROBABILITY.
C
C     EXP ... IS THE EXPONENTIAL FUNCTION.
C     ABS ... IS THE ABSOLUTE VALUE FUNCTION.
C
      DOUBLE PRECISION PC
C
      P  = .5
      IF (X.EQ.0.0) GOTO 20
      AX = ABS(X)
      PC = EXP(-((83.0*AX + 351.0)*AX + 562.0)/(703.0/AX + 165.0))
      P  = 1.0 - 0.5*PC
      IF (X.GT.0.0) GOTO 20
      P  = 1.0 - P
C
20    RETURN
      END
C
```

```
C     ***********************************************************
      SUBROUTINE WRITE1(IFOUT,NSITE,ALPHA)
C     ***********************************************************
C
C       THIS SUBROUTINE OUTPUTS THE RESULTS FROM ALL THE TREND
C     TESTS.
C
C     IFOUT .... OUTPUT UNIT NUMBER
C     NSITE .... NUMBER OF SITES.
C
C
      PARAMETER NYRS  = 30, NSEAS = 12, NTOT = 180, NSIT = 10,
     1      NTTOT = 16110, NYS = 2500, NA = 10, NE = 2
C
      COMMON /DATA/  NYEAR,        YEAR(NTOT),
     1        NSEASON,  SEASON(NTOT),
     2        NDATA(NSIT),   DATA(NTOT)
      COMMON /SORT/  SORTY(NTOT), SORTS(NTOT), SORTD(NTOT), NCR(NTOT)
      COMMON /SLOPE/ ZSLOPE(NSEAS,NSIT), SEASSL(NSEAS), SITESL(NSIT),
     1          YSSLOPE(NYS,NSEAS), NYSSLOPE(NSEAS)
      COMMON /ZST/   ZSTAT(NSEAS,NSIT), NSTAT(NSEAS,NSIT),
     1          SSTAT(NSEAS,NSIT)
      COMMON /HOMO/  TOTAL,        NTOTAL,        PTOTAL,
     1          HOMOGEN,     NHOMOGEN,     PHOMOGEN,
     2          SEASONAL,    NSEASONAL,    PSEASONAL,
     3          SITE,        NSITES,       PSITES,
     4          SITESEAS,    NSITESEAS,    PSITESEAS,
     5          TRENDS,      NTRENDS,      PTRENDS,
     6          ZSEASON(NSEAS),  PSEASON(NSEAS),
     7          ZSITE(NSIT),     PSITE(NSIT)
      COMMON /CI1/ ALP(NA), ZA(NA)
      COMMON /CI2/ CIMKS(NA,NE,NSEAS,NSIT), CISIS(NA,NE,NSIT),
     1        VSEA(NSEAS), NUME,       CISES(NA,NE,NSEAS)
      COMMON /WR1/ SENT(NSIT), PSENT(NSIT), ZKEN(NSIT), PKENZ(NSIT)
      COMMON /WR2/ NC(NTOT), NCRS(NTOT,NSIT), NCRC(NSIT)
C
      BYTE INFILE(80), OUTFILE(80), FMT(80), IFPRT
      CHARACTER AOUT*66, C1*5, C2*11, C3*12, C4*12, C5*10, C6*12
      CHARACTER DOUT*58, D1*8, D2*12, D3*12, D4*10, D5*12
      CHARACTER CNL*10, CNU*10
      EQUIVALENCE (AOUT(1:5), C1),  (AOUT(6:16),C2), (AOUT(17:28),C3),
     1       (AOUT(29:40),C4),(AOUT(43:52),C5), (AOUT(53:64),C6),
     2       (AOUT(33:42),CNL), (AOUT(57:66),CNU)
      EQUIVALENCE (DOUT(1:8), D1),  (DOUT(9:20),D2), (DOUT(21:32),D3),
     1       (DOUT(35:44),D4),  (DOUT(45:56),D5),
     2       (DOUT(25:34),CNL), (DOUT(49:58),CNU)
C
C     HOMOGENEITY STUFF
C
      IF (NTOTAL.GT.1)    WRITE(IFOUT,300)
     1            TOTAL,    NTOTAL,    PTOTAL
      IF (NHOMOGEN.NE.0)  WRITE(IFOUT,310)
     1            HOMOGEN,  NHOMOGEN,  PHOMOGEN
      IF (NSEASONAL.NE.0 .AND. NSEASONAL.NE.NHOMOGEN) WRITE(IFOUT,320)
     1            SEASONAL, NSEASONAL, PSEASONAL
      IF (NSITES.NE.0 .AND. NSITES.NE.NHOMOGEN)    WRITE(IFOUT,330)
     1            SITE,     NSITES,    PSITES
      IF (NSITESEAS.NE.0) WRITE(IFOUT,340)
     1            SITESEAS, NSITESEAS, PSITESEAS
      IF (NTOTAL.GT.1)    WRITE(IFOUT,350)
     1            TRENDS,   NTRENDS,   PTRENDS
      IF (PSEASONAL.GT.ALPHA.AND.PSITES.LT.ALPHA.AND.NSEASON.GT.1)
     1 WRITE(IFOUT,370) ((I,ZSITE(I),  PSITE(I)),I=1,NSITE)
C
      IF (PSEASONAL.LT.ALPHA.AND.PSITES.GT.ALPHA.AND.NSEASON.GT.1)
     1 WRITE(IFOUT,360) ((I,ZSEASON(I),PSEASON(I)),I=1,NSEASON)
C
C         SEN SLOPES AND CONFIDENCE LIMITS FOR EACH SEASON.
          WRITE(IFOUT,200)
200       FORMAT(//
```

```
     1    T26'SEN SLOPE'/T20'CONFIDENCE INTERVALS'//
     2    4X,'SEASON ',4X,'ALPHA',4X,'LOWER LIMIT',4X,'SLOPE',4X,
     3    'UPPER LIMIT')
C
         DO 260 K=1,NSEASON
          FM = 1
          DO 250 J=1,NUME
              D1  = '          '
              CNL = '            '
              CNU = '            '
              D3  = '              '
              D5  = '              '
              IF (FM.EQ.1) THEN
               ENCODE(8,210,D1) K
210            FORMAT(I8)
              ENDIF
C
              ENCODE(12,220,D2) ALP(J)
C
              CL = CISES(J,1,K)
              IF (CL.EQ.-99.99) THEN
               CNL = 'N TO SMALL'
              ELSE
               CNL = '            '
               ENCODE(12,220,D3) CL
220            FORMAT(F12.3)
              ENDIF
C
              ENCODE(10,230,D4) SEASSL(K)
230           FORMAT(F10.3)
C
              CU = CISES(J,2,K)
              IF (CU.EQ.-99.99) THEN
               CNU = 'N TO SMALL'
              ELSE
               CNU = '            '
               ENCODE(12,220,D5) CU
              ENDIF
C
              WRITE(IFOUT,240) DOUT
240           FORMAT(A)
              FM = 2
250       CONTINUE
          WRITE(IFOUT,*)
             FM = 1
260      CONTINUE
      ENDIF
C
300   FORMAT(1H1/'          HOMOGENEITY TEST RESULTS'//
     7    T48,'PROB. OF A'/
     1    '  SOURCE',T20,'CHI-SQUARE',T40,'DF',T47,
     1    'LARGER VALUE'/X,57('-')/
     1    ' TOTAL            ',T20,F9.5,T37,I5,T49,F7.3)
310   FORMAT( ' HOMOGENEITY    ',T20,F9.5,T37,I5,T49,F7.3)
320   FORMAT( '   SEASON       ',T20,F9.5,T37,I5,T49,F7.3)
330   FORMAT( '   STATION      ',T20,F9.5,T37,I5,T49,F7.3)
340   FORMAT( '   STATION-SEASON',T20,F9.5,T37,I5,T49,F7.3)
350   FORMAT( ' TREND          ',T20,F9.5,T37,I5,T49,F7.3)
360   FORMAT(///T16,'INDIVIDUAL SEASON TREND'/
     1 T41,'PROB. OF A',/
     2 4X,'SEASON',5X,'CHI-SQUARE',6X,'DF',6X,'LARGER VALUE'/
     3 <NSEASON>(I8,2X,F14.5,'        1 ',F14.3/))
370   FORMAT(///T15,'INDIVIDUAL STATION TREND'/T57,/
     1 T41,'PROB. OF A',T57,/
     2 4X,'STATION',4X,'CHI-SQUARE',6X,'DF',6X,'LARGER VALUE'/
     3 <NSITE>(I8,2X,F14.5,'        1 ',F14.3/))
C
      WRITE(IFOUT,380)
380   FORMAT(1H1)
C
```

```
C     SEASONAL KENDALL STATISTICS
C
      IF (NSEASON.GT.1) THEN
          WRITE(IFOUT,390)
      ENDIF
390   FORMAT(//T29,'PROB. OF EXCEEDING'/T29,'THE ABSOLUTE VALUE',
      1  /T12,'SEASONAL',T26,'OF THE KENDALL STATISTIC'/
      2  ' STATION',T12,'KENDALL',T24,'N',T30,'(TWO-TAILED TEST)')
C
      IF (NSEASON.GT.1) THEN
          DO 410 I=1,NSITE
           N = NCRC(I)
              WRITE(IFOUT,400) I, ZKEN(I), N, PKENZ(I)
400           FORMAT(I5,2X,F11.5,I6,4X,F12.3)
410       CONTINUE
C
C         SEASONAL KENDALL SLOPES
          WRITE(IFOUT,500)
500       FORMAT(//
      1    T19'SEASONAL-KENDALL SLOPE'/T20'CONFIDENCE INTERVALS'//
      2    4X,'STATION',4X,'ALPHA',4X,'LOWER LIMIT',4X,'SLOPE',4X,
      3    'UPPER LIMIT')
C
          DO 560 K=1,NSITE
           FM = 1
           DO 550 J=1,NUME
               D1  = '        '
               CNL = '            '
               CNU = '            '
               D3  = '              '
               D5  = '              '
               IF (FM.EQ.1) THEN
                ENCODE(8,510,D1) K
510             FORMAT(I8)
               ENDIF
C
               ENCODE(12,520,D2) ALP(J)
C
               CL = CISIS(J,1,K)
               IF (CL.EQ.-99.99) THEN
                CNL = 'N TO SMALL'
               ELSE
                CNL = '            '
                ENCODE(12,520,D3) CL
520             FORMAT(F12.3)
               ENDIF
C
               ENCODE(10,530,D4) SITESL(K)
530            FORMAT(F10.3)
C
               CU = CISIS(J,2,K)
               IF (CU.EQ.-99.99) THEN
                CNU = 'N TO SMALL'
               ELSE
                CNU = '            '
                ENCODE(12,520,D5) CU
               ENDIF
C
               WRITE(IFOUT,540) DOUT
540            FORMAT(A)
               FM = 2
550        CONTINUE
           WRITE(IFOUT,*)
              FM = 1
560       CONTINUE
      ENDIF
C
C     SEN STATISTICS
C
      IF (NSEASON.GT.1) THEN
          WRITE(IFOUT,600)
```

```
600       FORMAT(//T29,'PROB. OF EXCEEDING'/T29,
     1     'THE ABSOLUTE VALUE'/T27,'OF THE SEN T STATISTIC'/
     2     ' STATION',T13,'SEN T',T24,'N',T30,
     3                      '(TWO-TAILED TEST)'/)
C
          DO 630 I=1,NSITE
           N = NCRC(I)
           IF (SENT(I).EQ.0.0) THEN
                 WRITE(IFOUT,610) I
610           FORMAT(I5,'       MISSING VALUES IN DATA')
           ELSE
                 WRITE(IFOUT,620) I, SENT(I), N, PSENT(I)
620           FORMAT(I5,2X,F11.5,I6,4X,F12.3)
           ENDIF
630       CONTINUE
      ENDIF
C
C     INDIVIDUAL MANN KENDALL Z STATISTICS
C
      WRITE(IFOUT,640)
640   FORMAT(///T49,'PROB. OF EXCEEDING'/,T21,'MANN',T49,
     1    'THE ABSOLUTE VALUE'/T20,'KENDALL',
     1    T49,'OF THE Z STATISTIC',/,T23,'S',
     1    T34,'Z',T50,'(TWO-TAILED TEST)'/
     2    ' STATION  SEASON  STATISTIC  STATISTIC',
     3    '      N         IF  N > 10')
C
      N = NSEASON
      IF (N.EQ.0) N = 1
      DO 670 J=1,NSITE
      DO 670 I=1,N
          Z = ZSTAT(I,J)
          CALL PNORM(Z,PZ)
          IF (Z.GT.0) THEN
           PZ = 2*(1.0 - PZ)
          ELSE
           PZ = 2*PZ
          ENDIF
          NS = NSTAT(I,J)
          IF (I.EQ.1) THEN
           WRITE(IFOUT,650) J,I,SSTAT(I,J),Z,NS,PZ
650        FORMAT(/I5,I9,F11.2,F12.5,I7,2X,F14.3,6X)
          ELSE
           WRITE(IFOUT,660) I, SSTAT(I,J), Z, NS, PZ
660        FORMAT(5X,I9,F11.2,F12.5,I7,2X,F14.3,6X)
          ENDIF
670   CONTINUE
C
C     SEN SLOPES FOR MANN KENDALL TEST
C
      WRITE(IFOUT,700)
700   FORMAT( //T29'SEN SLOPE'/T24'CONFIDENCE INTERVALS'//,
     1    ' STATION',4X,'SEASON',5X,'ALPHA',4X,'LOWER LIMIT',
     2    4X,'SLOPE',4X,'UPPER LIMIT')
C
      DO 780 K=1,NSITE
          FM = 1
          DO 770 I=1,N
           DO 760 J=1,NUME
               C1  = '     '
               C2  = '           '
               CNL = '           '
               CNU = '           '
               C4  = '             '
               C6  = '             '
               IF (FM.EQ.1) THEN
                ENCODE(5,710,C1) K
710             FORMAT(I5)
                ENCODE(11,720,C2) I
720                FORMAT(I11)
               ENDIF
```

```
C
            IF (FM.EQ.2) ENCODE(11,720,C2) I
C
            ENCODE(12,730,C3) ALP(J)
C
            CL = CIMKS(J,1,I,K)
            IF (CL.EQ.-99.99) THEN
             CNL = 'N TO SMALL'
            ELSE
             CNL = '           '
             ENCODE(12,730,C4) CL
730             FORMAT(F12.3)
            ENDIF
C
            ENCODE(10,740,C5) ZSLOPE(I,K)
740         FORMAT(F10.3)
C
            CU = CIMKS(J,2,I,K)
            IF (CU.EQ.-99.99) THEN
             CNU = 'N TO SMALL'
            ELSE
             CNU = '
             ENCODE(12,730,C6) CU
            ENDIF
C
            WRITE(IFOUT,750) AOUT
750         FORMAT(A)
            FM = 3
760       CONTINUE
          FM = 2
          WRITE(IFOUT,*)
770      CONTINUE
         FM = 1
780   CONTINUE
C
      RETURN
      END
```

Symbols

B	Sign test statistic (18.1.1)
$C - c_0$	Dollars available for collecting and measuring samples, not including fixed overhead expenses (5.6.1)
c_h	Cost of collecting and measuring a population unit in stratum h (5.5)
$1 - f$	Finite population correction factor (4.2)
$f_h = n_h/N_h$	Proportion of N_h population units in stratum h that are measured (5.2)
$1 - f_h$	Finite population correction factor for stratum h (5.2)
$f_M = m/M$	Proportion of the M subunits selected for measurement (6.2.1)
$f_N = n/N$	Proportion of the N primary units selected for measurement (6.2.1)
F_r	Friedman's test statistic (18.1.3)
GM	Geometric mean (13.3.3)
GSE	Geometric standard error (13.4)
I	Total amount (inventory) of pollutant in the target population (4.2)
$\hat{I}$	Estimate of I
K_w	Kruskal-Wallis test statistic (18.2.2)
L	Number of strata (5.1)
M	Number of subunits in each primary unit (subsampling) (6.2.1)
N_h	Number of population units in stratum h (5.1)
n_h	Number of population units measured in stratum h (5.1)
N	Number of population units in the target population (4.1)
n	Number of population units selected for measurement; more generally, the number of measurements in a data set (4.2)

Note: The numbers in parentheses are the section numbers where the symbols are first mentioned.

n_s	Number of monitoring stations (4.5.1)
$N(\mu, \sigma^2)$	A normal (Gaussian) distribution with mean μ and variance σ^2 (11.1)
$N(0, 1)$	The standard normal distribution (11.1)
p_{x_c}	The proportion of the population that exceeds the value x_c (11.4)
R_v	Rank von Neumann test for serial correlation (11.13)
s^2	Variance of n measurements (4.2)
s_h^2	Variance of the n_h measurements in stratum h (5.2)
$s^2(\hat{I})$	Estimated variance of $\hat{I}$ (4.2)
s_w	Winsorized standard deviation (14.2.4)
$s(\bar{x})$	Estimated standard error of $\bar{x}$ (4.2)
$s^2(\bar{x})$	Estimate of $\mathrm{Var}(\bar{x})$ (4.2)
$s^2(\bar{x}_{st})$	Estimated variance of $\bar{x}_{\mathrm{st}}$ (5.2)
$s^2(\bar{x}_{\mathrm{lr}})$	Estimated variance of $\bar{x}_{\mathrm{lr}}$ (9.1.1)
s_y^2	Variance of n log-transformed data (12.1)
$t_{1-\alpha/2, n-1}$	$(1 - \alpha/2)$ quantile of the t distribution with $n - 1$ degrees of freedom (11.5.2); value that cuts off $(100\alpha/2)\%$ of the upper tail of the t distribution with $n - 1$ degrees of freedom (4.4.2)
$\mathrm{Var}(\hat{I})$	True variance of $\hat{I}$ (4.2)
$\mathrm{Var}(\bar{x})$	True variance of $\bar{x}$ (4.2)
$\mathrm{Var}(\bar{x}_{\mathrm{st}})$	True variance of $\bar{x}_{\mathrm{st}}$ (5.2)
W	The W statistic for testing that a data set is from a normal distribution (12.3.1)
$w_h = n_h/n$	Proportion of n measurements that were made in stratum h (5.1)
$W_h = N_h/N$	Proportion of population units in stratum h (5.1)
W_i	Proportion of all subunits that are in primary unit i (6.3.1)
W_{rs}	Wilcoxon rank sum test statistic (18.2.1)
$\bar{x}$	Arithmetic mean of n measurements (4.2)
$\bar{x}_h$	Arithmetic mean of the n_h measurements in stratum h (5.2)
$x_{[i]}$	The ith order statistic (ith largest value) of a data set (11.2)
x_i	The measurement on the ith population unit (4.2)
$\bar{x}_{\mathrm{lr}}$	Linear regression (double sampling) estimator of the population mean (9.1.1)
x_p	The pth quantile (percentile) of a distribution. That value, x_p, below which lies $100p\%$ of the population (11.1)

$\bar{x}_{st}$	Estimated mean of a stratified population (5.2)
$\bar{x}_w$	Winsorized mean (14.2.4)
$\bar{y}$	Arithmetic mean of *n* log-transformed data (12.1)
Y	D'Agostino's test statistic for testing that a data set is from a normal distribution (12.3.2)
$Z_{1-\alpha/2}$	$1 - \alpha/2$ quantile of the $N(0, 1)$ distribution (11.5.1); value that cuts off $(100\alpha/2)\%$ of the upper tail of an $N(0, 1)$ distribution (4.4.2)
Z_p	The *p*th quantile of the $N(0, 1)$ distribution (11.2)
$\Lambda(\mu_y, \sigma_y^2)$	A two-parameter lognormal distribution with parameters μ_y and σ_y^2, the mean and variance of the logarithms, respectively. (12.1)
$\Lambda(\mu_y, \sigma_y^2, \tau)$	A three-parameter lognormal distribution with parameters μ_y, σ_y^2, and τ (12.1)
μ	The true mean over all *N* units in the target population (4.2)
μ_h	True mean for stratum *h* (5.2)
μ_i	True mean for primary unit *i* (6.2.1)
μ_{ij}	True amount of pollutant present in the *j*th subunit of primary unit *i* (6.2.1)
μ_y	True mean of the logarithms of the population values (12.1)
$\eta = \mu/\sigma$	Population coefficient of variation (4.4.3)
$\rho_{ii'}$	Correlation between stations *i* and *i'* (4.5.1)
ρ_c	Average of all possible cross-correlations between monitoring stations (4.5.1)
ρ_l	True correlation between measurements *l* lags apart collected along a line in time or space (4.5.2)
$\psi_n(t)$	Infinite series used to estimate the mean and variance of a lognormal distribution (13.1.1)
σ^2	True variance of the *N* population units in the target population (4.2)
σ_y^2	True variance of the logarithms of the population values (12.1)

Glossary

Accuracy	A measure of the closeness of measurements to the true value (2.5.2)
Censored data set	Measurements for some population units are not available, for example, they are reported as "trace" or "not detected" (11.8)
$100(1 - \alpha)\%$ confidence interval on a population parameter	If the process of drawing n samples from the population is repeated many times and the $100(1 - \alpha)\%$ confidence interval computed each time, $100(1 - \alpha)\%$ of those intervals will include the population parameter value (11.5.2)
Measurement bias	Consistent under- or overestimation of the true values in population units (2.5.2)
Median	That value above which and below which half the population lies (13.3)
Nonparametric technique	One that does not depend for its validity upon the data being drawn from a specific distribution, such as the normal or lognormal. A distribution-free technique (11.9)
Outlier observation	An observation that does not conform to the pattern established by other observations in the data set (Hunt et al., 1981) (11.8)
Precision	A measure of the closeness of agreement among individual measurements (2.5.2)
Probability sampling	Use of a specific method of random selection of population units for measurement (3.3.2)

Note: The numbers in parentheses are the section numbers where the terms are first mentioned.

Random measurement uncertainty	Unpredictable deviation from the true value of a unit (2.5.2)
Random sampling error	Variation in an estimated quantity due to the random selection of environmental units for measurement (2.5.4)
Representative unit	One selected from the target population that in combination with other representative units will give an accurate picture of the phenomenon being studied (2.3)
Sampled population	Set of population units available for measurement (2.2)
Statistical bias	Discrepancy between the expected value of an estimator and the population parameter being estimated (2.5.3)
Target population	Set of N population units for which inferences will be made (2.2)
Trimmed mean	Arithmetic mean of the data remaining after a specified percent of the n data in both tails is discarded (14.2.3)

Bibliography

Abramowitz, M., and I. A. Stegun, 1964. *Handbook of Mathematical Functions With Formulas, Graphs and Mathematical Tables*, Applied Mathematics Series 55. National Bureau of Standards, U.S. Government Print Office, Washington, D.C.

Agterberg, F. P., 1974. *Geomathematics*. Elsevier, New York.

Aitchison, J., and J. A. C. Brown, 1969. *The Lognormal Distribution*. Cambridge University Press, Cambridge.

Akland, G. G., 1972. Design of sampling schedules, *Journal of the Air Pollution Control Association* **22**:264–266.

Albers, W., 1978*a*. One-sample rank tests under autoregressive dependence, *Annals of Statistics* **6**:836–845.

Albers, W., 1978*b*. Testing the mean of a normal population under dependence, *Annals of Statistics* **6**:1337–1344.

American Society for Testing and Materials, 1976. *ASTM Manual on Presentation of Data and Control Chart Analysis*, ASTM Special Technical Publication 15D. American Society for Testing and Materials, Philadephia.

American Society for Testing and Materials, 1984. *Annual Book of ASTM Standards Section 11, Water and Environmental Technology*, vol. 11.01. Designation: D4210-83, pp. 7–15.

Apt, K. E., 1976. Applicability of the Weibull distribution function to atmospheric radioactivity data, *Atmospheric Environment* **10**:777–781.

Barnett, V., and T. Lewis, 1978. *Outliers in Statistical Data*. Wiley, New York.

Bartels, R., 1982. The rank version of von Neumann's ratio test for randomness, *Journal of the American Statistical Association* **77**:40–46.

Beckman, R. J., and R. D. Cook, 1983. Outlier . . . . . . s, *Technometrics* **25**:119–149.

Bellhouse, D. R., and J. N. K. Rao, 1975. Systematic sampling in the presence of a trend, *Biometrika* **62**:694–697.

Berry, B. J. L., and A. M. Baker, 1968. Geographic sampling, in *Spatial Analysis*, B. J. L. Berry and D. F. Marble, eds. Prentice-Hall, Englewood Cliffs, N.J.

Berthouex, P. M., W. G. Hunter, and L. Pallesen, 1978. Monitoring sewage treatment plants: Some quality control aspects, *Journal of Quality Technology* **10**:139–149.

Berthouex, P. M., W. G. Hunter, and L. Pallesen, 1981. Wastewater treatment: A review of statistical applications, in *Environmetrics 81: Selected Papers*, SIAM-SIMS Conference Series No. 8. Society for Industrial and Applied Mathematics, Philadelphia, pp. 77–99.

Berthouex, P. M., W. G. Hunter, L. C. Pallesen, and C. Y. Shih, 1975. Modeling sewage treatment plant input BOD data, *Journal of the Environmental Engineering Division, Proceedings of the American Society of Civil Engineers*, vol. 101, no. EE1, Proceedings Paper 11132, pp. 127–138.

Biomedical Computer Programs P-Series, 1983. *BMDP Statistical Software. 1983 Printing and Additions*, W. J. Dixon, chief ed. University of California Press, Berkeley.

Blyth, C. R., and H. A. Still, 1983. Binomial confidence intervals, *American Statistical Association Journal* **78:**108–116.

Bowman, K. O., and L. R. Shenton, 1975. Omnibus test contours for departures from normality based on $\sqrt{b_1}$ and b_2, *Biometrika* **62:**243–250.

Box, G. E. P., and G. M. Jenkins, 1976. *Time Series Analysis: Forecasting and Control*, 2nd ed. Holden-Day, San Francisco.

Box, G. E. P., and G. C. Tiao, 1975. Intervention analysis with applications to economic and environmental problems, *Journal of the American Statistical Association* **70:**70–79.

Bradu, D., and Y. Mundlak, 1970. Estimation in lognormal linear models, *Journal of the American Statistical Association* **65:**198–211.

Bromenshenk, J. J., S. R. Carlson, J. C. Simpson, and J. M. Thomas, 1985. Pollution monitoring of Puget Sound with honey bees, *Science* **227:**632–634.

Brumelle, S., P. Nemetz, and D. Casey, 1984. Estimating means and variances: The comparative efficiency of composite and grab samples, *Environmental Monitoring and Assessment* **4:**81–84.

Burr, I. W., 1976. *Statistical Quality Control Methods*. Dekker, New York.

Carlson, R. F., A. J. A. MacCormick, and D. G. Watts, 1970. Application of linear random models to four annual streamflow series, *Water Resources Research* **6:**1070–1078.

Chamberlain, S. G., C. V. Beckers, G. P. Grimsrud, and R. D. Shull, 1974. Quantitative methods for preliminary design of water quality surveillance systems, *Water Resources Bulletin* **10:** 199–219.

Chambers, J. M., W. S. Cleveland, B. Kleiner, and P. A. Tukey, 1983. *Graphical Methods for Data Analysis*. Duxbury Press, Boston.

Chatfield, C., 1984. *The Analysis of Time Series: An Introduction*, 3rd ed. Chapman and Hall, London.

Clark, C., and B. A. Berven, 1984. Results of the Groundwater Monitoring Program Performed at the Former St. Louis Airport Storage Site For the Period of January 1981 through January 1983. National Technical Information Service, Springfield, Va., ORNL/TM-8879.

Clark, I., 1979. *Practical Geostatistics*. Applied Science Publishers, London.

Cochran, W. G., 1947. Some consequences when the assumptions for the analysis of variance are not satisfied, *Biometrics* **3:**22–38.

Cochran, W. G., 1963. *Sampling Techniques*, 2nd ed. Wiley, New York.

Cochran, W. G., 1977. *Sampling Techniques*, 3rd ed. Wiley, New York.

Cochran, W. G., F. Mosteller, and J. W. Tukey, 1954. Principles of sampling, *Journal of the American Statistical Association* **49:**13–35.

Cohen, A. C., Jr., 1959. Simplified estimators for the normal distribution when samples are single censored or truncated, *Technometrics* **1:**217–237.

Cohen, A. C., Jr., 1961. Tables for maximum likelihood estimates: Singly truncated and singly censored samples, *Technometrics* **3:**535–541.

Cohen, A. C., and B. J. Whitten, 1981. Estimation of lognormal distributions, *American Journal of Mathematical and Management Sciences* **1:**139–153.

Conover, W. J., 1980. *Practical Nonparameteric Statistics*, 2nd ed., Wiley, New York.

Cox, D. R., 1952. Estimation by double sampling, *Biometrika* **39:**217–227.

Crager, M. R., 1982. *Exponential Tail Quantile Estimators for Air Quality Data*, Parts I and II, Technical Reports No. 4 and 5. Society for Industrial and Applied Mathematics Reports, Bay Area Air Quality Management District, San Francisco.

Curran, T. C., 1978. *Screening Procedures for Ambient Air Quality Data*, U.S. Environmental Protection Agency Guideline Series, EPA-450/2-78-037. National Technical Information Service, Springfield, Va.

D'Agostino, R. B., 1971. An omnibus test of normality for moderate and large size samples, *Biometrika* **58:**341–348.

Das, A. C., 1950. Two-dimensional systematic sampling and the associated stratified and random sampling, *Sankhya* **10:**95–108.

David, H. A., 1981. *Order Statistics*, 2nd ed. Wiley, New York.

Davis, J. C., 1973. *Statistics and Data Analysis in Geology*. Wiley, New York.

Delfiner, P., and J. P. Delhomme, 1975. Optimal interpolations by kriging, in *Display and Analysis of Spatial Data*, J. C. Davis and M. J. McCullagh, eds. Wiley, New York, pp. 96–114.

Delhomme, J. P., 1978. Kriging in the hydrosciences, *Advances in Water Resources* **1:**251–266.

Delhomme, J. P., 1979. Spatial variability and uncertainty in groundwater flow parameters: A geostatistical approach, *Water Resources Research* **15:**269–280.

Deming, W. E., 1950. *Some Theory of Sampling*. Wiley, New York (available as a Dover reprint, 1966).

Deming, W. E., 1960. *Sample Design in Business Research*. Wiley, New York.

Dietz, E. J., and T. J. Killeen, 1981. A nonparametric multivariate test for monotone trend with pharmaceutical applications, *Journal of the American Statistical Association* **76:**169–174.

Dixon, W. J., 1953. Processing data for outliers, *Biometrics* **9:**74–89.

Dixon, W. J., and J. W. Tukey, 1968. Approximate behavior of the distribution of Winsorized *t* (Trimming/Winsorization 2), *Technometrics* **10:**83–98.

Doctor, P. G., and R. O. Gilbert, 1978. Two studies in variability for soil concentrations: With aliquot size and with distance, in *Selected Environmental Plutonium Research Reports of the Nevada Applied Ecology Group*, M. G. White and P. B. Dunaway, eds. U.S. Department of Energy, NVO-192, Las Vegas, pp. 405–449.

Eberhardt, L. L., and R. O. Gilbert, 1980. Statistics and sampling in transuranic studies, in *Transuranic Elements in the Environment*. National Technical Information Service, DOE/TIC-22800, Springfield, Va., pp. 173–186.

Elder, R. S., W. O. Thompson, and R. H. Myers, 1980. Properties of composite sampling procedures, *Technometrics* **22:**179–186.

Elsom, D. M., 1978. Spatial correlation analysis of air pollution data in an urban area, *Atmospheric Environment* **12:**1103–1107.

Environmental Protection Agency, 1980. *Upgrading Environmental Radiation Data*, EPA 520/1-80-012. Office of Radiation Programs, U.S. Environmental Protection Agency, Washington, D.C.

Environmental Protection Agency, 1982. *Handbook for Sampling and Sample Preservation of Water and Wastewater*, EPA-600/4-82-029. Environmental Monitoring and Support Laboratory, U.S. Environmental Protection Agency, Cincinnati.

Everitt, B. S., 1978. *Graphical Techniques for Multivariate Data*. North-Holland, New York.

Ewan, W. D., 1963. When and how to use Cu-Sum charts, *Technometrics* **5:**1–22.

Eynon, B. P., and P. Switzer, 1983. The variability of rainfall acidity, *Canadian Journal of Statistics* **11:**11–24.

Farrell, R., 1980. *Methods for Classifying Changes in Environmental Conditions*, Technical Report VRF-EPA7. 4-FR80-1. Vector Research Inc., Ann Arbor, Mich.

Feltz, H. R., W. T. Sayers, and H. P. Nicholson, 1971. National monitoring program for the assessment of pesticide residues in water, *Pesticides Monitoring Journal* **5:**54–62.

Ferrell, E. B., 1958. Control charts for log-normal universes, *Industrial Quality Control* **15:**4–6.

Filliben, J. J., 1975. The probability plot correlation coefficient test for normality, *Technometrics* **17:**111–117.

Finney, D. J., 1941. On the distribution of a variate whose logarithm is normally distributed, *Journal of the Royal Statistical Society* (suppl.), **7:**155–161.

Fisher, R. A., and F. Yates, 1974. *Statistical Tables for Biological, Agricultural and Medical Research*, 5th ed. Longman, London. (Previously published by Oliver & Boyd, Edinburgh.)

Fisz, M., 1963. *Probability Theory and Mathematical Statistics,* 3rd ed. Wiley, New York.

Flatman, G. T., 1984. Using geostatistics in assessing lead contamination near smelters, in *Environmental Sampling for Hazardous Wastes*, G. E. Schweitzer and J. A. Santolucito, eds., ACS Symposium Series 267. American Chemical Society, Washington, D.C., pp. 43–52.

Flatman, G. T., and A. A. Yfantis, 1984. Geostatistical strategy for soil sampling: The survey and the census, *Environmental Monitoring and Assessment* **4:**335–350.

Fuller, F. C., Jr., and C. P. Tsokos, 1971. Time series analysis of water pollution data, *Biometrics* **27:**1017–1034.

Gastwirth, J. L., and H. Rubin, 1971. Effect of dependence on the level of some one-sample tests, *Journal of the American Statistical Association* **66:**816–820.

Geigy, 1982. Geigy Scientific tables, in Vol. 2, Introduction to Statistics, *Statistical Tables and Mathematical Formulae*, 8th ed., C. Lentner, ed. Ciba-Geigy Corporation, West Caldwell, N.J.

Georgopoulos, P. G., and J. H. Seinfeld, 1982. Statistical distributions of air pollutant concentrations, *Environmental Science and Technology* **16:**401–416.

Gibbons, J. D., I. Olkin, and M. Sobel, 1977. *Selecting and Ordering Populations: A New Statistical Methodology*. Wiley, New York.

Gibra, I. N., 1975. Recent developments in control chart techniques, *Journal of Quality Technology* **7:**183–192.

Giesy, J. P., and J. G. Wiener, 1977. Frequency distributions of trace metal concentrations in five freshwater fishes, *Transactions of the American Fisheries Society* **196:**393–403.

Gilbert, R. O., 1977. Revised total amounts of ^{239,240}Pu in surface soil at safety-shot sites, in *Transuranics in Desert Ecosystems*, M. G. White, P. B. Dunaway, and D. L. Wireman, Eds. National Technical Information Services (NTIS), NVO-181, Springfield Va., pp. 423–429.

Gilbert, R. O., 1982. Some statistical aspects of finding hot spots and buried radioactivity in *TRAN-STAT: Statistics for Environmental Studies*, No. 19. Pacific Northwest Laboratory, Richland, Wash., PNL-SA-10274.

Gilbert, R. O., and P. G. Doctor, 1985. Determining the number and size of soil aliquots for assessing particulate contaminant concentrations, *Journal of Environmental Quality* **14:**286–292.

Gilbert, R. O., L. L. Eberhardt, E. B. Fowler, E. M. Romney, E. H. Essington, and J. E. Kinnear, 1975. Statistical analysis of $^{239-240}$Pu and ^{241}Am contamination of soil and vegetation on NAEG study sites, *The Radioecology of Plutonium and Other Transuranics in Desert Environments*, M. G. White and P. B. Dunaway, eds. U.S. Energy Research and Development Administration, NVO-153. Las Vegas, pp. 339–448.

Gilbert, R. O., and R. R. Kinnison, 1981. Statistical Methods for estimating the mean and variance from radionuclide data sets containing negative, unreported or less-than values, *Health Physics* **40:**377–390.

Gilbert, R. O., and J. C. Simpson, 1985. Kriging from estimating spatial pattern of contaminants: Potential and problems, *Environmental Monitoring and Assessment* **5:**113–135.

Gilliom, R. J., R. M. Hirsch, and E. J. Gilroy, 1984. Effect of censoring trace-level water-quality data on trend-detection capability, *Environmental Science and Technology* **18:**530–535.

Glass, G. V., P. D. Peckham, and J. R. Sanders, 1972. Consequences of failure to meet assumptions underlying the fixed effects analyses of variance and covariance, *Review of Educational Research* **42:**237–288.

Gnanadesikan, R., and J. R. Kettenring, 1972. Robust estimates, residuals, and outlier detection with multiresponse data, *Biometrics* **28:**81–124.

Green, R. H., 1968. Mortality and stability in a low density subtropical intertidal community, *Ecology* **49:**848–854.

Green, R. H., 1979. *Sampling Design and Statistical Methods for Environmental Biologists*. Wiley, New York.

Grivet, C. D., 1980. *Modeling and Analysis of Air Quality Data*, SIMS Technical Report No. 43. Department of Statistics, Stanford University, Stanford.

Gunnerson, C. G., 1966. Optimizing sampling intervals in tidal estuaries, *Journal of the Sanitary Engineering Division, Proceedings of the American Society of Civil Engineers* **92:**103–125.

Gunnerson, C. G., 1968. Optimizing sampling intervals, in *Proceedings IBM Scientific Computing Symposium Water and Air Resources Management*. IBM, White Plains, N.Y., pp. 115–139.

Hahn, G. J., and S. S. Shapiro, 1967. *Statistical Models in Engineering*. Wiley, New York.

Hakonson, T. E., and K. V. Bostick, 1976. The availability of environmental radioactivity to honey bee colonies at Los Alamos, *Journal of Environmental Quality* **5(3):**307–310.

Hakonson, T. E., and G. C. White, 1979. Effect of rototilling on the distribution of cesium-137 in trinity site soil, *Health Physics* **40:**735–739.

Hale, W. E., 1972. Sample size determination for the log-normal distribution, *Atmospheric Environment* **6:**419–422.

Handscome, C. M., and D. M. Elsom, 1982. Rationalization of the national survey of air pollution monitoring network of the United Kingdom using spatial correlation analysis: A case study of the greater London area, *Atmospheric Environment* **16:**1061–1070.

Hansen, M. H., W. N. Hurwitz, and W. G. Madow, 1953. *Sample Survey Methods and Theory. Volume 1. Methods and Applications*. Wiley, New York.

Harned, D. A., C. C. Daniel III, and J. K. Crawford, 1981. Methods of discharge compensation as an aid to the evaluation of water quality trends, *Water Resources Research* **17:**1389–1400.

Harter, H. L., and A. H. Moore, 1966. Local-maximum-likelihood estimation of the parameters of three-parameter lognormal populations from complete and censored samples, *Journal of the American Statistical Association* **61:**842–851.

Hawkins, D. M., 1974. The detection of errors in multivariate data using principal components, *Journal of the American Statistical Association* **69:**340–344.

Hawkins, D. M., 1980. *Identification of Outliers*. Chapman and Hall, London.

Healy, M. J. R., 1968. Multivariate normal plotting, *Applied Statistics* **17:**157–161.

Heien, D. M., 1968. A note on log-linear regression, *Journal of the American Statistical Association* **63:**1034–1038.

Hill, M., and W. J. Dixon, 1982. Robustness in real life: A study of clinical laboratoy data, *Biometrics* **38:**377–396.

Hipel, K. W., W. C. Lennox, T. E. Unny, and A. I. McLeod, 1975. Intervention analysis in water resources, *Water Resources Research* **11:**855–861.

Hipel, K. W., A. I. McLeod, and W. C. Lennox, 1977*a*. Advances in Box-Jenkins modeling. Part 1. Model construction, *Water Resources Research* **13:**567–575.

Hipel, K. W., A. I. McLeod, and W. C. Lennox, 1977*b*. Advances in Box-Jenkins modeling. Part 2. Applications, *Water Resources Research* **13:**577–586.

Hirsch, R. M., J. R. Slack, and R. A. Smith, 1982. Techniques of trend analysis for monthly water quality data, *Water Resources Research* **18:**107–121.

Hirtzel, C. S., J. E. Quon, and R. B. Corotis, 1982*a*. Mean of autocorrelated air quality measurements, *Journal of the Environmental Engineering Division, Proceedings of the American Society of Civil Engineers* **108:**488–501.

Hirtzel, C. S. R. B. Corotis, and J. E. Quon, 1982*b*. Estimating the maximum value of autocorrelated air quality measurements, *Atmospheric Environment* **16:**2603–2608.

Hirtzel, C. S., and J. E. Quon, 1979. Statistical dependence of hourly carbon monoxide measurements, *Journal of the Air Pollution Control Association* **29:**161–163.

Hoaglin, D. C., F. Mosteller, and J. W. Tukey, 1983. *Understanding Robust and Exploratory Data Analysis*. Wiley, New York.

Holland, D. M., and T. Fitz-Simons, 1982. Fitting statistical distributions to air quality data by the maximum likelihood method, *Atmospheric Environment* **16:**1071–1076.

Hollander, M., and D. A. Wolfe, 1973. *Nonparametric Statistical Methods*. Wiley, New York.

Holoway, C. F., J. P. Witherspoon, H. W. Dickson, P. M. Lantz, and T. Wright, 1981. *Monitoring for Compliance with Decommissioning Termination Survey Criteria*. Oak Ridge National Laboratory, Oak Ridge, Tenn., NUREG/CR-2082, ORNL/HASRD-95.

Horton, J. H., J. C. Corey, D. C. Adriano, and J. E. Pinder III, 1980. Distribution of surface-deposited plutonium in soil after cultivation, *Health Physics* **38:**697–699.

Hsu, Der-Ann, and J. S. Hunter, 1976. Time series analysis and forecasting for air pollution concentrations with seasonal variations, in *Proceedings of the EPA Conference on Environmental Modeling and Simulation*, PB-257 142. National Technical Information Service, Springfield, Va., pp. 673–677.

Hughes, J. P., and D. P. Lettenmaier, 1981. Data requirements for kriging: Estimation and network design, *Water Resources Research* **17:**1641–1650.

Hunt, W. F., Jr., G. Akland, W. Cox, T. Curran, N. Frank, S. Goranson, P. Ross, H. Sauls, and J. Suggs, 1981. *U.S. Environmental Protection Agency Intra-Agency Task Force Report on Air Quality Indicators*, EPA-450/4-81-015. Environmental Protection Agency, National Technical Information Service, Springfield, Va.

Hunt, W. F., Jr., J. B. Clark, and S. K. Goranson, 1978. The Shewhart control chart test: A recommended procedure for screening 24 hour air pollution measurements, *Journal of the Air Pollution Control Association* **28:**508–510.

Hunter, J. S., 1977. Incorporating uncertainty into environmental regulations, in *Environmental Monitoring*. National Academy of Sciences, Washington, D.C., pp. 139–153.

Iachan, R., 1982. Systematic sampling: A critical review, *International Statistical Review* **50:**293–302.

Iman, R. L., 1982. Graphs for use with the Lilliefors test for normal and exponential distributions, *The American Statistician* **36:**109–112.

Iman, R. L., D. Quade, and D. A. Alexander, 1975. Exact probability levels for the Kruskal-Wallis test, in *Selected Tables in Mathematical Statistics*, vol. III. American Mathematical Society, Providence, R.I.

International Mathematical and Statistical Library, *IMSL Library Reference Manual*, vol. 1. IMSL, Houston.

Jensen. A. L., 1973. Statistical analysis of biological data from preoperational-post-operational industrial water quality monitoring, *Water Research* **7:**1331–1347.

Jessen, R. J., 1978. *Statistical Survey Techniques*. Wiley, New York.

Johnson, N. L., and S. Kotz, 1970*a*. *Continuous Univariate Distributions—1*. Houghton Mifflin, Boston.

Johnson, N. L., and S. Kotz, 1970*b*. *Continuous Univariate Distributions—2*. Houghton Mifflin, Boston.

Johnson, T., 1979. A comparison of the two-parameter Weibull and Lognormal distributions fitted to ambient ozone data, in *Proceedings, Quality Assurance in Air Pollution Measurements*. Air Pollution Control Association, New Orleans, pp. 312–321.

Journel, A. G., 1984. New ways of assessing spatial distributions of pollutants, in *Environmental Sampling for Hazardous Wastes*, G. E. Schweitzer and J. A. Santolucito, eds. ACS Symposium Series 267, American Chemical Society, Washington, D.C., pp. 109–118.

Journel, A. G., and CH. J. Huijbregts, 1978. *Mining Geostatistics*. Academic Press, New York.

Keith, L. H., W. Crummett, J. Deegan, Jr., R. A. Libby, J. K. Taylor, and G. Wentler, 1983. Principles of environmental analysis, *Analytical Chemistry* **55:**2210–2218.

Kendall, M. G., 1975. *Rank Correlation Methods*, 4th ed. Charles Griffin, London.

Kendall, M. G., and A. Stuart, 1961. *The Advanced Theory of Statistics*, vol. 2. Charles Griffin, London.

King, J. R., 1971. *Probability Charts for Decision Making*. Industrial Press, New York.

Kinnison, R. R. 1985. *Applied Extreme Value Statistics*. Battelle Press, Columbus, Ohio.

Kish, L., 1965. *Survey Sampling*. Wiley, New York.

Koch, G. S., Jr., and R. F. Link, 1980. *Statistical Analyses of Geological Data*, vols. I, II. Dover, New York.

Königer, W., 1983. A remark on AS177. Expected normal order statistics (exact and approximate), *Applied Statistics* **32:**223–224.

Koop, J. C., 1971. On Splitting a systematic sample for variance estimation, *Annals of Mathematical Statistics* **42:**1084–1087.

Krumbein, W. C., and F. A. Graybill., 1965. *An Introduction to Statistical Models in Geology*. McGraw-Hill, New York.

Kurtz, S. E., and D. E. Fields, 1983*a*. *An Analysis/Plot Generation Code with Significance Levels Computed Using Kolmogorov-Smirnov Statistics Valid for Both Large and Small Samples*, ORNL-5967. Oak Ridge National Laboratory, Oak Ridge, Tenn. Available from National Technical Information Service, Springfield, Va.

Kurtz, S. E., and D. E. Fields, 1983*b*. *An Algorithm for Computing Significance Levels Using the Kolmogorov-Smirnov Statistic and Valid for Both Large and Small Samples*, ORNL-5819. Oak Ridge National Laboratory, Oak Ridge, Tenn. Available from National Technical Information Service, Springfield, Va.

Kushner, E. J., 1976. On determining the statistical parameters for pollution concentration from a truncated data set, *Atmospheric Environment* **10:**975–979.

Lahiri, D. B., 1951. A method for sample selection providing unbiased ratio estimates, *Bulletin of the International Statistical Institute* **33:**133–140.

Land, C. E., 1971. Confidence intervals for linear functions of the normal mean and variance, *Annals of Mathematical Statistics* **42:**1187–1205.

Land, C. E., 1975. Tables of confidence limits for linear functions of the normal mean and variance, in *Selected Tables in Mathematical Statistics*, vol. III. American Mathematical Society, Providence, R.I., pp. 385–419.

Landwehr, J. M., 1978. Some properties of the geometric mean and its use in water quality standards, *Water Resources Research* **14:**467–473.

Langford, R. H., 1978. National monitoring of water quality, in *Establishment of Water Quality Monitoring Programs*, L. G. Everett and K. D. Schmidt, eds. American Water Resources Association Technical Publication Series TPS79-1, Minneapolis, pp. 1–6.

Lee, L., and R. G. Krutchkoff, 1980. Mean and variance of partially-truncated distributions, *Biometrics* **36:**531–536.

Lehmann, E. L., 1975. *Nonparametrics: Statistical Methods Based on Ranks*. Holden-Day, San Francisco.

Lettenmaier, D. P., 1975. *Design of Monitoring Systems for Detection of Trends in Stream Quality*. C. W. Harris Hydraulics Laboratory, Department of Civil Engineering, University of Washington, Seattle, Wash. Technical Report. No. 39, National Technical Information Service, PB-272 814.

Lettenmaier, D. P., 1976. Detection of trends in water quality data from records with dependent observations, *Water Resources Research* **12:**1037–1046.

Lettenmaier, D. P., 1977. *Detection of Trends in Stream Quality: Monitoring Network*

Design and Data Analysis. C. W. Harris Hydraulics Laboratory, Department of Civil Engineering, University of Washington, Seattle, Wash. Technical Report. No. 51, National Technical Information Service, PB-285 960.

Lettenmaier, D. P., 1978. Design considerations for ambient stream quality monitoring, *Water Resources Bulletin* **14:**884–902.

Lettenmaier, D. P., L. L. Conquest, and J. P. Hughes, 1982. *Routine Streams and Rivers Water Quality Trend Monitoring Review*, C. W. Harris Hydraulics Laboratory, Department of Civil Engineering, University of Washington, Seattle, Wash., Technical Report No. 75.

Lettenmaier, D. P., and L. C. Murray, 1977. *Design of Nonradiological Aquatic Sampling Programs for Nuclear Power Plant Impact Assessment Using Intervention Analysis*. Technical Report UW-NRC-6, College of Fisheries, University of Washington, Seattle, Wash.

Liebetrau, A. M., 1979. Water quality sampling: Some statistical considerations, *Water Resources Research* **15:**1717–1725.

Liggett, W., 1984. Detecting elevated contamination by comparisons with background, in *Environmental Sampling for Hazardous Wastes*, G. E. Schweitzer and J. A. Santolucito, eds. ACS Symposium Series 267, American Chemical Society, Washington, D.C., pp. 119–128.

Lilliefors, H. W., 1967. On the Kolmogorov-Smirnov test for normality with mean and variance unknown, *Journal of the American Statistical Association* **62:**399–402.

Lilliefors, H. W., 1969. Correction to the paper "On the Kolmogorov-Smirnov test for normality with mean and variance unknown," *Journal of the American Statistical Association* **64:**1702.

Lindgren, B. W., 1976. *Statistical Theory*, 3rd ed. MacMillan, New York.

Loftis, J. C., and R. C. Ward., 1979. *Regulatory Water Quality Monitoring Networks—Statistical and Economic Considerations*. U.S. Environmental Protection Agency, Las Vegas, EPA-600/4-79-055.

Loftis, J. C., and R. C. Ward., 1980*a*. Sampling frequency selection for regulatory water quality monitoring, *Water Resources Bulletin* **16:**501–507.

Loftis, J. C., and R. C. Ward., 1980*b*. Water quality monitoring—some practical sampling frequency considerations, *Environmental Management* **4:**521–526.

Loftis, J. C., and R. C. Ward, and G. M. Smillie, 1983. Statistical models for water quality regulation, *Journal of the Water Pollution Control Federation* **55:**1098–1104.

Looney, S. W., and T. R. Gulledge, 1985. Use of the correlation coefficient with normal probability plots, *The American Statistician* **39:**75–79.

Lorenzen, T. J., 1980. Determining statistical characteristics of a vehicle emissions audit procedure, *Technometrics* **22:**483–494.

McBratney, A. B., and R. Webster, 1981. The design of optimal sampling schemes for local estimation and mapping of regionalized variables. Part 2. Program and examples, *Computers and Geosciences* **7:**335–365.

McBratney, A. B., R. Webster, and T. M. Burgess, 1981. The design of optimal sampling schemes for local estimation and mapping of regionalized variables. Part 1. Theory and method, *Computers and Geosciences* **7:**331–334.

McCleary, R., and R. A. Hay, Jr., 1980. *Applied Time Series Analysis for the Social Sciences*. Sage, Beverly Hills, Calif.

McCollister, G. M., and K. R. Wilson, 1975. Linear stochastic models for forecasting daily maxima and hourly concentrations of air pollutants, *Atmospheric Environment* **9:**417–423.

McKerchar, A. I., and J. W. Delleur, 1974. Application of seasonal parametric linear stochastic models to monthly flow data, *Water Resources Research* **10:**246–255.

McLendon, H. R., 1975. Soil monitoring for plutonium at the Savannah River plant, *Health Physics* **28:**347–354.

McMichael, F. C., and J. S. Hunter, 1972. Stochastic modeling of temperature and flow in rivers, *Water Resources Research* **8:**87–98.

Mage, D. T., 1975. Data analysis by use of univariate probability models, in *Symposium on Air Pollution Control*. Ministry of Health and Social Affairs, Seoul, Korea, Paper No. 1-G/75.

Mage, D. T., 1980. An explicit solution for S_B parameters using four percentile points, *Technometrics* **22**:247–252.

Mage, D. T., 1981. A review of the application of probability models for describing aerometric data,'' in *Environmetrics 81: Selected Papers*, SIAM-SIMS Conference Series No. 8. Society for Industrial and Applied Mathematics, Philadelphia, pp. 42–51.

Mage, D. T., 1982. An objective graphical method for testing normal distributional assumptions using probability plots, *The American Statistician* **36**:116–120.

Mage, D. T., and W. R. Ott, 1978. Refinements of the lognormal probability model for analysis of aerometric data, *Journal of the Air Pollution Control Association* **28**:796–798.

Mage, D. T., and W. R. Ott, 1984. An evaluation of the methods of fractiles, moments and maximum likelihood for estimating parameters when sampling air quality data from a stationary lognormal distribution, *Atmospheric Environment* **18**:163–171.

Mann, H. B., 1945. Non-Parametric tests against trend, *Econometrica* **13**:245–259.

Marasinghe, M. G., 1985. A multistage procedure for detecting several outliers in linear regression, *Technometrics* **27**:395–400.

Marsalek, J., 1975. Sampling techniques in urban runoff quality studies, in *Water Quality Parameters*. American Society for Testing and Materials, STP 573, pp. 526–542.

Matalas, N. C., and W. B. Langbein, 1962. Information content of the mean, *Journal of Geophysical Research* **67**:3441–3448.

Miesch, A. T., 1976. *Sampling Designs for Geochemical Surveys—Syllabus for a Short Course*. U.S. Department of the Interior, Geological Survey, Open-file Report 76–772, Denver.

Montgomery, D. C., and L. A. Johnson, 1976. *Forecasting and Time Series Analysis*. McGraw-Hill, New York.

Montgomery, H. A. C., and I. C. Hart, 1974. The design of sampling programmes for rivers and effluents, *Water Pollution Control (London)* **73**:77–101.

Montgomery, R. H., and K. H. Reckhow, 1984. Techniques for detecting trends in lake water quality, *Water Resources Bulletin* **20**:43–52.

Moore, S. F., G. C. Dandy, and R. J. DeLucia, 1976. Describing variance with a simple water quality model and hypothetical sampling programs, *Water Resources Research* **12**:795–804.

Morrison, J., 1958. The lognormal distribution in quality control, *Applied Statistics* **7**:160–172.

Mosteller, F., and R. E. K. Rourke, 1973. *Sturdy Statistics*. Addison-Wesley, Reading, Mass.

Mosteller, F., and J. W. Tukey, 1977. *Data Analysis and Regression, A Second Course in Statistics*. Addison-Wesley, Reading, Mass.

Munn, R. E., 1981. The estimation and interpretation of air quality trends, including some implications for network design, *Environmental Monitoring and Assessment* **1**:49–58.

Murthy, M. N., 1967. *Sampling Theory and Methods*. Statistical Publishing Society, Calcutta, India.

National Academy of Sciences, 1977. *Environmental Monitoring*, Analytical Studies for the U.S. Environmental Protection Agency, vol. IV. National Academy of Sciences, Washington, D.C.

Nehls, G. J., and G. G. Akland, 1973. Procedures for handling aerometric data, *Journal of the Air Pollution Control Association* **23**:180–184.

Nelson, A. C., Jr., D. W. Armentrout, and T. R. Johnson, 1980. *Validation of Air Monitoring Data*, EPA-600/4-80-030. U.S. Environmental Protection Agency, National Technical Information Service, Springfield, Va.

Nelson, J. D., and R. C. Ward, 1981. Statistical considerations and sampling techniques for ground-water quality monitoring, *Ground Water* **19:**617–625.

Noll, K. E., and T. L. Miller, 1977. *Air Monitoring Survey Design*. Ann Arbor Science, Ann Arbor, Mich.

Odeh, R. E., D. B. Owen, Z. W. Birnbaum, and L. Fisher, 1977. *Pocket Book of Statistical Tables*. Dekker, New York.

Olea, R. A., 1984. *Systematic Sampling of Spatial Functions*, Series on Spatial Analysis No. 7. Kansas Geological Survey, University of Kansas, Lawrence, Kan.

Olsson, D. M., and L. S. Nelson, 1975. The Nelder-Mead simplex procedure for function minimization, *Technometrics* **17:**45–51.

Ott, W., and R. Eliassen, 1973. A survey technique for determining the representativeness of urban air monitoring stations with respect to carbon monoxide, *Journal of the Air Pollution Control Association* **23:**685–690.

Ott, W. R., and D. T. Mage, 1976. A general purpose univariate probability model for environmental data analysis, *Computers and Operations Research* **3:**209–216.

Owen, D. B., 1962. *Handbook of Statistical Tables*. Addison-Wesley, Palo Alto, Calif.

Page, E. S., 1961. Cumulative sum charts, *Technometrics* **3:**1–9.

Page, E. S., 1963. Controlling the standard deviation by cusums and warning lines, *Technometrics* **5:**307–315.

Parkhurst, D. F., 1984. Optimal sampling geometry for hazardous waste sites, *Environmental Science Technology* **18:**521–523.

Pearson, E. S., and H. O. Hartley, 1966. *Biometrika Tables for Statisticians*, vol. 1, 3rd ed. Cambridge University Press, London.

Pielou, E. C., 1977. *Mathematical Ecology*. Wiley, New York.

Pinder, J. E., III, and M. H. Smith, 1975. Frequency distributions of radiocesium concentrations in soil and biota, in *Mineral Cycling in Southeastern Ecosystems*, F. B. Howell, J. B. Gentry, and M. H. Smith, eds. CONF-740513, National Technical Information Service, Springfield, Va., pp. 107–125.

Price, K. R., R. O. Gilbert, and K. A. Gano, 1981. *Americium-241 in Surface Soil Associated with the Hanford Site and Vicinity*. Pacific Northwest Laboratory, Richland, Wash., PNL-3731.

Provost, L. P., 1984. Statistical methods in environmental sampling, in *Environmental Sampling for Hazardous Wastes*, G. E. Schweitzer and J. A. Santolucito, eds., ACS Symposium Series 267. American Chemical Society, Washington, D.C., pp. 79–96.

Quenouille, M. H., 1949. Problems in plane sampling, *Annals of Mathematical Statistics* **20:**355–375.

Rabosky, J. G. and D. L. Koraido, 1973. Gaging and sampling industrial wastewaters, *Chemical Engineering* **80:**111–120.

Rajagopal, R., 1984. Optimal sampling strategies for source identification in environmental episodes, *Environmental Monitoring and Assessment* **4:**1–14.

Rand Corporation, 1955. *A Million Random Digits*. Free Press, Glencoe, ILL.

Reckhow, K. H., and S. C. Chapra, 1983. *Engineering Approaches for Lake Management. volume I. Data Analysis and Empirical Modeling*. Butterworth, Boston.

Reinsel, G., G. C. Tiao, and R. Lewis, 1981. Statistical analysis of stratospheric ozone data for trend detection, in *Environmetrics 81: Selected Papers*, SIAM-SIMS Conference Series No. 8. Society for Industrial and Applied Mathematics, Philadelphia, pp. 215–235.

Reinsel, G., G. C. Tiao, M. N. Wang, R. Lewis, and D. Nychka, 1981. Statistical analysis of stratospheric ozone data for the detection of trends, *Atmospheric Environment* **15:**1569–1577.

Rhodes, R. C., 1981. Much ado about next to nothing, or what to do with measurements below the detection limit, *Environmetrics 81: Selected Papers*, SIAM-SIMS Conference Series No. 8. Society for Industrial and Applied Mathematics, Philadelphia, pp. 157–162.

Rickert, D. A., W. G. Hines, and S. W. McKenzie, 1976. *Methodology for River-Quality Assessment with Application to the Willamette River Basin, Oregon*, Geological Survey Circular 715-M. U.S. Geological Survey, Arlington, Va.

Rodriguez-Iturbe, I., and J. M. Mejia, 1974. The design of rainfall networks in time and space, *Water Resources Research* **10:**713–728.

Rohde, C. A., 1976. Composite sampling, *Biometrics* **32:**273–282.

Rohlf, F. J., 1975. Generalization of the gap test for the detection of multivariate outliers, *Biometrics* **31:**93–102.

Rosner, B., 1983. Percentage points for a generalized ESD many-outlier procedure, *Technometrics* **25:**165–172.

Rosner, B., 1975. On the detection of many outliers, *Technometrics* **17:**221–227.

Roy, R., and J. Pellerin, 1982. On long term air quality trends and intervention analysis, *Atmospheric Environment* **16**:161–169.

Royston, J. P., 1982*a*. An extension of Shapiro and Wilk's W test for normality to large samples, *Applied Statistics* **31:**115–124.

Royston, J. P., 1982*b*. The *W* test for normality, Algorithm AS181, *Applied Statistics* **31:**176–180.

Royston, J. P., 1982*c*. Expected normal order statistics (exact and approximate), Algorithm AS177, *Applied Statistics* **31:**161–165.

Royston, J. P., 1983. The *W* test for normality, a correction, *Applied Statistics* **32:**224.

Ryan, T. A., Jr., B. L. Joiner, and B. F. Ryan, 1982. *Minitab Reference Manual*. Statistics Department, The Pennsylvania State University, University Park, Pa.

Saltzman, B. E., 1972. Simplified methods for statistical interpretation of monitoring data, *Journal of the Air Pollution Control Association* **22:**90–95.

Sanders, T. G., and D. D. Adrian, 1978. Sampling frequency for river quality monitoring, *Water Resources Research* **14:**569–576.

Schaeffer, D. J., and K. G. Janardan, 1978. Theoretical comparisons of grab and composite sampling programs, *Biometrical Journal* **20:**215–227.

Schaeffer, D. J., H. W. Kerster, and K. G. Janardan, 1980. Grab versus composite sampling: A primer for the manager and engineer, *Environmental Management* **4:**157–163.

Schaeffer, D. J., H. W. Kerster, D. R. Bauer, K. Rees, and S. McCormick, 1983. Composite samples overestimate waste loads, *Journal of the Water Pollution Control Federation* **55:**1387–1392.

Scheffé, H., 1959. *The Analysis of Variance*. Wiley, New York.

Schubert, J., A. Brodsky, and S. Tyler, 1967. The log-normal function as a stochastic model of the distribution of strontium-90 and other fission products in humans, *Health Physics* **13:**1187–1204.

Schweitzer, G. E., 1982. Risk assessment near uncontrolled hazardous waste sites: Role of monitoring data, *Environmental Monitoring and Assessment* **2:**15–32.

Schweitzer, G. E., and S. C. Black, 1985. Monitoring statistics, *Environmental Science and Technology* **19:**1026–1030.

Sen, P. K., 1963. On the properties of U-statistics when the observations are not independent. Part 1. Estimation of non-serial parameters in some stationary stochastic process, *Calcutta Statistical Association Bullitin* **12:**69–92.

Sen, P. K., 1965. Some non-parametric tests for m-dependent time series, *Journal of the American Statistical Association* **60:**134–147.

Sen, P. K., 1968*a*. On a class of aligned rank order tests in two-way layouts, *Annals of Mathematical Statistics* **39:**1115–1124.

Sen, P. K., 1968*b*. Estimates of the regression coefficient based on Kendall's tau, *Journal of the American Statistical Association* **63:**1379–1389.

Shapiro, S. S., and R. S. Francia, 1972. An approximate analysis of variance test for normality, *Journal of the American Statistical Association* **67:**215–216.

Shapiro, S. S., and M. B. Wilk, 1965. An analysis of variance test for normality (complete samples), *Biometrika* **52**:591–611.

Sharp, W. E., 1970. Stream order as a measure of sample source uncertainty, *Water Resources Research* **6**:919–926.

Sharp, W. E., 1971. A topologically optimum water-sampling plan for rivers and streams, *Water Resources Research* **7**:1641–1646.

Shewhart, W. A., 1931. *Economic Control of Quality of Manufactured Product*. Van Nostrand, Princeton, N.J.

Sichel, H. S., 1952. New Methods in the statistical evaluation of mine sampling data, *Transactions Institute of Mining and Metallurgy* **61**:261–288.

Sichel, H. S., 1966. The estimation of means and associated confidence limits for small samples from lognormal populations, in *Proceedings of the Symposium on Mathematical Statistics and Computer Applications in Ore Valuation*. South African Institute of Mining and Metallurgy, Johannesburg, pp. 106–123.

Simonoff, J. S., 1982. A comparison of robust methods and detection of outliers techniques when estimating a location parameter, in *Computer Science and Statistics: Proceedings of the 14th Symposium on the Interface*, K. W. Heiner, R. S. Sacher, and J. W. Wilkinson, eds. Springer-Verlag, New York, pp. 278–281.

Singer, D. A., 1972. ELIPGRID: A Fortran IV program for calculating the probability of success in locating elliptical targets with square, rectangular and hexagonal grids, *Geocom Programs* **4**:1–16.

Singer, D. A., 1975. Relative Efficiencies of square and triangular grids in the search for elliptically shaped resource targets, *Journal of Research of the U.S. Geological Survey* **3**(2):163–167.

Skalski, J. R., and D. H. McKenzie, 1982. A design for aquatic monitoring programs, *Journal of Environmental Management* **14**:237–251.

Slonim, M. J., 1957. Sampling in a nutshell, *Journal of the American Statistical Association* **52**:143–161.

Smith, W., 1984. Design of efficient environmental surveys over time, in *Statistics in the Environmental Sciences*, American Society for Testing and Materials, STP 845, S. M. Gertz and M. D. London, eds. American Society for Testing and Materials, Philadelphia, pp. 90–97.

Smith, R. A., R. M. Hirsch, and J. R. Slack, 1982. *A Study of Trends in Total Phosphorus Measurements at NASQAN Stations*. U.S. Geological Survey Water-Supply Paper 2190, U.S. Geological Survey, Alexandria, Va.

Snedecor, G. W., and W. G. Cochran, 1967. *Statistical Methods*, 6th ed. Iowa State University Press, Ames, Iowa.

Snedecor, G. W., and W. G. Cochran, 1980. *Statistical Methods*, 7th ed. Iowa State University Press, Ames, Iowa.

Sokal, R. R., and R. J. Rohlf, 1981. *Biometry*, 2nd ed. W. H. Freeman, San Francisco.

Sophocleous, M., 1983. Groundwater observation network design for the Kansas groundwater management districts, U.S.A., *Journal of Hydrology* **61**:371–389.

Sophocleous, M., J. E. Paschetto, and R. A. Olea, 1982. Ground-water network design for Northwest Kansas, using the theory of regionalized variables, *Ground Water* **20**:48–58.

Sparr, T. M., and D. J. Schaezler, 1974. Spectral analyses techniques for evaluating historical water quality records, in *Water Resources Instrumentation: Proceedings of the International Seminar and Exposition*, vol. 2, R. J. Krizek, ed. Ann Arbor Science, Ann Arbor, Mich., pp. 271–297.

Statistical Analysis System, 1982. *SAS Users Guide: Basics*. SAS Institute Cary, N.C.

Statistical Analysis System, 1985. *SAS Program Products*. SAS Institute, Cary, N.C.

Statistical Package for the Social Sciences, 1985. SPSS, Inc., Chicago.

Stefansky, W., 1972. Rejecting outliers in factorial designs, *Technometrics* **14**:469–478.

Sukhatme, P. V., and B. V. Sukhatme, 1970. *Sampling Theory of Surveys with Applications*, 2nd ed. Iowa State University Press, Ames, Iowa.

Tanur, J. M., F. Mosteller, W. H. Kruskal, R. F. Link, R. S. Pieters, and G. R. Rising, 1972. *Statistics: A Guide to the Unknown*. Holden-Day, San Francisco.

Tenenbein, A., 1971. A double sampling scheme for estimating from binomial data with misclassifications: Sample size determination, *Biometrics* **27:**935–944.

Tenenbein, A., 1974. Sample size determination for the regression estimate of the mean, *Biometrics* **30:**709–716.

Theil, H., 1950. A rank-invariant method of linear and polynomial regression analysis, Part 3, in *Proceedings of Koninalijke Nederlandse Akademie van Wetenschatpen A* **53:**1397–1412.

Thöni, H., 1969. A table for estimating the mean of a lognormal distribution, *Journal of the American Statistical Association* **64:**632–636.

Thornton, K. W., R. H. Kennedy, A. D. Magoun, and G. E. Saul, 1982. Reservoir water quality sampling design, *Water Resources Bulletin* **18:**471–480.

Tiku, M. L., 1968. Estimating the parameters of log-normal distribution from censored samples, *Journal of the American Statistical Association* **63:**134–140.

van Belle, G., and J. P. Hughes, 1983. Monitoring for water quality: Fixed stations vs intensive surveys, *Journal of Water Pollution Control Federation* **55:**400–404.

van Belle, G., and J. P. Hughes, 1984. Nonparametric tests for trend in water quality, *Water Resources Research* **20:**127–136.

Vardeman, S., and H. T. David, 1984. Statistics for quality and productivity. A new graduate-level statistics course, *The American Statistician* **38:**235–243.

Vaughan, W. J., and C. S. Russell, 1983. Monitoring point sources of pollution: Answers and more questions from statistical quality control, *The American Statistician* **37:**476–487.

Velleman, P. F., and D. C. Hoaglin, 1981. *Applications, Basics, and Computing of Exploratory Data Analysis*. Duxbury Press, Boston.

Verly, G., M. David, A. G. Journel, and A. Marechal, 1984. *Geostatistics for Natural Resources Characterization*. Parts 1, 2. NATO ASI Series C: Mathematical and Physical Sciences. vol. 122. Reidel, Boston.

Wallis, J. R., and P. E. O'Connell, 1972. Small sample estimation of ρ_1, *Water Resources Research* **8:**707–712.

Ward, R. C., J. C. Loftis, K. S. Nielsen, and R. D. Anderson, 1979. Statistical evaluation of sampling frequencies in monitoring networks, *Journal of the Water Pollution Control Federation* **51:**2292–2300.

Wetherill, G. B., 1977. *Sampling Inspection and Quality Control*, 2nd ed. Chapman and Hall, New York.

Whitten, E. H. T., 1975. The practical use of trend-surface analyses in the geological sciences, in *Display and Analysis of Spatial Data*, J. C. Davis and M. J. McCullagh, eds. Wiley, New York, pp. 282–297.

Williams, W. H., 1978. *A Sampler on Sampling*. Wiley, New York.

Wolter, K. M., 1984. An investigation of some estimators of variance for systematic sampling, *Journal of the American Statistical Association* **79:**781–790.

Yates, F., 1948. Systematic sampling, *Philosophical Transactions of the Royal Society of London*, series A, **241:**345–377.

Yates, F., 1981. *Sampling Methods for Censuses and Surveys*, 4th ed. MacMillan, New York.

Zeldin, M. D., and W. S. Meisel, 1978. *Use of Meteorological Data in Air Quality Trend Analysis*. U.S. Environmental Protection Agency, Research Triangle Park, National Technical Information Service, Springfield, Va., EPA-450/3-78-024.

Zinger, A., 1980. Variance estimation in partially systematic sampling, *Journal of the American Statistical Association* **75:**206–211.

Zirschky, J., and R. O. Gilbert, 1984. Detecting hot spots at hazardous-waste sites, *Chemical Engineering* **91:**97–100.

Index